An introduction to

Practical Hydraulic System Maintenance

Steve Skinner and Martin Cuthbert

WEBTEC

Published by Webtec Products Ltd

Nuffield Road, St. Ives, Cambridgeshire PE27 3LZ, UK

Tel: +44 (0)1480 397400

Email: marketing@webtec.com

Web: www.webtec.com

Ordering information for quantity sales: Special discounts are available on quantity purchases by corporations, associations, and others. For details, contact the publisher at the address above.

ISBN 978-1-9164485-0-6

TSK6944 – Issue B – 02/19

Printed on demand by Amazon Corp. USA

Designed by Webtec Products Ltd

Typeset by Helius

CONTENTS

FOREWORD

John R. Savage

Director, National Fluid Power Centre, UK

The necessary knowledge, skills and role competency of today's workforce to manage and maintain hydraulic systems far exceeds that of a decade ago. With the interface of electronics and control, today's hydraulic systems and components can outperform those of the past. However, as systems become more complex, their guaranteed future will depend on the investment in people to ensure that the education and training they receive enable them to perform effectively and safely in the workplace. To achieve through the knowledge and skills of people, we must lay down a strong foundation and a common-sense approach to learning that ensures results.

I congratulate the authors on producing this outstanding book, which reflects their many years of experience as both practising engineers and professional trainers. The book is easy to read, easy to follow and down to earth, and follows a learning progression route that is easy to understand. It is ideal for advanced apprentices, craftsmen, technicians and engineers in every sector where hydraulics are used.

This book will provide added value to those attending introductory level and Stage 1 courses at the National Fluid Power Centre, and we will clearly promote this book. Well done to all concerned.

Chris Buxton

CEO, British Fluid Power Association

It is a cliché to say that we are living in a changing world but it is truer now than it has ever been! The rate of technological change is exponential; Moore's law has now been exceeded and the convergence of technologies through platforms such as Industry 4.0 has led the pundits to capture its impact by recognising what has become the Fourth Industrial Revolution! Despite these changes and the resulting plethora of more highly skilled and, most significantly, more intellectually stimulating jobs, UK manufacturing is still dogged by chronic skills shortages. Since 2006 the single biggest obstacle to business growth in the UK has been a lack of suitably qualified employees with the right qualifications and a good work ethic. The statistics for how many engineers will be required over the next three years in order to simply 'stand still' are widely publicised and are – to say the least – worrying.

Against this background, the need for good-quality, sector-specific training, and the associated reference texts and books in support of that training, is paramount.

The fluid power sector is particularly short of good-quality reference and training material, so as the chief executive of the industry trade association I am especially pleased to be able to make my contribution to supporting this most excellent text by Martin Cuthbert and Steve Skinner.

Whether a new student in the field of hydraulics or a seasoned engineer in need of being reminded of a particular technical concept, this book will have a place on every fluid power engineer's shelf. Written in an easily digested style, it follows a logical path of learning such that each chapter builds upon the former in such a way that it tells an enticing story. From the British Fluid Power Association perspective, I would put it into the category of 'recommended reading' for all students and operatives in the fluid power sector.

Ian Morris FBFPA CEng FIMechE

Former CEO of the British Fluid Power Association

Francis Bacon is commonly attributed with the first coining of the expression 'Knowledge is power'. Although similar phrases have been found in early Arabic documents, for the purposes of this foreword I will simply assume that it is a valued expression that has clearly been around for a while.

However,' knowledge' and 'understanding' are not necessarily the same thing, and we all know people who may have great knowledge about, but very little understanding of, a particular subject. It is the assimilation of those two nouns that makes the true artisan in any subject, none more so perhaps than in the world of musical interpretation, and it is that which separates the world-class concert pianist from the rest.

The book you have just opened should be read by all in the fluid power industry, whether one has 40 years of experience or 4 months. It is probably the clearest depiction of the most important facts and techniques that I have seen in a long time. The book explores the technical aspects of the industry accurately, and with the comfort of knowing that it has been written by two authors both with long and respected experience in the world of fluid power.

We all need to ensure that the 'understanding' part of our 'knowledge' is constantly being topped up, and the clarity of the illustrations and the easy to follow text in this book makes it an essential read in this process. The Institution of Mechanical Engineers has been involved in the development of all aspects of engineering, not only in the UK but around the world, for many years, and *An Introduction to Practical Hydraulic System Maintenance* will add much to this task.

May I congratulate Steve Skinner and Martin Cuthbert for producing this excellent guide, and for the hours of dedicated work which has brought it to fruition.

Rob Bartlett FIAM

Director & CEO, British Valve & Actuator Association Ltd

This highly informative book is ideal for anyone new to fluid power hydraulics. It will provide a solid foundation from which to start your journey into hydraulics and hydraulic system maintenance. It is a must have for anyone who needs to brush up on hydraulics!

Rob Oliver

Chief Executive, Construction Equipment Association

www.thecea.org.uk

Hydraulic systems are at the heart of construction equipment. Without well-maintained hydraulics the machine will fail, accidents happen and the job site stop. This important new book is an invaluable guide to maintaining a machine's hydraulic system at peak performance. The Construction Equipment Association (CEA) is very pleased to endorse *An Introduction to Practical Hydraulic System Maintenance* and welcomes the book as an essential component of the hydraulic technician's tool kit. CEA members comprise leading construction equipment OEMs and the sector's supply chain, including manufacturers from the hydraulics sector, including Webtec.

Qixin Wang

General Manager, Guangzhou Xinou Machinery Co. Ltd

Current President, Guangzhou Huangpu Construction Machinery Industry Association

Hydraulic technology has seen increasingly wide application in many industrial sectors. This is thanks to its irreplaceable role in areas in which high power is required or strict space constraints exist, specialist areas or those areas in which accessibility to electricity is very poor (e.g. wind, ocean and solar energy equipment, construction machinery, marine equipment, aerospace equipment and robotics). Hydraulic equipment plays an important role in the system in which it is contained.

Hydraulic systems are integrated systems that combine mechanical, electrical and hydraulic systems, and feature complex structures and high precision. Hydraulic technology involves disciplines such as mechanical engineering, electrical engineering, fluid mechanics and control engineering. In addition, there is a current trend for the networking of intelligent hydraulic equipment. Therefore, it is difficult for maintenance staff dealing with hydraulic equipment to systematically acquire the required expertise in hydraulic systems.

The factors involved in hydraulic failure are complex and the fault points are hidden. The cause of the generation and development of a failure is characterised by time-based variation, randomness, overlapping and redundancy, and so it is difficult to diagnose and repair hydraulic faults. The failure of a hydraulic component affects not only the output of the device itself but also the entire production system. In particular, the failure of equipment in an automatic production line often leads to a breakdown of the entire line, entailing huge economic losses. Thus the quality of the maintenance of a hydraulic system is key to reliable system operation. Each link in a system must be made scientifically and reasonably so as to reduce the failure rate, ensure the normal operation of the hydraulic system and thus enable full effectiveness of the system to be achieved.

Guangzhou Xinou Machinery Co. Ltd has long been dedicated to the development and manufacture of various types of hydraulic testing systems. Such testing systems enable hydraulic equipment maintenance personnel to accurately detect faulty hydraulic components and systems and eliminate equipment failures.

An Introduction to Practical Hydraulic System Maintenance systematically introduces the main aspects, basic methods and technical points of hydraulic system maintenance. It covers hydraulic elements and systems, transmission media, oil cleanliness and temperature control, power and electronic controls, connectors, system safety, installation and maintenance, and troubleshooting. The book is up to date on the technology, strong in practicability, rich in content, clear in organisation and contains a wealth of diagrams and tips.

I believe this book will be of great use to hydraulic equipment maintenance personnel, and will be popular with readers. I would like to extend my warm congratulations to the authors Steve Skinner and Martin Cuthbert on the publication of this book.

INTRODUCTION

Why is Webtec sponsoring this book?

For Webtec, as a global hydraulic measurement and control manufacturer with over 50 years' experience, we are often surprised by the relative inexperience of fluid power hydraulics demonstrated in industry today. For example, a service or design engineer might be handed the job to maintain or update a hydraulic system, often when a more experienced person retires, without receiving any formal training. When the person charged with maintaining the hydraulics is not fully conversant with fluid power, we feel duty bound to help these technicians and engineers get to grips with the skills they need to carry out the 'hydraulics' part of their job.

We looked around for a book or training resource we could supply to end-users we meet all around the world, and found that, while there are lots of resources, most aren't readily available via Amazon or a similar web-based shop, and unless you enrol on a training course you can't access them at all. Those that are available tend to focus more on system design than maintenance. This book came about to address these issues and provide a readily available starting point for the 'novice hydraulic engineer', to encourage everyone to then go on to take a practical hydraulics course using the resources listed in the Resources section at the end of this book. In this way we hope we can help educate the next generation in how to work safely with hydraulics and to better understand the role of hydraulics in an increasingly interconnected world.

Using the royalties to encourage future fluid power engineers

Hydraulics and fluid power, like most disciplines in engineering, is constantly short of smart, motivated and skilled people to help advance technology and build the hydraulic systems of the future. We feel the answer to this skills shortage lies in attracting younger people into the industry, by demonstrating the fun, flexibility and usefulness of fluid power through games and challenges for primary and secondary school students. Webtec is committing all royalties from this book to charities aimed at inspiring the next generation of engineers to get involved in fluid power, like Primary Engineer, which runs the UK Fluid Power Challenge, and is similar to the Fluid Power Challenge run by the National Fluid Power Association (NFPA) in the USA. If you would like to recommend a charity that you feel would meet these criteria, please contact us at marketing@webtec.com.

We hope you enjoy the book.

To keep up to date with hydraulic maintenance training tips and resources, please visit www.webtec.com/education

Martin Cuthbert and Steve Skinner

November 2018

ACKNOWLEDGEMENTS

It has taken over three years to put this book together and wouldn't have been possible without the help of many contributors. We wanted the book to be very visual and represent real-life solutions. This has only been possible with help of the many companies that have granted us permission to use their images: thank you to them all. The project has only become a reality through Steve's expertise, and we are indebted to him for sharing his extensive knowledge and amazing illustrations. Special thanks must also go to the guest authors who have freely shared the wisdom of their collective 200+ years' experience in Chapter 10 and to Alan Hitchcox of *Hydraulics and Pneumatics Magazine* for his contribution to Chapter 9. Initially, we were worried no one would be willing to contribute; in the end we ran out of space and had to leave some material out. We are very grateful for the positive words of support from the many trainers and industry associations who work with fluid power, as you have been the inspiration that has kept us going. Lastly, a special thanks to the graphic design, editing and production team who have helped bring this book to life.

Martin Cuthbert

CHAPTER

1 THE BASIC PRINCIPLES OF HYDRAULIC FLUID POWER

POWER TRANSMISSION AND CONTROL

For many thousands of years, mankind relied on muscle power to perform tasks such as building shelters, digging for minerals and felling trees. Then, as farming developed, animal power was harnessed to perform some of the heavier-duty tasks of ploughing fields or moving heavy objects. Next, wind and water power opened up more possibilities for activities such as milling flour and draining fields, so the tasks that could be carried out grew larger in scale. Then along came the Industrial Revolution, when steam power provided many new possibilities for powering machinery, which allowed humans to concentrate more on the control of the machines.

The steam engines of the Industrial Revolution were superseded by today's more common sources of power, namely the **internal combustion engine** and the **electric motor**.

Traditionally, machines that have to move around, such as earth-moving machinery and agricultural vehicles, have relied on the diesel engine to provide their motive power. The electric motor, on the other hand, has provided the main source of power for stationary machines, such as presses, plastics machinery and machine tools. Some applications, fork-lift trucks for example, may use either power source, depending on their intended area of use. As **hybrid technology** advances, a combination of the internal combustion engine and electric motors will most likely become increasingly commonplace.

FURTHER READING

For a further insight into the history of hydraulic power transmission, see one of the author's other publications: *Hydraulic Fluid Power – A Historical Timeline* (ISBN 978-1-291-67689-1)

Furthermore, humans now tend to be less directly involved in the control of machines. As the requirements for fast, precise control have increased, first **analogue electronics** and then **digital electronics** have taken over the control responsibility from the human eye and hand for many machine functions.

Connection of the power source (or **prime mover** as it is sometimes called) to the machine it is powering still requires some form of transmission mechanism, because rarely will the output of a diesel engine or electric motor exactly match the requirements of the machine.

In simple applications, such as a desk fan or power drill, an electric motor may be used to directly drive the machine, but these situations are relatively few. Traditionally, electric motors have been single-speed, single-direction drives, although **variable-speed electric drives** are now becoming more common.

Most diesel engines are also single-direction drives, but their drive speed can be varied, although only over a relatively narrow speed range. A machine may, however, require a bi-directional drive over a wide range of infinitely variable speeds. It may also require a **linear** rather than **rotary** movement with the capability to limit the maximum torque or force output. A machine may also have idling periods, when no movement is required, but frequent stopping and starting of the prime mover may not be practical.

> **DEFINITION**
>
> A **prime mover** is a device that powers a machine. In the case of hydraulic power transmission systems it is the component that drives the pump and is normally an electric motor or diesel engine. A diesel engine may drive the pump either directly or through a gearbox known as a **power take-off (PTO)**.

It is often the case that several different functions all have to be powered from a single power source, so some means of dividing the drive power is required. In some applications the available power has to be shared, but with some functions having priority over others. For example, the prime mover on a vehicle may have to share the available power between the vehicle's transmission (which drives it along) and auxiliary functions (such as a boom and bucket arrangement for handling a payload), while at the same time ensuring that the vehicle's brakes and steering have priority over all other functions.

There are many ways in which power can be transmitted and converted by mechanical means. Most of us are familiar with gearboxes and clutches from driving our cars or motorcycles. However, less obviously, deep within the engine the linear, up-and-down movement of the pistons is converted into the rotary movement of the drive shaft by connecting-rods and cranks. Also, lead screws can convert the rotary motion of an electric motor into the linear movement required by a machine tool, for example, and rack and pinions have been used for many years to convert linear motion into rotary motion, or vice versa.

The 'solid' gears, cams, clutches, lead screws and other components of these mechanical or electro-mechanical transmissions have to be manufactured to close tolerances and assembled carefully to avoid excessive binding or backlash. Very often they also have to be lubricated to reduce wear between components, which can limit their performance or service life.

Hydraulics and pneumatics, known together as **fluid power**, differ from mechanical power transmissions by using a flexible medium to transmit power, namely fluids. This is usually some form of oil for hydraulic systems and air for pneumatic equipment.

> **DEFINITION**
>
> **Hydrodynamic** refers to the transmission of power using a fluid's momentum. A typical example is the torque converter or fluid coupling used in the transmission of an automatic gearbox car.
>
> **Hydrostatic** refers to the transmission of power using the flow and pressure of a fluid. An example is the transmission of power in a car power-steering system.

Hydraulic technology can be subdivided into **hydrodynamic** and **hydrostatic** systems. Hydrodynamic systems rely on the momentum of a fluid to transmit power smoothly through the vanes of a turbine. The most common application is the torque converter of the automatic transmissions used in vehicles. Contrary to what the name suggests, the fluid in hydrostatic systems can also be moving or flowing, but in this case the transmission of force or torque is achieved by virtue of the fluid's **pressure** rather than its momentum.

Figure 1.1 illustrates the power transmission options available, but it is hydrostatic systems that will be covered in this book.

There are many similarities between the two sister technologies of hydraulics and pneumatics, but there are also some important differences. The most obvious difference is in the properties of the fluid itself. The air employed in pneumatic systems is a gas, which is easily compressed. However, the oil or other liquid used in hydraulic systems is virtually incompressible (Fig. 1.2).

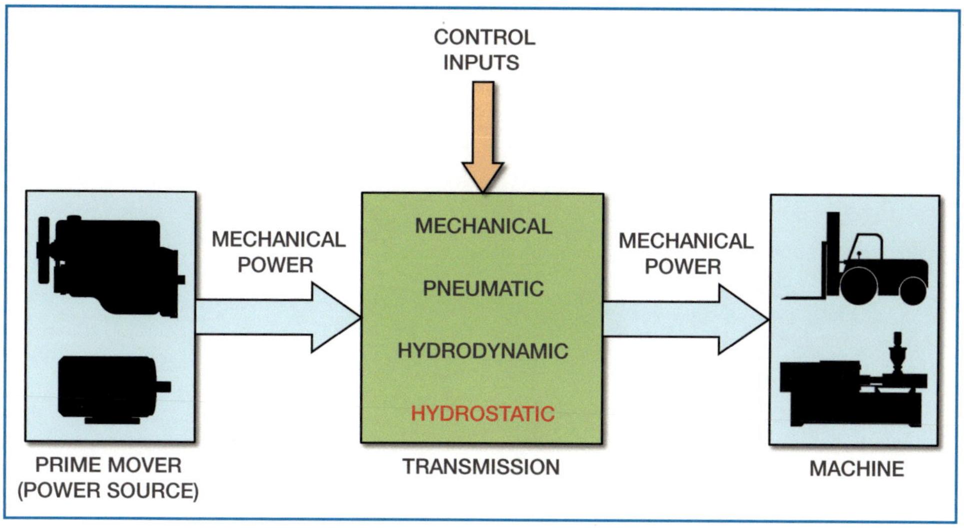

Fig. 1.1 Transmission and control of power

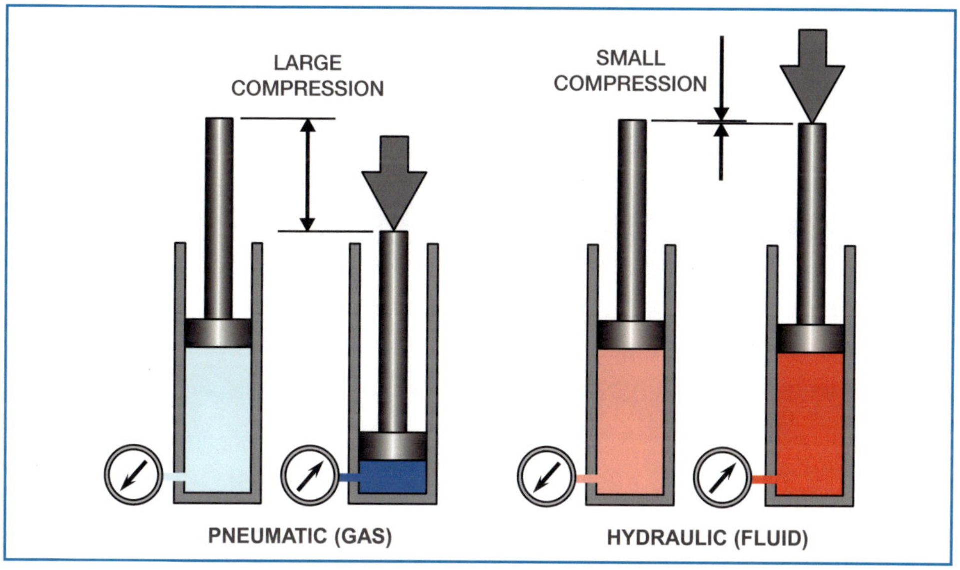

Fig. 1.2 Fluids are (virtually) incompressible

The very small amount by which liquids compress under pressure can be ignored in many applications. In other cases, however, the compressed volume of fluid can be significant (typically in machinery that uses large volumes of fluid at high pressure), and this must be taken into account in the design of the hydraulic system to avoid unwanted shocks detracting from its performance.

The gas invariably used in pneumatic systems is air. As air has a high **compressibility**, the pressure used in pneumatic systems is very low compared with that used in hydraulic systems. The amount of power that can be practically transmitted in a pneumatic system is, therefore, also correspondingly less. So, in general, pneumatic systems are confined to relatively light-duty applications compared with the heavy-duty tasks that can be performed by hydraulic systems.

A pick-and-place robot used to assemble components onto an electronic circuit board is an ideal application for pneumatics, whereas hydraulics is a natural choice for a 10,000 tonne forging press. The advantages and disadvantages of transmitting power hydraulically are summarised in Fig. 1.3.

POINT OF INTEREST

Currently, the largest hydraulic press in the world is an 80,000 ton die forging press in China that stands 10 stories tall

POINT OF INTEREST

As a rule of thumb, hydraulic fluid compresses by approximately 0.5% of its volume for every 70 bar (1000 psi) pressure. The presence of any air in the fluid will increase its compressibility and may cause an erratic or 'spongy' movement of actuators.

ADVANTAGES	DISADVANTAGES
1. Power density (large amounts of power from small components)	1. Generally less efficient than mechanical transmission
2. Capability to provide linear or rotary output movement	2. Requirement for fluid cleanliness
3. Simple bi-directional output	3. Fluid leakage may create pollution, fire risk or health hazard
4. Infinitely variable speed control	
5. Force/torque, speed and power limiting easily achieved	
6. Multiple outputs from single-input power source	
7. Manual, electrical or electronic control inputs	
8. Self-lubricating	
9. Ability to cope with difficult working environments	
10. Relatively simple, well-understood technology	

Fig. 1.3 Advantages and disadvantages of hydraulic power transmission

TRANSMITTING MOVEMENT

The fact that a liquid is (virtually) incompressible but at the same time able to take up the shape of its container is the principle upon which hydraulic fluid power is used to transmit movement. Consider the situation shown in Fig. 1.4, where two **cylinders** are connected together by means of a pipe. Each cylinder is fitted with a **piston**, which is free to move up and down and is perfectly sealed against the walls of the cylinder. If the cylinders and the interconnecting pipe are filled with fluid, then pushing down on one (input) piston will displace fluid along the pipe and cause the other (output) piston to rise. If both cylinders have the same dimensions, the distance moved by one will be exactly the same as the distance moved by the other.

POINT OF INTEREST

If the fluid is incompressible, the volume of fluid pushed out of the input cylinder is exactly equal to that pushed into the output cylinder. For cylinders of equal size, therefore, the amount of output movement will be the same as the amount of input movement, irrespective of the distance between the two cylinders.

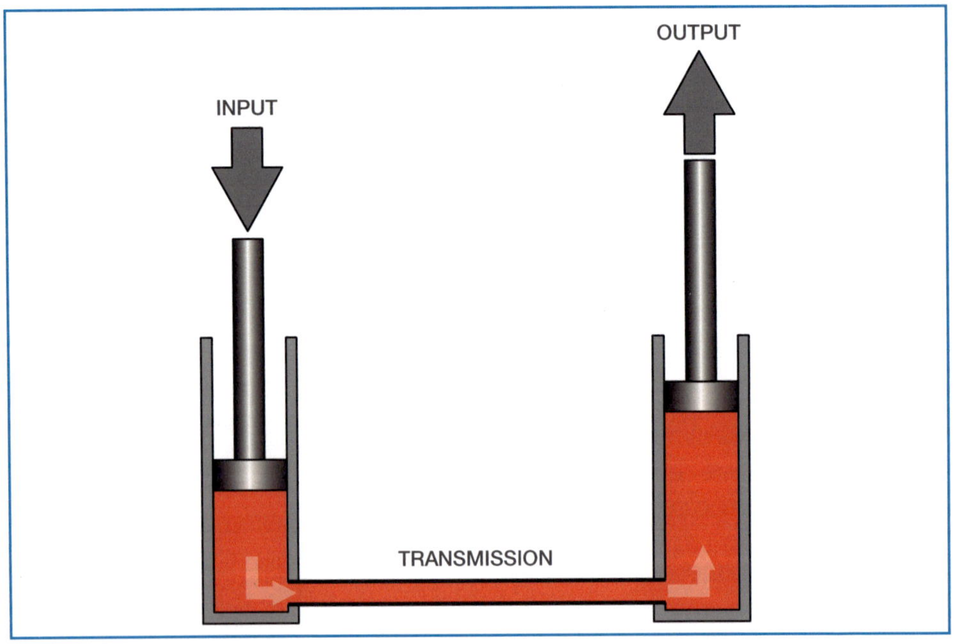

Fig. 1.4 Transmitting movement

An arrangement as simple as this is therefore capable of transmitting movement from one place to another. In theory, there is no limit to how far apart the input and output cylinders can be, and the route taken by the interconnecting pipe can be as tortuous as it needs to be. That is, it can follow walls and ceilings or be attached to convenient parts of the machinery as required. Furthermore, the output piston can be in a different orientation relative to the input piston (e.g. the piston movement can be 'side to side' rather than 'up and down').

If the movement of the output cylinder means that it has to pivot as its piston extends and retracts, then a **flexible mounting** can be used to attach the cylinder to the machine and a **flexible connector** can be incorporated in the pipe (normally in the form of a reinforced rubber **hose**).

If the output movement needs to be a rotation rather than a linear up-and-down or side-to-side movement, a hydraulic motor could take the place of the output cylinder, thus converting the input linear movement to an output rotary movement. Finally, multiple outputs can be connected to the input cylinder if necessary, so that the input power can be shared between several output devices and distributed to wherever it is required, as illustrated in Fig. 1.5.

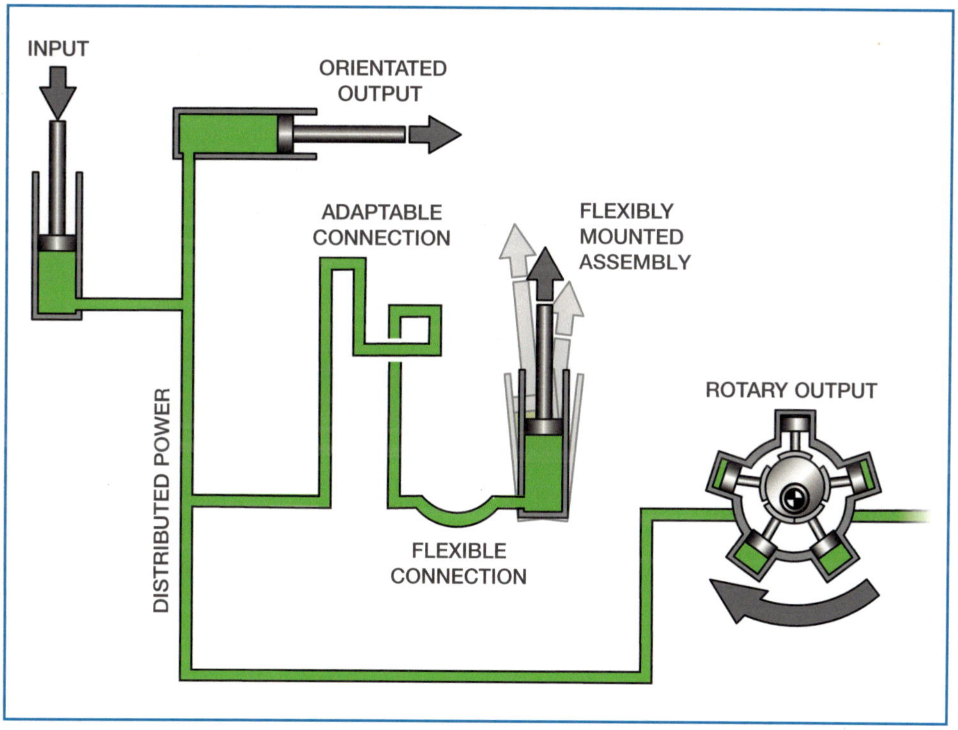

Fig. 1.5 Transmission, conversion and distribution of input power

Closing off the opposite side of a cylinder, sealing the rod and adding an extra connection means that the piston can now be pushed hydraulically in both directions (**double acting**) (Fig. 1.6) and does not have to rely on gravity or any other means to retract it. The same also applies to hydraulic motors, which can normally be driven in either direction.

Semi-rotary actuators also provide a rotary output movement but only over a certain arc (maybe 90° or 180°) rather than the continuous rotation of a motor. A typical application is to open and close a butterfly valve in a large pipeline.

POINT OF INTEREST

One of the key advantages of hydraulic transmission is its flexibility to be able to provide linear or rotary outputs from a single input source, with few restrictions on the interconnections from input to output.

DEFINITION

Hydraulic actuator is the general term for a device that converts hydraulic power back into mechanical movement. The two most common types are linear actuators (or cylinders) and rotary actuators (or motors).

POINT OF INTEREST

Double-acting cylinders require fluid seals on both the piston and the piston rod. They are generally the most common type of cylinder used in hydraulic systems.

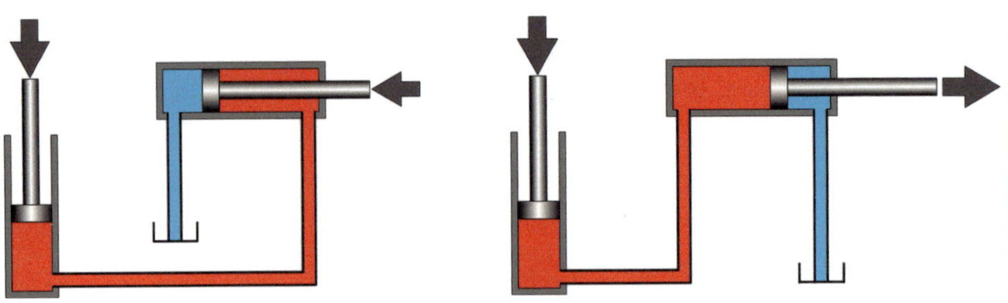

Fig. 1.6 Double-acting cylinder

It is now possible to start thinking of the two cylinders and pistons as the basis of a simple hydraulic system, with the left-hand cylinder acting as a **pump** and the right-hand one as the actuator (Fig. 1.7). In order to make the operation of the pump more convenient, a hand lever can be attached to the piston rod.

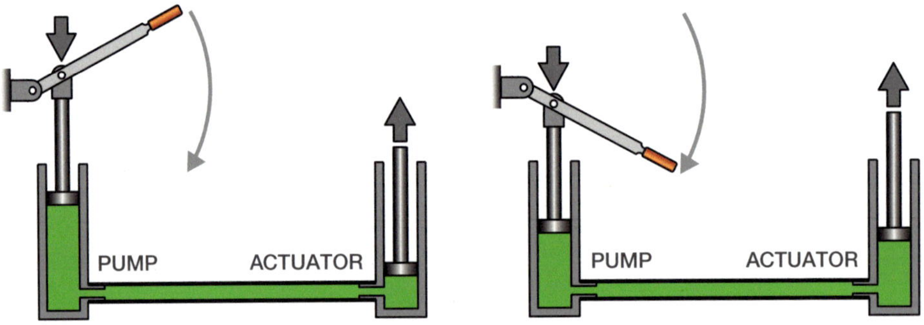

Fig. 1.7 Pump and actuator

Pushing the hand lever down will in turn push the cylinder piston down, thus forcing fluid out of the left-hand cylinder, along the pipe and into the right-hand cylinder. As described earlier, if both cylinders have the same dimensions, the piston of the right-hand cylinder will move upwards by an amount equal to the downward movement of the left-hand pump piston.

GENERATING PRESSURE

If the output piston has no resistance (i.e. it is not moving any load), the fluid can be pushed out of the pumping cylinder very easily. The only force that is required on the hand lever is that required to overcome the friction of the pistons and the friction of the fluid flowing through the passageways of the interconnection.

Any force that acts on the fluid creates a 'pressure' in the fluid. If the force is small, the pressure created in the fluid will also be small. If the output cylinder meets some resistance to its movement however, such as when jacking up a vehicle, the force required on the hand lever of the pump will increase and a higher pressure will be generated within the fluid (Fig. 1.8).

The pressure that is created within the fluid is dependent on both the resistance to movement (how heavy the load is) and the area of the piston over which the load

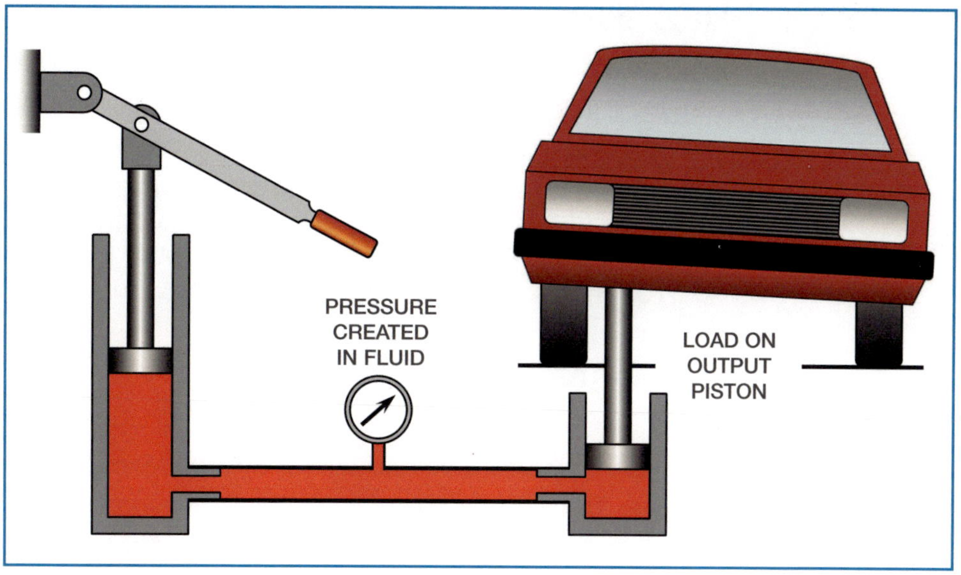

Fig. 1.8 Generation of pressure

acts. Therefore, pressure can be thought of as the distribution of force (or weight) over an area, which can be expressed mathematically as

$$\text{Pressure} = \frac{\text{Force}}{\text{Area}}$$

To further illustrate the concept of pressure, consider a metal cone placed on a surface with the point facing upwards. The whole weight of the cone is distributed over the large surface area of the bottom of the cone, so the pressure created at this surface is low. But if the cone is placed the other way up, so that it is balanced on its point, its weight is distributed over a much smaller area, thus creating a higher pressure. The weight of the cone is the same in both cases, but the area over which the weight acts is different and hence the pressure on the surface will be different (Fig. 1.9). The same situation applies to a hydraulic cylinder – a small-diameter cylinder supporting or moving a load will require a higher pressure than a large-diameter cylinder supporting or moving the same load.

POINT OF INTEREST

Although a large-diameter cylinder will require a lower pressure to move a certain load, the volume of fluid required to lift the load a certain distance will be greater than with a small-diameter cylinder. Designers of hydraulic systems therefore have to find the best compromise between using large volumes of fluid at low pressure and smaller volumes at higher pressure.

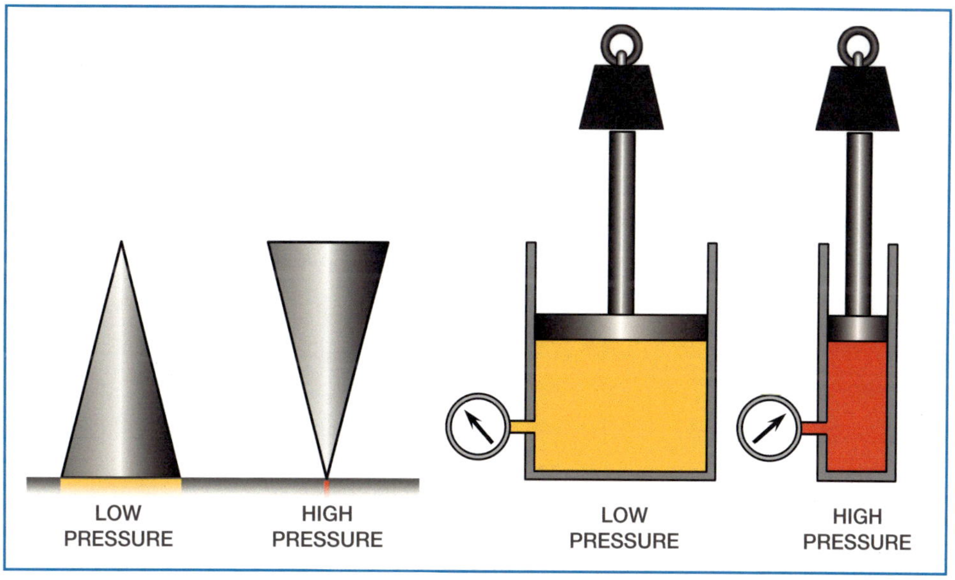

Fig. 1.9 Pressure is the distribution of force or weight over an area

POINT OF INTEREST

SI is the abbreviation for the French name Système International d'Unités – called in English the **International System of Units**.

DEFINITION

1 kilopascal (kPa) = 1000 Pa

1 megapascal (MPa) = 1,000,000 Pa

1 bar = 100,000 Pa = 100 kPa = 0.1 MPa

UNITS OF PRESSURE

How is pressure measured and calculated? Unfortunately, there is no standard unit of measurement that is used in every country around the world. Most countries have adopted the metric, or **SI system**, of measurement, but in others (primarily the USA) a system of units based on imperial measurements is still in common use.

The official SI unit of pressure is the **pascal** (symbol Pa), which is defined as a force of 1 **newton** (N) acting on an area of 1 square metre (m^2):

$$1 \text{ pascal (Pa)} = \frac{1 \text{ newton (N)}}{1 \text{ square metre } (m^2)}$$

A force of 1 N is approximately equivalent to the weight of a small apple, so 1 Pa is a very small amount of pressure. Therefore, the pascal is not a practical unit for hydraulic technology, where working pressures up to 40,000,000 Pa are not uncommon. More convenient units to use for hydraulic systems are **kilopascals** (kPa) and **megapascals** (MPa), which are equivalent to 1000 and 1,000,000 Pa, respectively.

However, when these units were first introduced, hydraulic engineers argued that the use of two different units would be confusing, and potentially dangerous if they were mixed up. They argued in favour of a common unit lying between kPa and MPa in size that would be called the **bar**. The definition of 1 bar is 100,000 Pa (10^5 Pa), so just one unit of pressure can be used for all hydraulic systems, which typically operate in the range up to 400 or 500 bar.

Many countries and equipment manufacturers (especially in Europe) have adopted the bar as the standard unit of pressure for hydraulic applications (although it is not an official SI unit). However, other countries still insist on using the official units of kPa and MPa.

Fluid pressure in bar can be determined from

$$1 \text{ bar} = \frac{1 \text{ newton} \times 10}{1 \text{ mm}^2}$$

Note the use here of the more convenient units of area, namely square millimetres (mm^2), rather than square metres (m^2).

Before we go further, it is worth looking at the difference between weight and mass. The **mass** of an object is a measure of how much matter it contains (measured in kilograms). The **weight** of an object is the force exerted on it by gravity (measured in newtons). The mass of an object will be the same on the Earth and on Mars, but its weight will be higher on the Earth (where gravity is stronger) than on Mars (where gravity is weaker). On Earth the acceleration due to gravity is about 9.81 m/s^2, so an object with a mass of 1 kg will weigh 9.81 N (its mass multiplied by gravitational acceleration). Unfortunately, confusion often arises due to the fact that the general public uses the kilogram as its common unit of weight.

An approximate calculation of pressure can be carried out by determining the load or weight in kilograms-force (kgf), where 1 kilogram (mass) weighs 1 kilogram (force), and determining the piston area in square centimetres (cm²):

$$1 \text{ bar} \approx \frac{1 \text{ kilogram (kgf)}}{1 \text{ cm}^2}$$

(Note: the symbol ≈ denotes 'approximately equal to'.) Convenient as this formula may be, it should be noted that it is an approximation, and kilogram-force and centimetres are not recommended SI units.

DEFINITION

1 bar = 14.5 psi

In the imperial system, the standard unit of pressure is the **pound per square inch (psi)**, and the equation for the determination of pressure is

$$1 \text{ psi} = \frac{1 \text{ pound (lbf)}}{1 \text{ in}^2}$$

Note that the imperial system uses the same unit (pound (lb)) for both mass and force (or weight), whereas in the SI system mass is measured in kilograms and force (or weight) is measured in newtons (where 1 kg mass weighs 9.81 N) (Fig. 1.10).

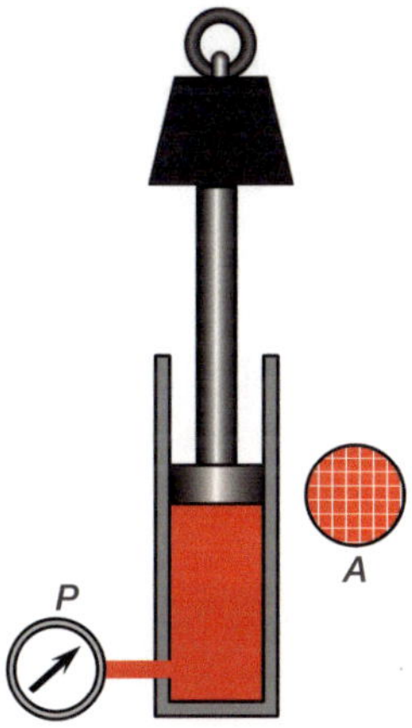

Metric SI units:

$$P(\text{Pa}) = \frac{F(\text{N})}{A(\text{m}^2)}$$

Practical units:

$$P(\text{bar}) = \frac{F(\text{N}) \times 10}{A(\text{mm}^2)}$$

Imperial units:

$$P(\text{psi}) = \frac{F(\text{lbf})}{A(\text{in}^2)}$$

Fig. 1.10 Pressure calculation

In some situations a distinction has to be made between **gauge pressure** and **absolute pressure**. Absolute pressure is measured relative to a perfect vacuum, which is taken as 0 bar, whereas gauge pressure is measured using the surrounding atmospheric pressure as the reference. Relative to a perfect vacuum, atmospheric pressure is approximately 1 bar (14.5 psi), although it varies both with atmospheric conditions and altitude above sea level. Pressure gauges measure the pressure relative to atmospheric pressure, hence the term 'gauge pressure'.

FORCE MULTIPLICATION

The French philosopher and mathematician Blaise Pascal formulated the principle that when pressure is created in a fluid it is transmitted undiminished in all directions and acts with equal force on equal areas at right angles to the area. He went on to propose a machine that would multiply a force by using two pistons of different areas.

Applying Pascal's principle to the arrangement shown in Fig. 1.11, if the right-hand piston has an area equal to 5 times the area of the left-hand piston and the pressure

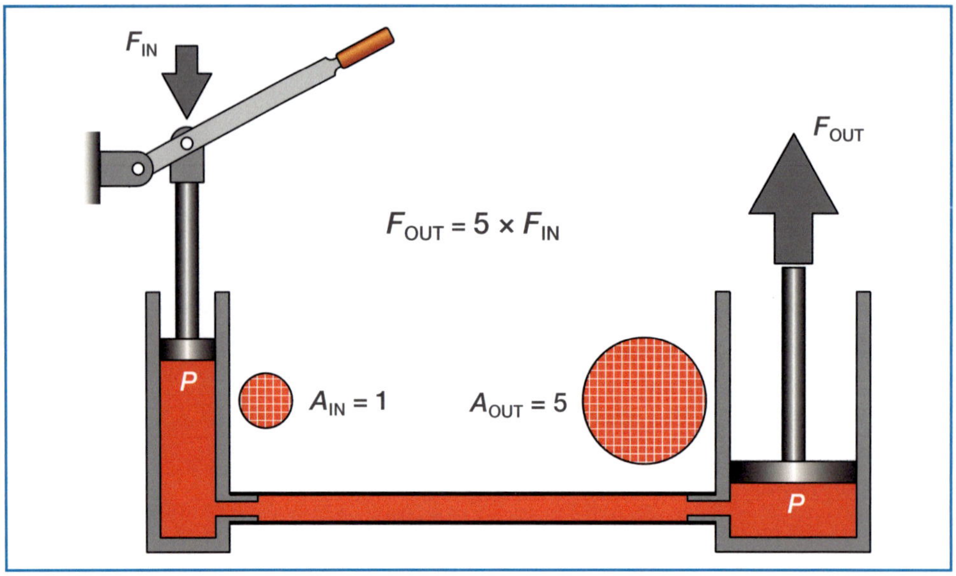

Fig. 1.11 Multiplication of force

is the same underneath both pistons, the force acting on the right-hand piston (F_{OUT}) will be 5 times as large as that on the left-hand piston (F_{IN}). So, if the left-hand cylinder is considered to be the pump and the right-hand cylinder the actuator, it can be seen that the force generated by the actuator will be 5 times as great as that applied to the pump piston.

In fact, this hydraulic multiplication of force is acting in the same way as the mechanical lever attached to the pump piston. If the lever arm ratio used on the pump piston is also 5 to 1, the total multiplication of force between the end of the pump handle and the actuator is 25 times (5 times from the mechanical lever multiplied by 5 times from the piston area ratio).

So, as well as transmitting the force from the pump handle to the piston of the actuator cylinder, the system is also multiplying the input force to provide a higher force at the actuator. This means that a relatively small input force on the pump handle can generate enough force to, for example, lift up a car in order to change a wheel. However, although the piston force has been multiplied by a factor of 5, the amount of movement achieved from the actuator will only be 1/5 of the input movement of the pump piston. This is because, although the volume of fluid pushed out of the left-hand cylinder is the same as the volume pushed into the right-hand cylinder, the larger area of the actuator piston means that the amount of movement required to accommodate that volume is less than the movement of the pump piston. So having completed a full stroke of the pump piston, the car will have hardly moved at all (Fig. 1.12).

POINT OF INTEREST

The multiplication of force that can be obtained in a hydraulic system can be thought of as **hydraulic leverage**.

Neglecting the effects of friction or other losses, the work input will be the same as the work output:

$$\text{Input force} \times \text{Input distance} = \text{Output force} \times \text{Output distance}$$

Furthermore, as soon as the pump handle is released, the weight of the car will push the fluid back into the pump cylinder again, and the car will fall back to its initial position. To make the system more useful it is necessary to enable the pump piston to make multiple strokes, which can be achieved by adding in two **non-return valves**

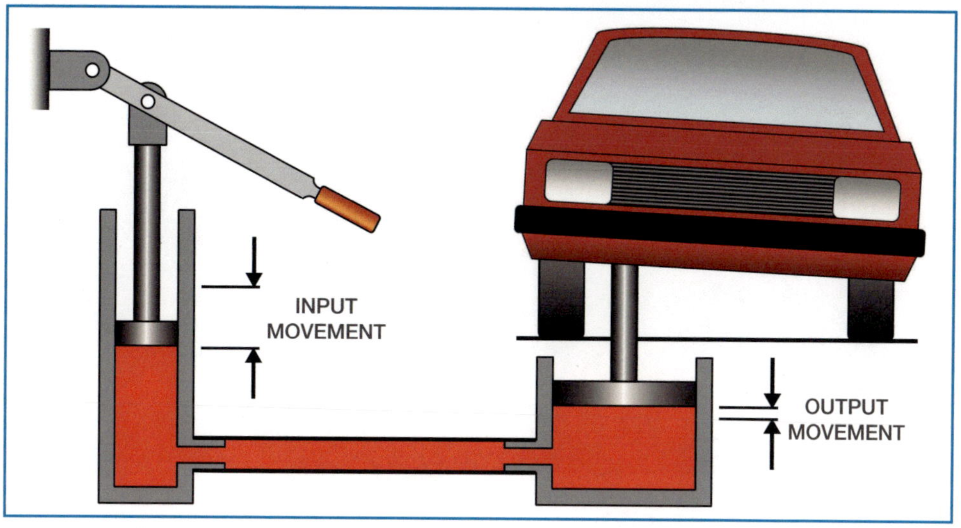

Fig. 1.12 Reduction of movement

and a fluid **reservoir**. As their name suggests, non-return valves allow fluid to pass through them in one direction but prevent fluid passing in the reverse direction.

In Fig. 1.13, valve A allows fluid to flow downwards (from the reservoir) and valve B allows fluid to flow from left to right (i.e. from the pump to the actuator). Having completed one pumping stroke and lifted the car up a small distance, the pump lever can be raised, allowing the pump cylinder to be refilled with fluid from the reservoir but holding the fluid in the actuator cylinder to prevent the car from dropping. This cycle can then be repeated as often as necessary in order to lift the car to the required height.

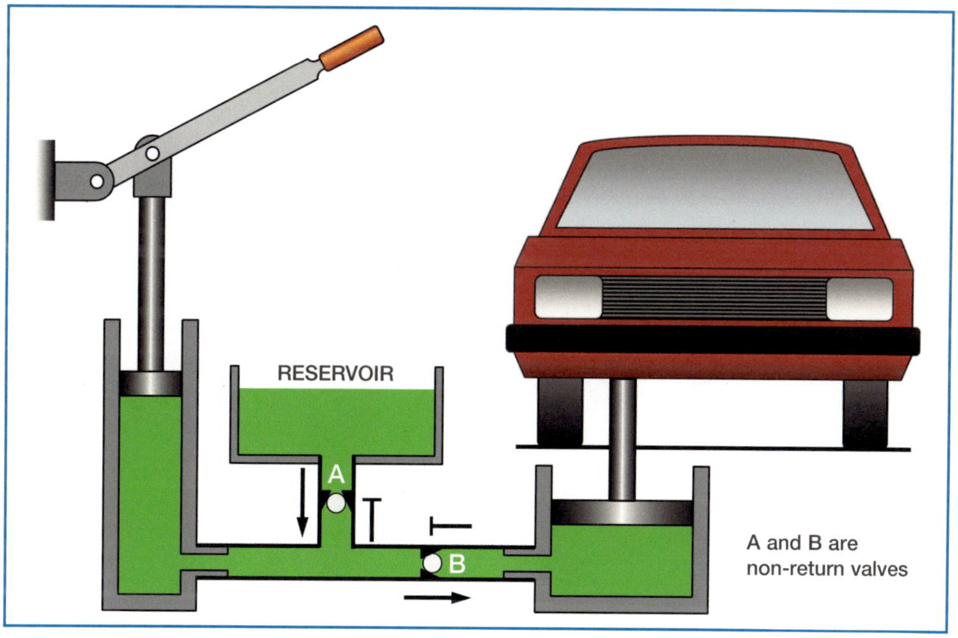

Fig. 1.13 Non-return valves

When it is required to lower the car back onto the ground, fluid must be allowed out of the actuator cylinder back into the reservoir. This requires the addition of a small **release valve** (simply an open or closed valve), which remains closed during the lifting of the car. In this application it is not necessary to pump the piston down, as gravity will act to lower the car and retract the piston once the release valve has been opened (Fig. 1.14).

POINT OF INTEREST

Non-return valves are also referred to as **check valves** or, sometimes, **one-way valves**.

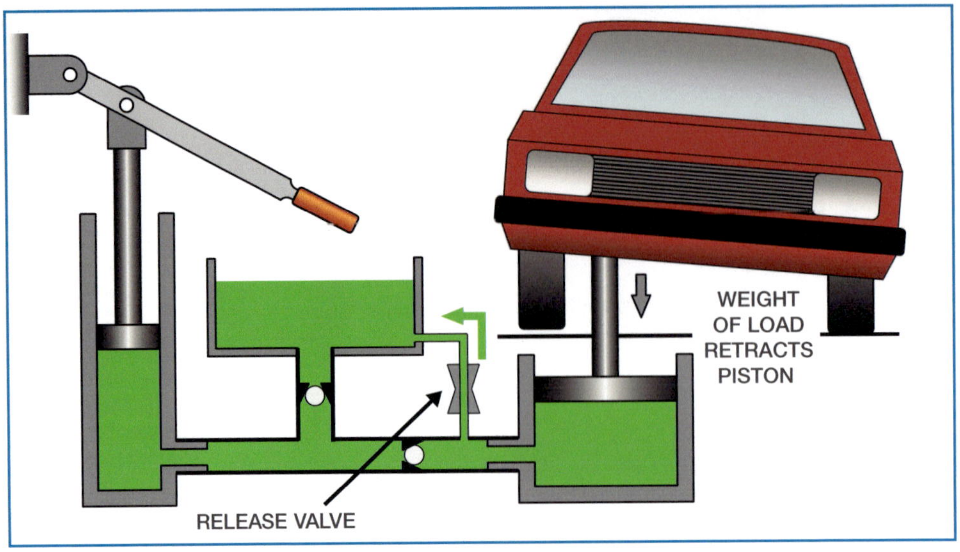

Fig. 1.14 Gravity lowering

If it is required to lift the whole car off the ground a sufficient distance for someone to work underneath it, in theory the same principle could be used. However, to lift the full weight of the car a larger actuator cylinder is required for the same amount of force input on the pump handle. This in turn means that each stroke of the pump will lift the car only a tiny amount, and therefore the system requires a lot of pumping to lift the car to the required height (probably about an hour for a reasonably fit mechanic).

As the amount of power a human being can produce over a sustained period is limited, a more practical solution is to use a power-driven pump, in this case probably using an electric motor as the prime source of power. Within reason, the electric motor can then be as large as necessary to lift the car at the required speed, which might take typically around 10 or 15 seconds.

POINT OF INTEREST

Human beings can create a power output of up to approximately 1 kilowatt (kW) (1.3 horsepower (hp)) for short periods of time (although trained athletes may create more). However, their sustained output over an hour or more is limited to approximately 100 W (0.13 hp).

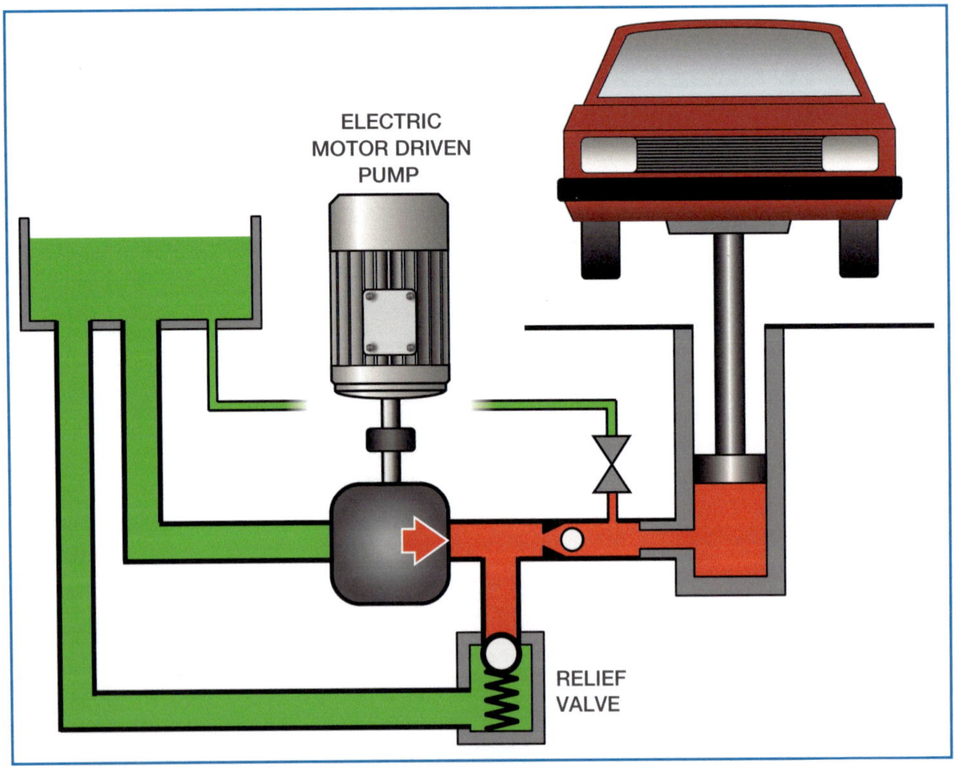

Fig. 1.15 Power-driven pump plus relief valve

When using a power-driven system, and even a hand-operated system in many cases, a **relief valve** is required in the system to prevent a dangerous build-up of pressure should a jam occur or should the operator try to lift a vehicle that is too heavy for the machine (Fig. 1.15). This is described in more detail in Chapter 2.

FLUID FLOW RATE

Compared with a manually operated system, a power-driven pump can produce a higher flow rate of fluid at the pressure required to lift the car. Flow rate is defined as the volume of fluid that is created by, or passes through, a component in a given period of time. The official SI unit of flow (cubic metres per second (m^3/s)) is not a practical unit for typical hydraulic systems, so the more convenient unit of litres per minute (L/min) is the one most commonly used:

$$1\,\text{L/min} = 0.0000167\,\text{m}^3/\text{s}$$

In the USA, the normal unit of flow rate is the **gallon per minute (gpm)**, although it should be noted that this is the American gallon, not the imperial gallon which was used in the UK for many years. The imperial gallon is equal to approximately 1.2 US gallons.

For very low flow rates (such as those used to define leakage) the units in the SI system are millilitres per minute (ml/min) or cubic centimetres per minute (cm^3/min), which are numerically the same, and the American standard unit is cubic inches per minute (in^3/min).

The amount of fluid flowing into an actuator in a given period of time (i.e. the flow rate) and the physical size of the actuator determine how quickly the actuator will move (Fig. 1.16). This is the case for both linear actuators (cylinders) and rotary actuators (motors).

POINT OF INTEREST

The use of upper-case 'L' to represent litres is often preferred to avoid confusion with the number '1'.

DEFINITION

1 cubic metre = 1000 litres

1 litre = 1000 millilitres
= 1000 cubic centimetres

1 US gallon = 3.785 litres

1 imperial gallon = 4.55 litres

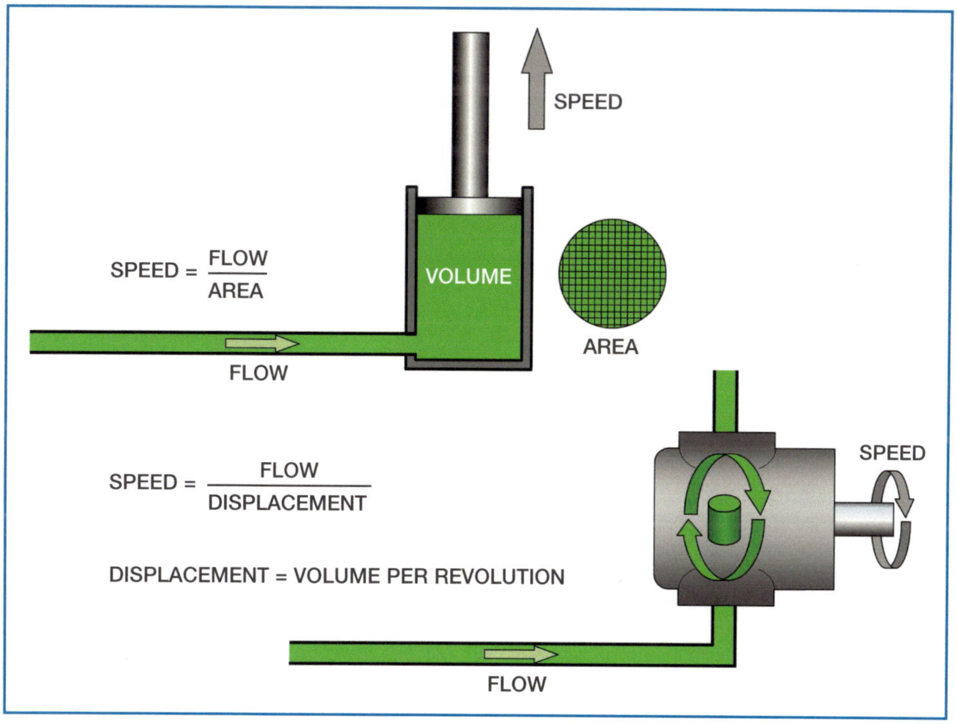

Fig. 1.16 Flow rate and actuator size determine speed

DEFINITION

The **displacement** of a hydraulic motor is defined as the theoretical volume of fluid required to turn the shaft one revolution (i.e. it ignores any leakage that may occur inside the motor). It is normally quoted in terms of cubic centimetres per revolution (cm^3/rev) or cubic inches per revolution (in^3/rev).

In certain cases, hydraulic motors can mechanically alter their effective size (**displacement**), either between two fixed values or by infinitely varying their displacement between minimum and maximum values. In the case of **variable-displacement motors** the output speed of the motor can change even though the fluid flow rate remains the same. An increased speed will, however, result in a proportionate reduction in torque output, and vice versa. This is a useful feature when a hydraulic motor is used to drive the wheels of a vehicle both on-road and off-road, for example. The motor can be used in high-speed/low-torque mode for on-road and low-speed/high-torque mode for off-road driving.

FLOW, PRESSURE AND POWER

In a hydraulic system the pump is normally the primary source of flow. Contrary to what is stated in many hydraulic texts, the pump is also the component that generates pressure in a hydraulic system. However, in order to do so it needs to have something to push against (i.e. some flow resistance). A pump producing flow with no restriction on its outlet (Fig. 1.17a) will not generate any significant pressure on its outlet port, and therefore not much effort is required to drive it. A similar situation arises when pumping up a bicycle tyre from flat – the effort required is quite small to start with when there is very little pressure within the tyre.

However, if the flow from the pump meets some resistance, for example by having to lift or turn a load (Fig. 1.17b), by having to push a poppet open against a spring in a relief valve (Fig. 1.17c), by compressing a gas in a device known as an accumulator (Fig. 1.17d), or simply by having to squeeze through a narrow opening or passageway in a valve (Fig. 1.17e), then pressure will build up at the pump outlet. Therefore, the best way to describe the situation is 'the system provides the resistance that enables the pump to generate pressure'.

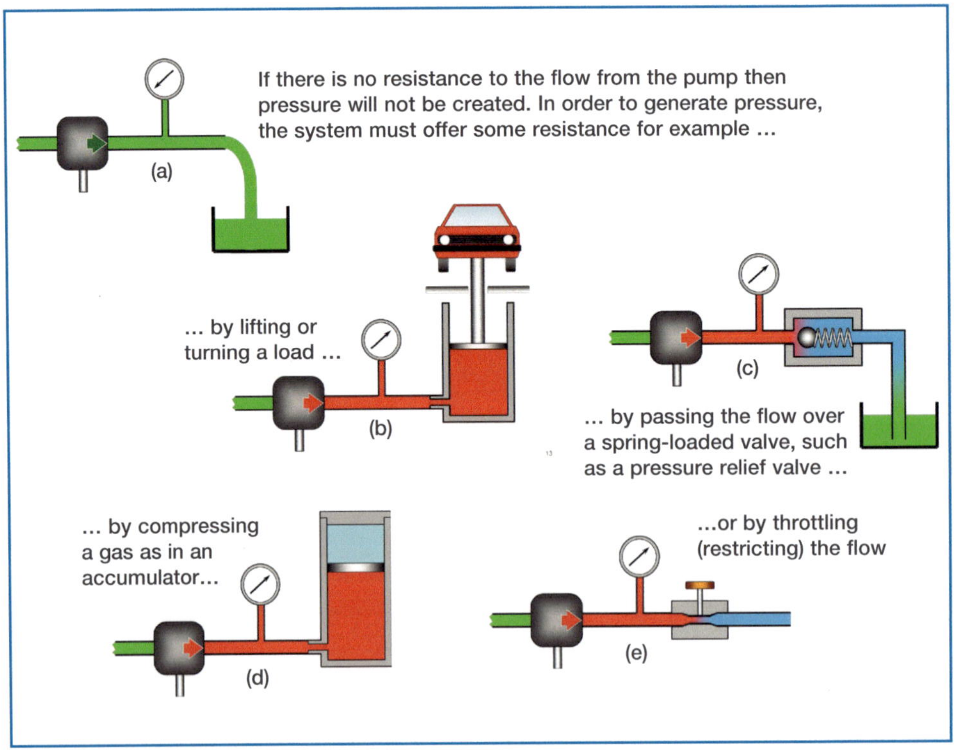

Fig. 1.17 Generating pressure in the system

As mentioned above in relation to the bicycle tyre, as the pressure on the pump outlet increases, the greater the pumping effort required. By 'effort' we mean power (or the rate of doing work) in mechanical terminology. So it follows that the power required to drive a pump is dependent not only on how much flow the pump produces but also on the pressure generated at the pump outlet. In SI units the convenient unit of power is the **kilowatt (kW)**. The basic formula for calculating pump drive power is

$$\text{Power (kW)} = \frac{\text{Flow (L/min)} \times \text{Pressure (bar)}}{600}$$

where 600 is a conversion factor to keep the units consistent.

This formula really calculates the amount of hydraulic power emerging from the pump, which if the pump is 100% efficient is also the amount of mechanical power required to drive the pump. In reality, however, the pump will have friction between its moving parts and will also have some internal leakage, so its overall efficiency would typically be closer to 80–85%. Incorporating this fact in the formula gives

$$\text{Pump drive power (kW)} = \frac{\text{Flow (L/min)} \times \text{Pressure (bar)}}{600 \times \eta}$$

where η is the **overall pump efficiency**, typically 0.85 or less (Fig. 1.18).

In the USA, power is measured in **horsepower (hp)** and the relevant formula is

$$\text{Pump drive power (hp)} = \frac{\text{Flow (gpm)} \times \text{Pressure (psi)}}{1715 \times \eta}$$

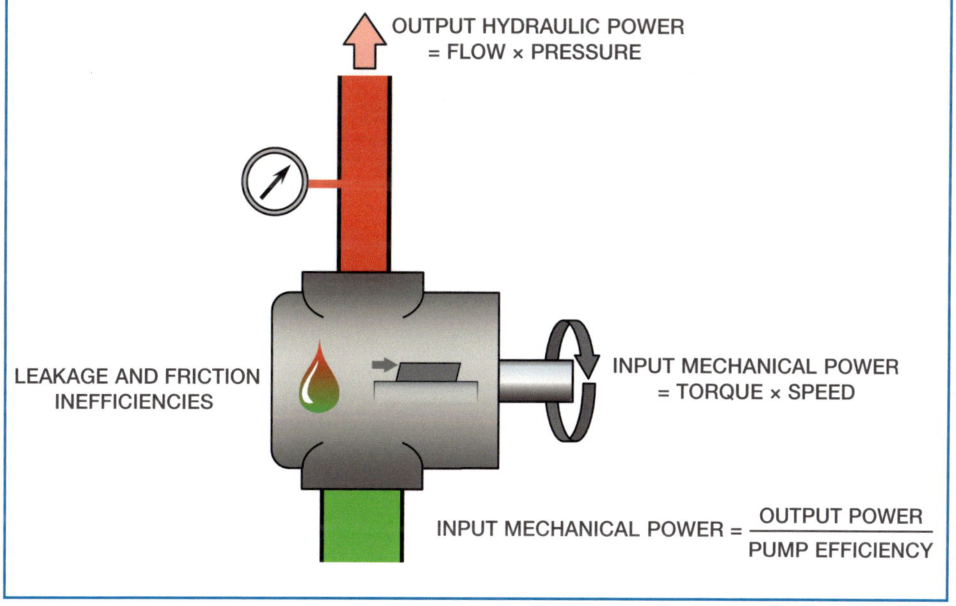

Fig. 1.18 Pump drive power

When the flow from the pump is lifting or turning a load, then useful mechanical work is being done. This means that power is being transmitted through the hydraulic system to operate the machine, with perhaps only a relatively small amount being lost along the way due to inefficiencies in the hydraulic components. However, when the flow performs no useful work, such as when it passes across a relief valve or

DEFINITION

The **pressure rating** of a pump (often quoted in manufacturers' catalogues) is not the pressure that the pump will automatically create, as this will be determined by the flow resistance. Rather, it is the maximum pressure that the pump should be allowed to generate in order to avoid mechanical damage to the pump components.

is throttled by a restriction, a significant amount of power loss can occur within the hydraulic system itself.

POINT OF INTEREST

An approximate rule of thumb is that for every 17.5 bar pressure drop across a component (throttle valve, relief valve, etc.) an oil-based hydraulic fluid will increase in temperature by approximately 1°C (1°F per 140 psi).

It is a fundamental law of physics that energy can neither be created nor destroyed, so the power that is 'lost' in the hydraulic system is simply turned into heat. Basically, all the mechanical power that is used to drive the shaft of the hydraulic pump has to be accounted for somehow or another – either in the form of mechanical power output from the cylinder rod or motor shaft, or in the form of waste heat (caused by fluid friction, pressure drops and leakage). The only exception to this is when hydraulic energy is stored temporarily in an accumulator, as will be described later.

Like many things in life, when offered alternative routes the flow always takes the path of least resistance (Fig. 1.19). So if the flow has access to two or more relief valves or two or more actuators, whichever valve or actuator route requires the least amount of pressure will be the one through which the flow passes, until such time as that route is closed off (such as by a cylinder reaching the end of its stroke).

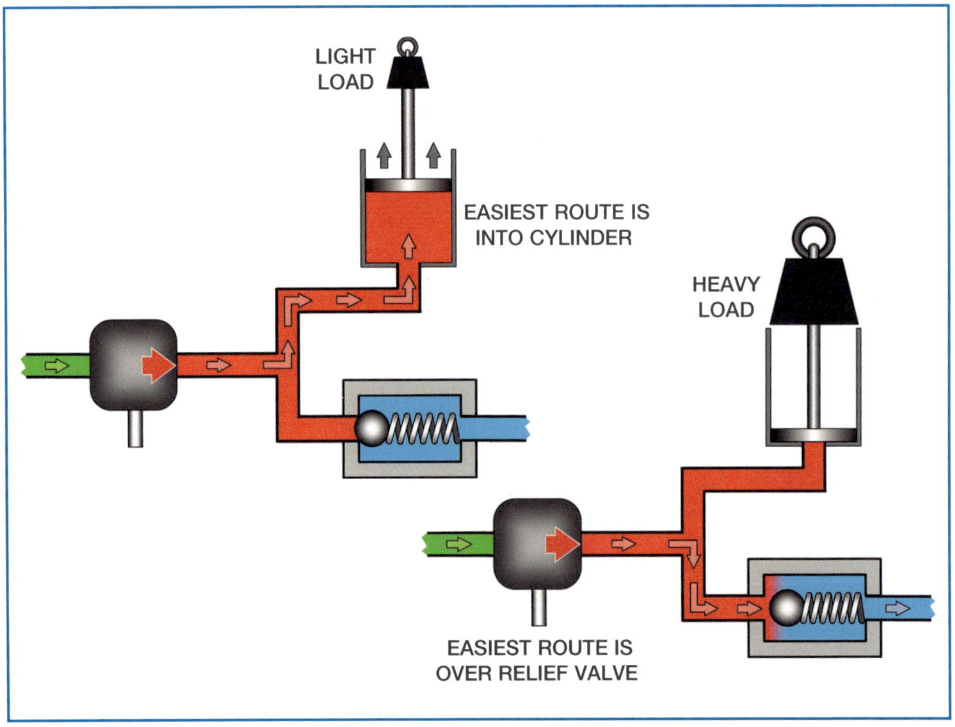

Fig. 1.19 Flow takes the easiest route

The relationship between the flow passing through a restriction and the pressure difference across it is, unfortunately, not a simple one. It is dependent not only on the size of the restriction (the cross-sectional area of the restriction) but also on its shape and the characteristics of the fluid.

The pressure drop across a sharp-edged restriction (such as an orifice plug) is proportional to the square of the flow rate. This means that to double the flow through a restriction requires four times the pressure difference across it. Where the restriction is not sharp-edged (such as through pipes, passageways in manifold blocks and valves), the relationship between the pressure drop and the flow rate is more complicated and depends very much on whether the flow is laminar, turbulent or something in between.

Laminar, or streamline, flow is characterised by an orderly flow profile parallel to the walls of the passageway (Fig. 1.20). In turbulent flow, eddies break up the regular flow profile and the flow tends to 'bounce around' within the passageway. As would be expected, a turbulent flow situation tends to create a higher pressure drop than does a laminar one. Therefore, where pressure drops need to be minimised, swept pipework bends and gently curved flexible hoses will prove better than sharp elbow joints or right-angled drillings in manifold blocks. Pressure drops through valves and other components can normally be estimated from the manufacturer's catalogue data, although corrections will often have to be made for the type and temperature of the fluid being used.

POINT OF INTEREST

There is no one point at which a flow suddenly changes from laminar to turbulent, and in many cases the flow will be partly laminar and partly turbulent. So, although pressure drops through a system can be estimated, precise determination is normally very difficult.

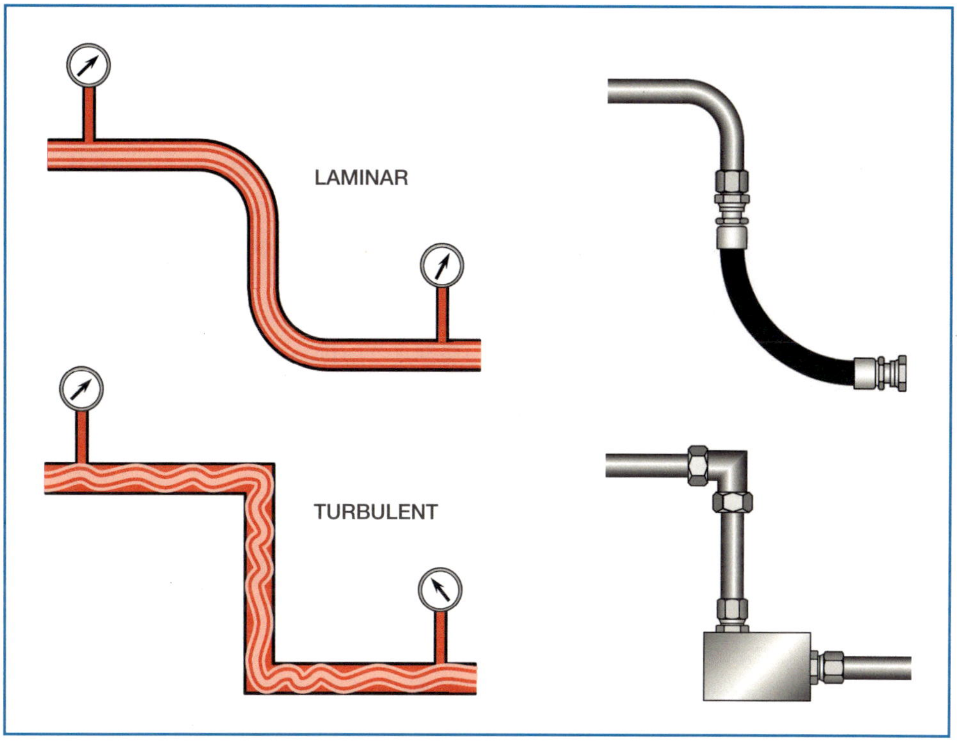

Fig. 1.20 Laminar and turbulent flow

Pressure in a fluid is also created simply by the weight of the fluid itself, and this is normally known as the static head of the fluid (Fig. 1.21). For example, the water pressure in the sea increases with depth, which is the phenomenon that limits the depth at which divers can safely operate or that submersibles can descend to. The same is also true with the fluid in hydraulic systems, although normally the heights or depths involved are much less.

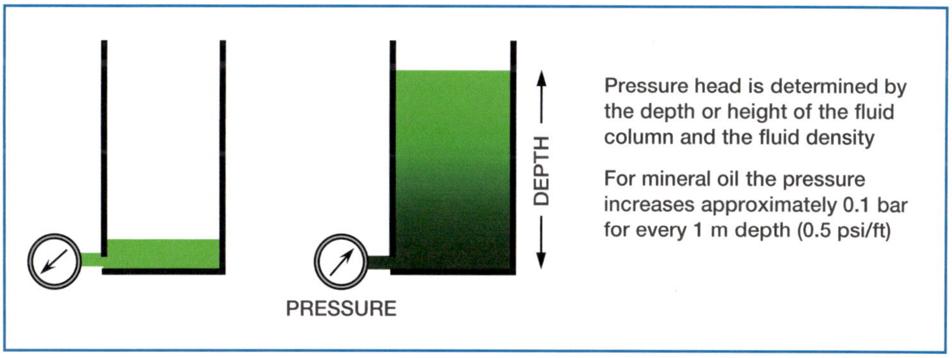

Fig. 1.21 Static head pressure

In most applications, where the difference in the height of the machinery is only a few metres, the variation in pressure due to the static head of the fluid can generally be ignored. The rule of thumb figure is a pressure variation of approximately 0.1 bar for every metre difference in height (0.5 psi/ft). However, there is one important part of a hydraulic system where the static head may have to be considered, which is on the inlet side of the pump. This is discussed below.

CAVITATION AND AERATION

If a hydraulic pump is mounted above the level of the fluid in the reservoir, the fluid has to flow 'uphill' to get into the pump. In order for this to happen, a negative pressure must be created at the pump inlet port. A 'negative' pressure is a pressure lower than the pressure on the fluid surface, which in normal situations is atmospheric pressure (approximately 1 bar (14.7 psi)).

At normal atmospheric pressure a hydraulic fluid will naturally dissolve a certain amount of air, but if the fluid pressure is reduced (e.g. in the inlet port of a pump), some of the air will come out of solution and form bubbles in the fluid. A similar situation occurs when removing the top from a bottle of sparkling water (although in this case the bubbles will be mainly carbon dioxide).

In addition, as the pressure of the hydraulic fluid is reduced, the fluid will tend to vaporise, creating more bubbles – this time bubbles of fluid vapour. The bubbles then travel through the pump from the low-pressure inlet port to the high-pressure outlet port, where the air bubbles dissolve back into solution and the vapour bubbles condense back to liquid form.

This condensation process (known as **cavitation**) is very violent and can cause extremely high pressures at the points in the fluid where the vapour bubbles collapse (Fig. 1.22). The pressures created are high enough to fatigue the surface of metal components, causing noticeable erosion and contamination particles.

POINT OF INTEREST

Atmospheric pressure is nothing more than the static head of the air in the atmosphere pushing down on the Earth's surface. It will thus vary slightly from one location to another and also reduce with increasing altitude above sea level.

WARNING

The maximum negative pressure (or vacuum) allowable at the pump inlet port will be stated by the manufacturer and should not be exceeded under any working conditions (e.g. even for cold starts). In fact, with certain types of pumps and/or fluids, no vacuum at all is allowed (i.e. the inlet pressure must be at least atmospheric pressure).

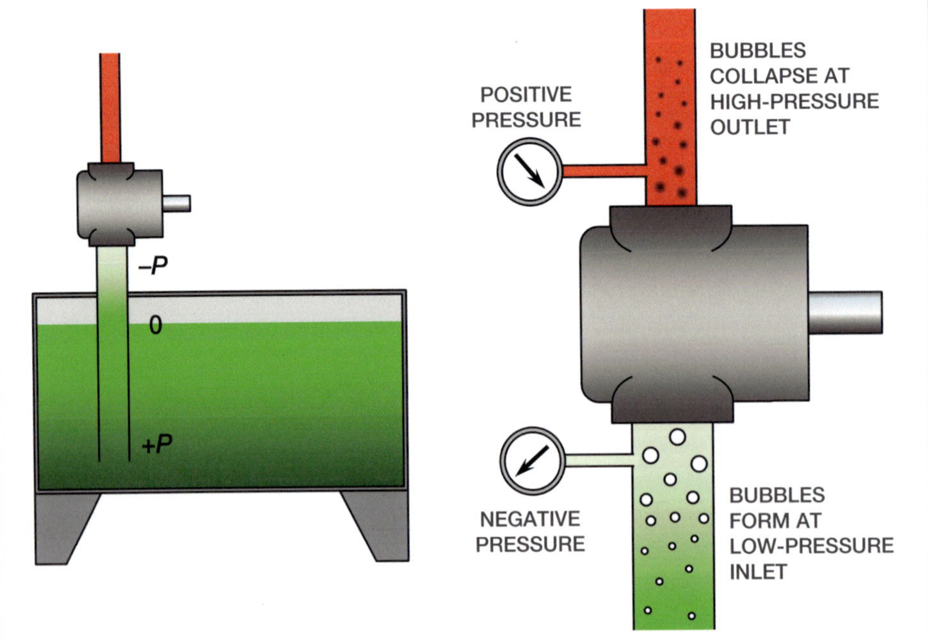

Fig. 1.22 Pump inlet cavitation

Severe cavitation can destroy a pump in a relatively short period of time, but even mild cavitation, occurring perhaps at start-up when the fluid is cold, will reduce the service life of the pump. Cavitation is characterised by noise at the pump outlet that is often described as sounding like small stones passing through the pump. Certain pumps and certain fluids are more susceptible to cavitation than others, so the manufacturer's recommendations should always be followed with regard to the maximum allowable vacuum on the pump inlet port and maximum pump drive speeds.

The inlet connection to a pump is very important, and every effort needs to be made to reduce the pressure drop in the inlet line (e.g. by increasing the pipe size or by reducing the number of bends). A reservoir mounted above the level of the pump will also reduce the risk of cavitation occurring.

A closely related phenomenon to cavitation is aeration, which is where a leak on the suction side of the pump allows air to be drawn into the fluid. This is more likely to occur in applications where the pump is mounted above the reservoir, such that the inlet fluid will be at a pressure slightly lower than atmospheric. As with cavitation, bubbles of air are drawn into the pump and travel round to the high-pressure outlet port, where some of the bubbles dissolve and some are compressed into much smaller volumes, causing similar damage and symptoms to those of cavitation. Aeration is also likely to cause frothing of the fluid in the reservoir when the pressure is released, and may also cause **dieseling** of the fluid.

Dieseling occurs with mineral oil fluids when air bubbles are compressed to high pressures, such as at the outlet port of a pump. The bubbles of air heat up rapidly as they are compressed and, as they are surrounded by a flammable fluid, a micro-explosion occurs (in much the same way as in the compression ignition effect in a diesel engine). Effectively, the fluid 'burns away' internally. It becomes acidic, darkens in colour and develops a characteristic smell. The lubrication properties of the fluid deteriorate as a result, and so, as with cavitation, dieseling is something that needs to be eliminated from the system through good maintenance procedures.

TYPICAL APPLICATIONS OF HYDRAULIC FLUID POWER

Hydraulic fluid power is used just about anywhere on land, sea or in the air (and even in space).

Construction machines, such as the excavator shown in Fig. 1.23, use hydraulics for all the machines' main functions. While hydraulic cylinders operate the boom, arm and bucket, hydraulic motors are commonly used to swing the body of the machine and also to drive the tracks, wheels and auxiliary attachments. Material-handling vehicles such as fork-lift trucks and telehandlers are another common application of hydraulics, where precise control combined with a high load capability are the main requirements.

Industrial applications, such as plastics and die-casting machines, again make use of the rapid and accurate control capability of hydraulic power, while presses benefit from the almost limitless amount of force that can be both generated and precisely controlled.

TOP TIP

A vacuum gauge (or test point) on the inlet port of a pump is useful for determining if cavitation is occurring within a pump.

DEFINITION

Cavitation damage is caused primarily by the violent collapse of fluid vapour bubbles at the outlet (high-pressure) port of a pump. The main cause of cavitation is a restricted inlet flow to the pump.

Aeration damage is caused by air bubbles dissolving or collapsing at the outlet port of a pump. The main causes of aeration are air leaks in the suction line or foam in the fluid reservoir.

TOP TIP

Hydraulic reservoirs on mobile vehicles (particularly off-road vehicles) may be susceptible to frothing of the fluid caused by the vehicle's movement. Suitable **baffles** may therefore be required within the reservoir to prevent this causing problems.

Fig. 1.23 Typical applications for hydraulic power transmission

In recent years, wind turbine generators have emerged as a new application for hydraulics, where again the precision of the hydraulic control is used to vary the blade pitch and apply the brakes when necessary. Many other civil engineering projects, such as flood barriers, swing bridges and cranes, exploit the capability of hydraulic systems to operate in harsh environments and in all weathers. Hydraulic systems are also used to power and control many theme park attractions and simulator rides.

Despite being a technology that can be traced back well over 200 years, hydraulic fluid power is a modern, up-to-the-minute technology that is ideally suited to interface with sophisticated digital electronic controls. However, like any technology, to be successfully applied it needs to be well understood so that if things do go wrong the fault can be diagnosed and rectified as quickly and accurately as possible.

FURTHER READING

For further information, including an updated list of training centres, white papers, free of charge checklists, videos and calculation apps, go to www.webtec.com/education

CHAPTER

2 HYDRAULIC COMPONENTS AND SYSTEMS

A BASIC INDUSTRIAL HYDRAULIC SYSTEM

Suppose that having inspected a car, as described in Chapter 1, we have come to the realisation that it is beyond economic repair. Having removed any saleable parts and hazardous materials and fluids, the remaining shell is sent off to the car crusher for recycling (Fig. 2.1).

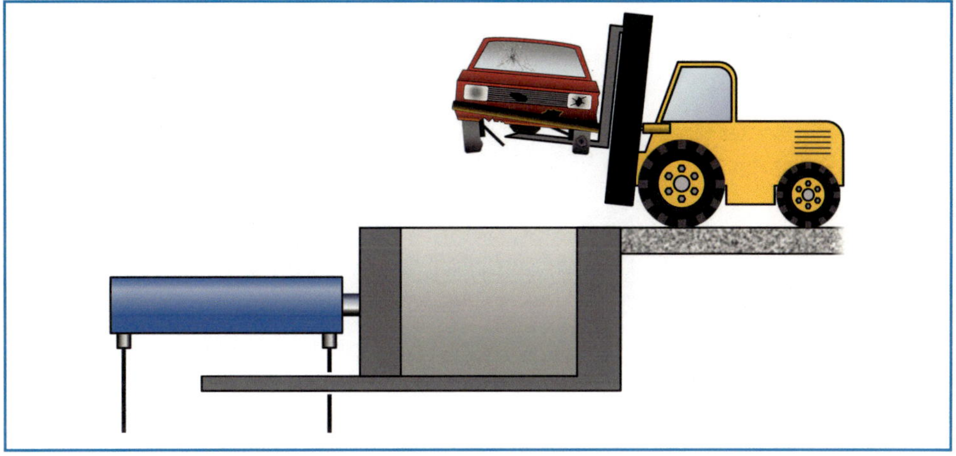

Fig. 2.1 Car crusher

The mechanical force required to crush the body of a car is obviously considerably greater than that required to simply lift it off the ground. So, although it would be theoretically feasible to operate a car crusher with a simple hand pump, in practice a power-operated pump will inevitably be required. However, starting at the actuator end of the system, the first hydraulic component required is a hydraulic cylinder to create the force output.

CYLINDERS

Although they are available in many different forms, hydraulic cylinders are relatively simple components from the point of view of their operation. As shown in Fig. 2.2, the cylinder consists of a cylindrical tube with a cap at either end. Inside the tube is a closely fitting piston connected to a piston rod that protrudes from one end cap

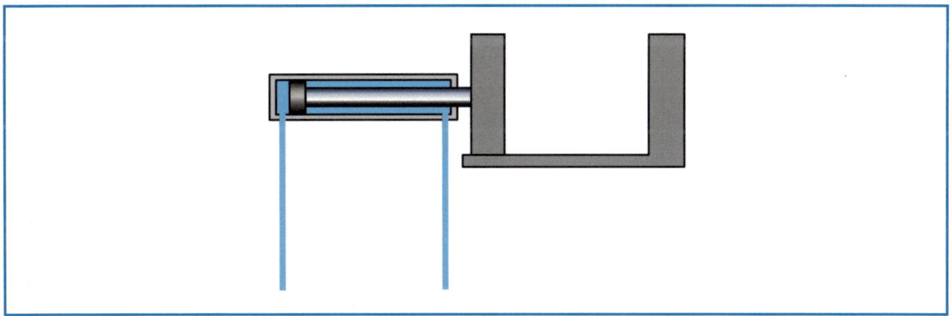

Fig. 2.2 Linear actuator (cylinder)

> **DEFINITION**
>
> **Industrial** (or stationary) systems consist of machinery that does not normally move around. Typically this applies to factory machinery such as presses, moulding machines and machine tools.
>
> Machinery designed to move around during its normal operation is classified as **mobile** machinery. Typical examples are fork-lift trucks, agricultural machines and earth-moving equipment.

of the cylinder. A port at each end of the cylinder enables fluid to enter or exit the cylinder, depending on which way the piston is moving.

The two sides of the cylinder are usually referred to as the **full-bore** or 'cap' end and the **annulus** or 'rod' end. The terms 'full bore' and 'annulus' refer to the area of the piston over which the fluid pressure will act. On the left-hand side in Fig. 2.3 the full piston area is exposed to the fluid pressure, whereas on the right-hand side only a ring-shaped (or annulus-shaped) area has pressure acting on it. The area difference between the two sides is equal to the cross-sectional area of the piston rod. When designing a hydraulic system, engineers use graphical symbols to represent components as shown in Fig. 2.3 and the following illustrations. Symbols and circuit diagrams are discussed in more detail later in this chapter.

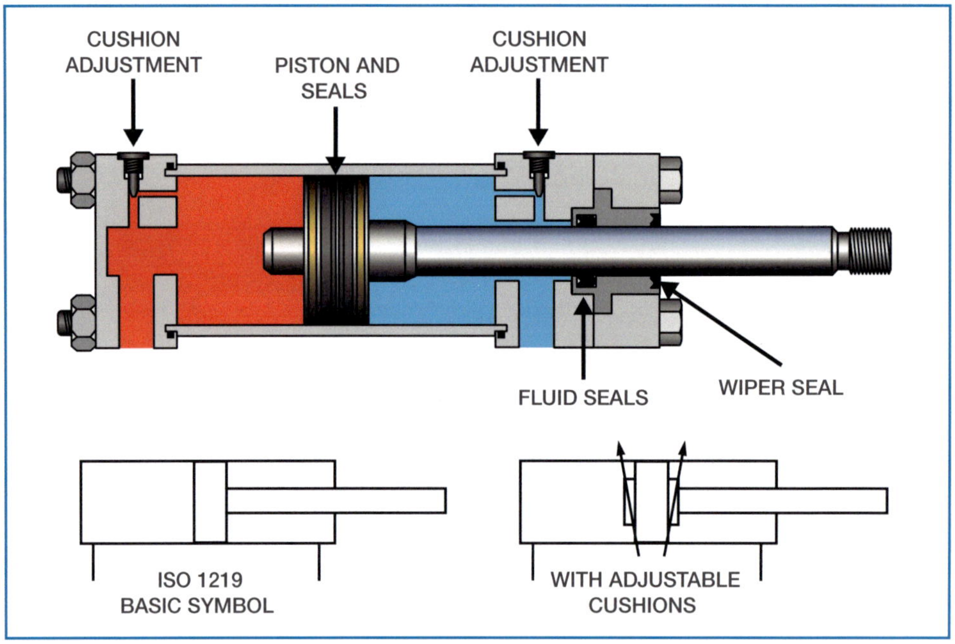

Fig. 2.3 Tie-rod cylinder (Image courtesy of Eaton Corp.)

The cylinder piston incorporates seals to positively seal each side of the piston from the other and prevent leakage of fluid. Normally the seals are made of an elastomeric (flexible) material such as synthetic rubber, and are constructed such that their sealing effect increases as pressure is applied to them. For example, a simple V-section seal, or **chevron seal**, tends to seal more effectively when pressure is applied to the inside of the 'V'. However, many different types of seal are used in practice, depending on the type of fluid to be used, the working pressure, the operational speed, the working temperature of the cylinder, and other possible requirements such as low friction. Also, two sets of back-to-back seals are often used on a cylinder piston to provide effective sealing, whichever side is under pressure.

Rod seals are also incorporated in the front cap to prevent fluid leaking out of the annulus side of the cylinder when pressurised, along with **wiper seals**, which are designed to prevent contamination being drawn into the cylinder when the piston rod retracts.

Cushions can be incorporated in one or both ends of the cylinder. They consist of tapered spigots or collars, which gradually restrict the exhaust flow from the cylinder as it approaches each end of its stroke. Adjustable needle valves then control the final creep speed to the fully extended or fully retracted position. As their name suggests, cushions soften the shock that might otherwise occur when a fast-acting cylinder moving a heavy load reaches the physical end of its stroke.

As cylinders are a 'dead end' from the fluid's point of view, any air in the fluid tends to collect in the cylinder, so **air bleed screws** are often incorporated to enable the air to be bled out.

There are many different types of cylinder construction for different applications (Fig. 2.4). For relatively light-duty applications (e.g. on agricultural machinery) and where serviceability is not of prime importance, **welded construction cylinders** are a popular, low-cost choice. The standard choice for many industrial applications is the **tie-rod cylinder**, in which the end caps are attached to the cylinder barrel by means of four or more tie-rods that run the length of the cylinder. As well as providing flexibility in its manufacture, the tie-rod cylinder is also relatively simple to service, (e.g. changing piston or rod seals).

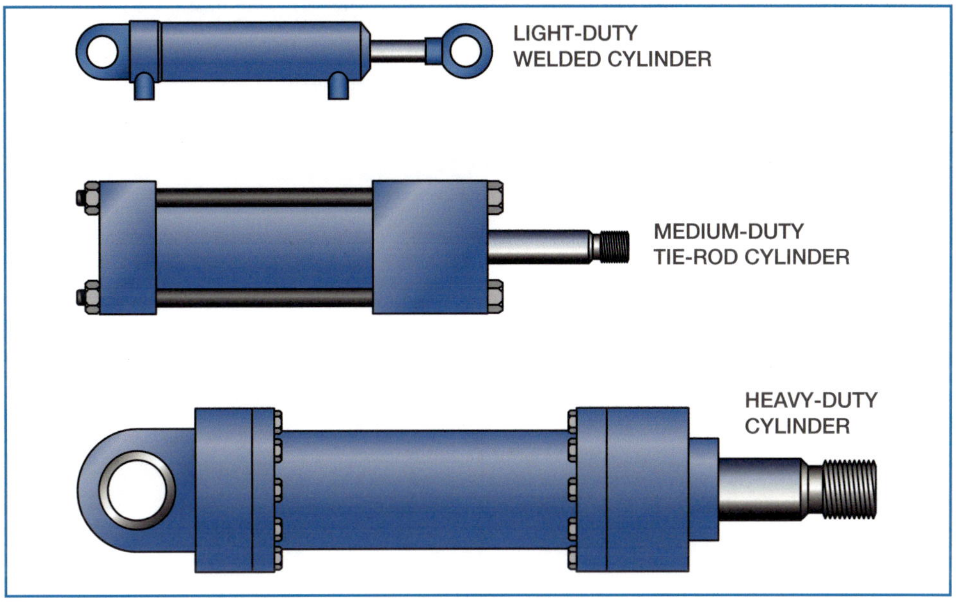

Fig. 2.4 Cylinder types

Heavy-duty mobile and industrial applications tend to use cylinders built specifically for a single application, and frequently use bolted-on end caps to provide a construction that is rugged but still serviceable.

There are also many other options that can be built into cylinders. For example:

- **Double-rod cylinders** have a piston rod at each end and equal areas on either side of the piston. Such cylinders are used for applications that require symmetry (e.g. vehicle steering).
- **Limit switches** or **position transducers** can be built into the cylinder to provide either a signal at a certain piston position or a continuously variable signal proportional to the piston stroke. The latter type are often used in closed-loop position-controlled applications where positioning of a load to a high degree of accuracy is required (e.g. plastic blow moulding machines, material testing machines and industrial robots).

TOP TIP

Air in a hydraulic fluid tends to create an erratic or 'spongy' movement of an actuator and can also cause foaming in the reservoir. Whenever maintenance work is carried out on a system involving the removal of components, it may be necessary to bleed air from the system in order to ensure satisfactory operation.

- **Gaiters** can be used to prevent contamination coming into contact with the cylinder rod when extended.
- **Stop tubes** are used to prevent the cylinder piston travelling to the very front of the cylinder. This provides additional resistance to the bending of long-stroke cylinders when fully extended.
- Numerous mounting configurations are available, depending on how the cylinder must be attached to the machine functions it will operate.

For an application such as a car crusher, heavy-duty cylinders would be the most likely choice, being sized according to the force output required at the desired working pressure of the hydraulic system. When selecting suitable components for a specified level of force output, a hydraulic system designer has a choice of either a small cylinder operating at high pressure or a large cylinder operating at a lower pressure, and any number of combinations in between. A low-pressure system will require physically larger components, while the components in a high-pressure system have to be rated for the higher pressure. In practice, therefore, the decision is often made on the basis of user preference, past experience or component availability. For example, until comparatively recently, aircraft hydraulic systems always operated at 210 bar (3000 psi) by tradition.

TOP TIP

Be aware that, when measuring the flow rate from a cylinder, because of the difference in area between the full-bore and annulus sides of the cylinder the exhaust flow rate from the full-bore side when retracting will be greater than the ingoing flow rate to the annulus side.

FIXED-DISPLACEMENT PUMPS

Having chosen a suitable cylinder, the next component required is a pump of some description to operate it (Fig. 2.5).

DEFINITION

Nearly all pumps used in hydraulic systems can be classified as **positive-displacement pumps**. This means that very little slippage can occur from outlet to inlet within the pump (unlike a centrifugal pump for example). For each turn of the shaft a certain volume of fluid will be pushed out, so care must be taken to ensure that the pump outlet is never completely blocked, otherwise damage may occur.

This is the hydraulic component that has probably the greatest number of options from which to choose, including:

- size (displacement and physical dimensions)
- maximum pressure rating
- maximum drive speed
- type of construction
- fluid compatibility
- noise level
- serviceability
- efficiency
- life expectancy
- cost.

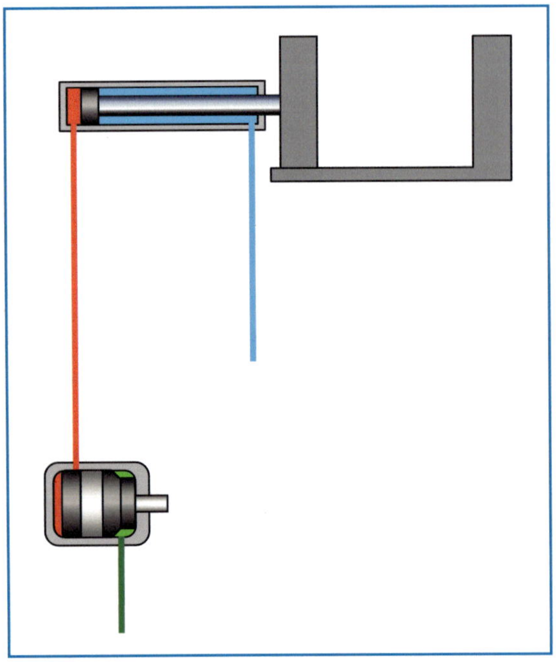

Fig. 2.5 Pump

Over the years, many different types of pump construction have been developed. The three main types in use today, however, are gear, vane and piston pumps.

External gear pumps

After the hand pump, probably the simplest hydraulic pump is the external gear pump, which has just two moving parts, namely two intermeshing gear wheels.

One gear wheel (the top one in Fig. 2.6) is driven round by the pump drive shaft and the other wheel (the bottom one in Fig. 2.6) rotates because it is in mesh with the driven gear. On either side of the two gears are closely fitting side plates, which are normally pressurised against the side faces of the gears to minimise the leakage path. Both gears then rotate (normally on plain bearings) inside a housing, which again is fitted closely around the tips of the gear teeth as they rotate.

POINT OF INTEREST

Most gear pumps use spur gears, although some are available with helical gear teeth, which tend to reduce their noise level.

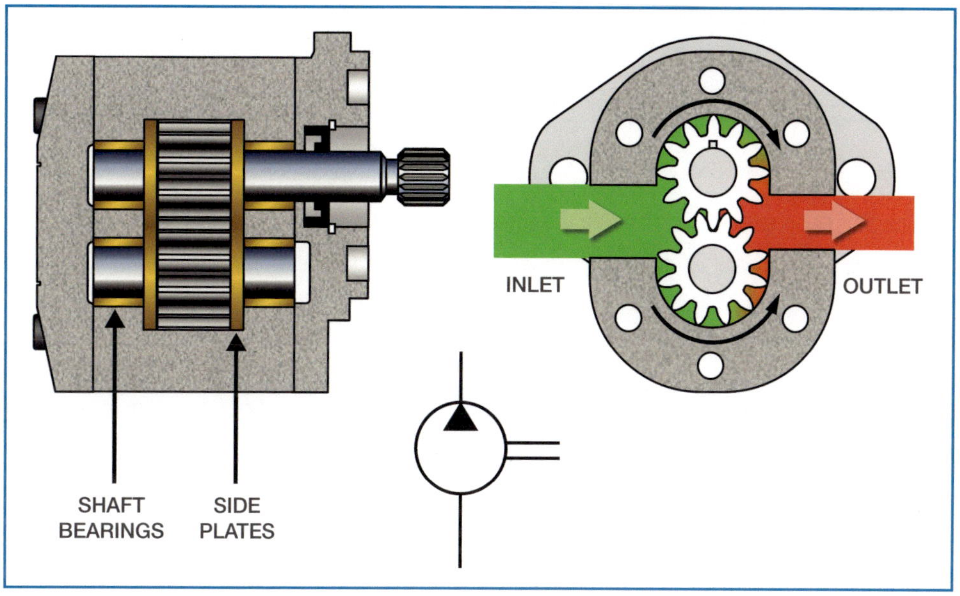

Fig. 2.6 External gear pump

On the inlet side of the pump the fluid fills the spaces between adjacent gear teeth, the pump housing and the side plates. It is then carried round the top and bottom of the two gears to the outlet port, where the gear teeth mesh, forcing the fluid out of the spaces between the gear teeth, thereby creating a flow of fluid from the outlet port. The amount of flow will be determined by the physical size of the pump components (diameter and width) and also the drive speed of the pump shaft.

Typical characteristics of external gear pumps can be summarised as follows:

- they are more tolerant of poorer fluid contamination levels
- failures tend to be a gradual loss of efficiency rather than a sudden, catastrophic failure
- they are compact and lightweight
- it is simple to configure multiple pumps
- they are readily available
- they are inexpensive
- the gears are side-loaded by out-of-balance pressure forces, thus limiting indirect drive possibilities (e.g. belt drive)
- they are normally not economical to repair
- they are generally noisier than some other types of pumps.

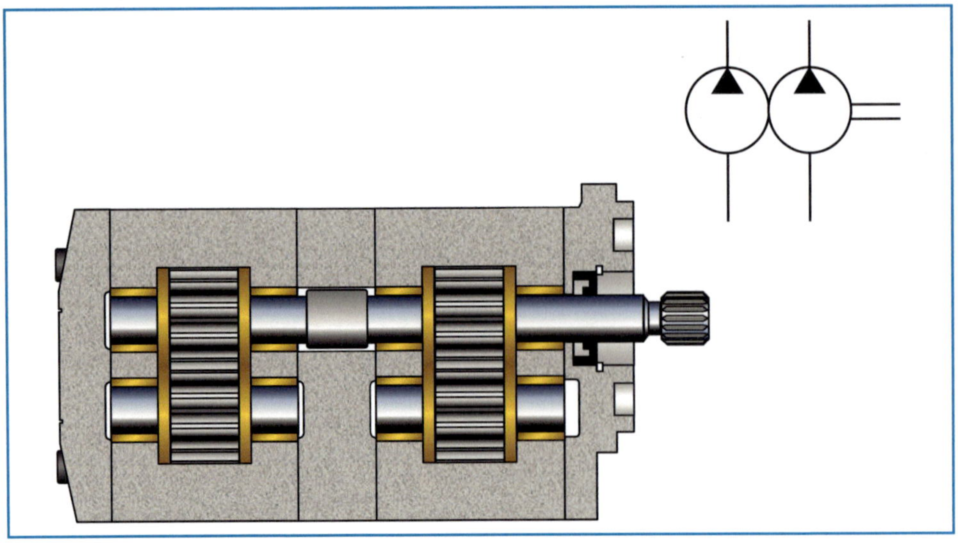

Fig. 2.7 Double gear pump

Figure 2.7 illustrates a double gear pump arrangement where the shaft is extended to drive a second set of gears, thus providing two independent pumps driven by a common shaft. Provided the shaft has sufficient torsional strength, this principle can be extended to three, four or even more sections relatively easily.

External gear pumps are ideal for low- to medium-pressure applications, especially those that do not operate continuously, such as agricultural equipment, fork-lift trucks and man-lifts. Although there are many simpler machines in industrial applications that use gear pumps, they are more commonly found in mobile applications, for which their characteristics are ideally suited.

POINT OF INTEREST

Multiple pumps can be used to separate a hydraulic system into independent circuits. For example, one circuit on a vehicle could be used for the steering and a second for the auxiliary functions. This ensures that the steering cannot be 'robbed' of flow by the other functions.

Internal gear pumps

The internal gear pump operates on a similar principle to that just described but has one external gear and one internal gear, which are separated by a crescent-shaped segment (Fig. 2.8). The inner gear is driven by the pump shaft and meshes with

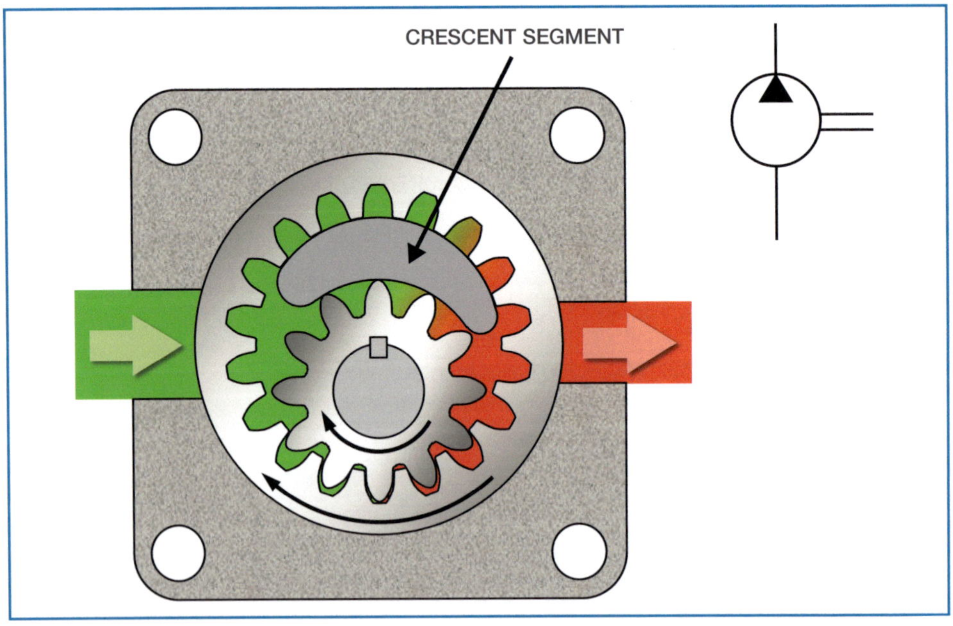

Fig. 2.8 Internal gear pump

the outer gear to drive it round within the pump housing, but this time the gears rotate in the same direction. As before, fluid entering the inlet port is carried round in the spaces between the gears to the outlet port area, where the meshing of the gears forces the fluid out of the pump. Although internal and external gear pumps are similar in principle, the characteristics of internal gear pumps tend to differ as follows:

- they are generally much quieter than external gear pumps
- they are available at higher pressure ratings than external gear pumps
- they are normally constructed for continuous, industrial applications
- they are more expensive than external gear pumps.

Other characteristics, such as the possibility of constructing multiple pump configurations, are similar to those of external gear pumps. Internal gear pumps are available for low-, medium- and high-pressure systems, and are commonly used in applications where low noise is an important consideration (e.g. die-casting machines, metal-working machinery and ship-borne systems).

Vane pumps

Vane pumps derive their name from a series of sliding vanes (typically 10 or 12) fitted into a rotor that is driven round by the pump drive shaft (Fig. 2.9). The rotor and vanes rotate within a cam ring, which is approximately elliptical in shape. As with the gear pump, side plates are pressure loaded against the sides of the rotor and vanes to ensure a minimal leakage path.

As the pump is driven up to speed, centrifugal action throws the vanes out of their slots so that the tips of the vanes follow the profile of the cam ring. The vanes slide in and out of their slots twice per revolution. As each vane passes an inlet port, the space between the rotor and the cam ring is increasing, so fluid is drawn in to fill the space. A quarter of a revolution later, the vane is passing an outlet port, where

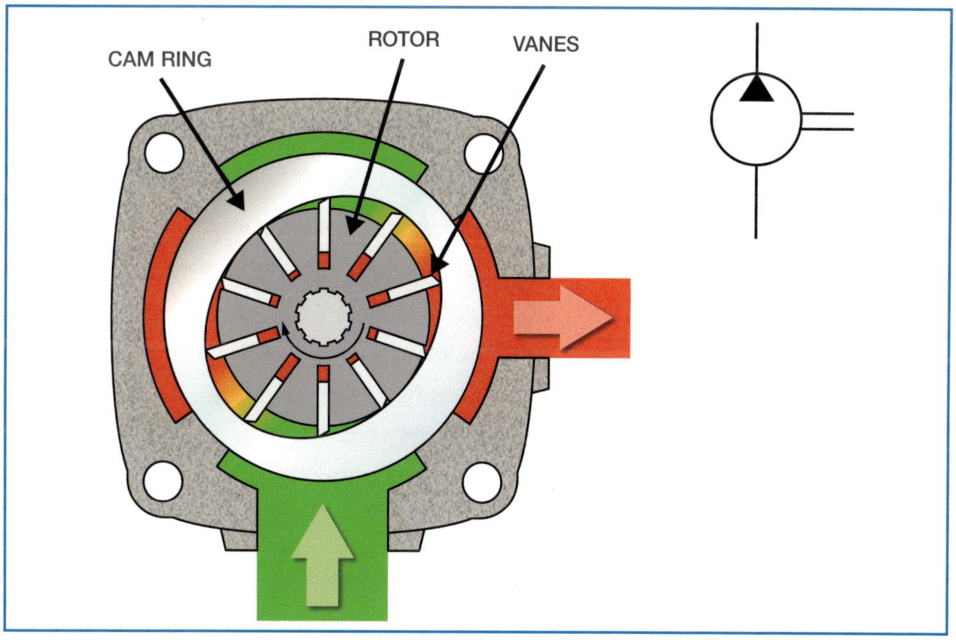

Fig. 2.9 Balanced vane pump

WARNING

Vane pumps can often be assembled for either right-hand or left-hand rotation. If a pump has been stripped down, therefore, it is important to reassemble it correctly for the drive rotation being used. Right-hand rotation is defined as clockwise, when looking at the shaft end of the pump.

the ring-to-rotor clearance is reducing, so fluid is squeezed out of the port. Exactly the same process then occurs on the second half of the revolution. Therefore, there are two pumping actions for each turn of the shaft. In practice, the two inlet ports and the two outlet ports are connected together within the body of the pump, thus providing single inlet and outlet connections.

As the pump starts to deliver fluid and generate pressure on the outlet port, the outlet pressure is connected via the side plates to the slots underneath the vanes. This hydraulically biases the vanes outwards, ensuring that the vane tips remain in contact with the cam ring. The reason for using the elliptical cam ring with two inlets and two outlets is to balance the pressures on either side of the rotor. Therefore, unlike gear pumps, the vane pump is pressure balanced and generates no side load on the shaft or bearings.

Smaller sizes of vane pump have a simple construction where the cam ring is normally the central section of the pump body (Fig. 2.10).

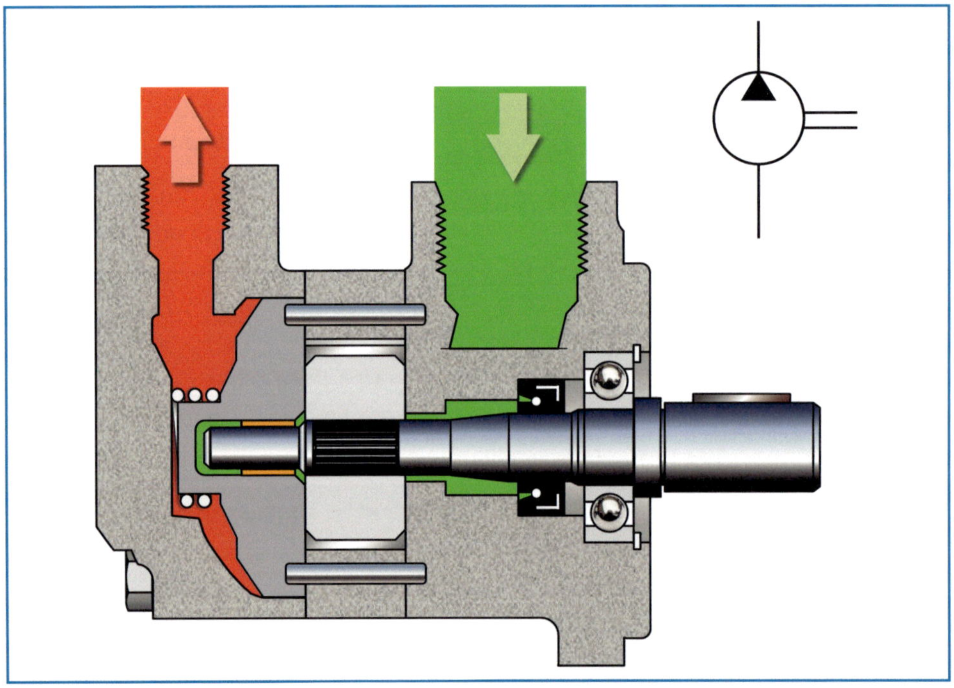

Fig. 2.10 Small-series vane pump (Image courtesy of Eaton Corp.)

Larger vane pumps (Fig. 2.11), which are often rated to higher output pressures, tend to contain small vane inserts or pins to reduce the area exposed to the outlet pressure under the vane. This ensures that the vanes are still biased outwards but not with too high a force, which would cause rapid wear on the vane tip. Such pumps also incorporate the ring, rotor, vanes and side plates as a self-contained sub-assembly, which is usually referred to as a pump cartridge. This makes the pump very serviceable, as a worn pump normally requires only a cartridge change. This is a relatively quick and simple process, which can often be carried out with the pump still in place.

As with gear pumps, double and triple vane pumps are readily available (Fig. 2.12), and with a through-shaft option it is possible to build different combinations of multiple pump assemblies.

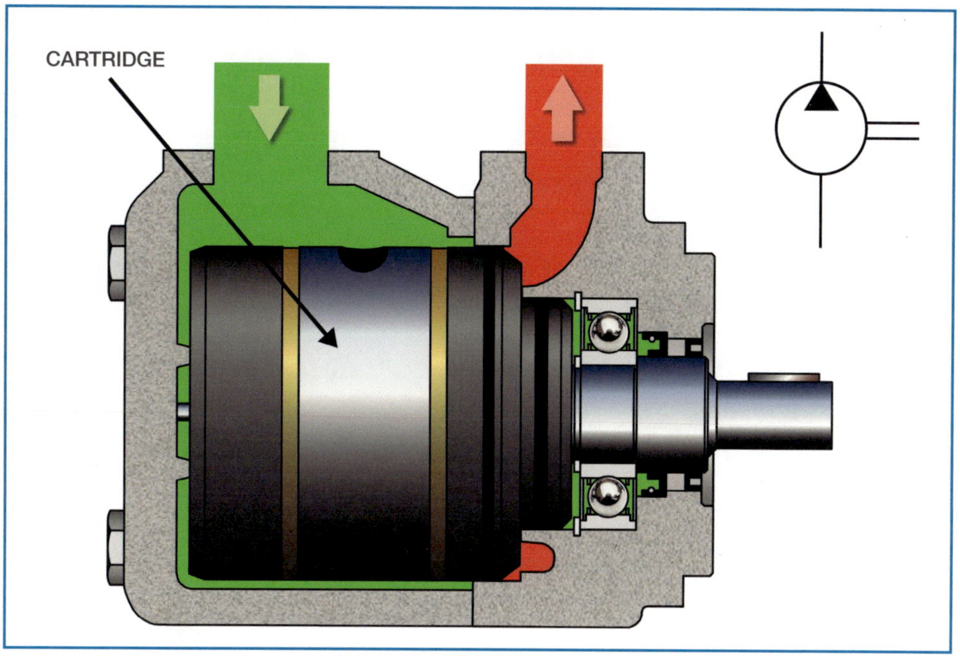

Fig. 2.11 Large-series vane pump (Image courtesy of Eaton Corp.)

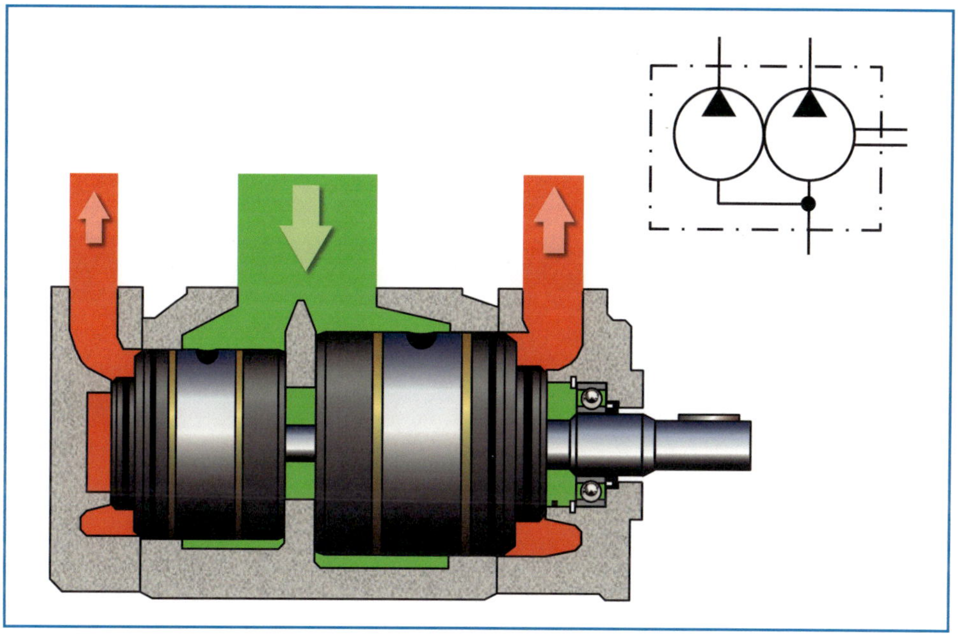

Fig. 2.12 Double vane pump (Image courtesy of Eaton Corp.)

Typical characteristics of vane pumps are:

- the gradual transition from inlet to outlet pressure provides a relatively low-pulsation, quiet pump
- the internal-pressure-balanced design provides a comparatively long service life
- cartridge-design vane pumps are easily serviced in the field
- the direction of pump rotation can be changed easily
- they are less tolerant of poor inlet conditions or contaminated fluid than are gear pumps
- they are more expensive than gear pumps.

WARNING

Pump cartridges in a double pump are normally arranged back to back, so it is important that their drive rotation is assembled accordingly.

Vane pumps are a popular choice for medium-pressure applications in both industrial and mobile applications. They have been used extensively in the power steering of both passenger and commercial vehicles for many years. They are also commonly found on construction and utility vehicles, baling presses and plastics machinery.

Piston pumps

There are several different designs of piston pumps, the axial and bent-axis designs being the two most common.

The **axial piston pump** consists of a number of sliding pistons (typically nine) fitted into bores in a rotating cylinder block (Fig. 2.13). The spherical front end of each piston has a **shoe** or **slipper** swaged over it, which is held in contact with an angled **swashplate**. At the opposite end of the cylinder block is a stationary **valve plate**, which incorporates kidney-shaped inlet and outlet ports.

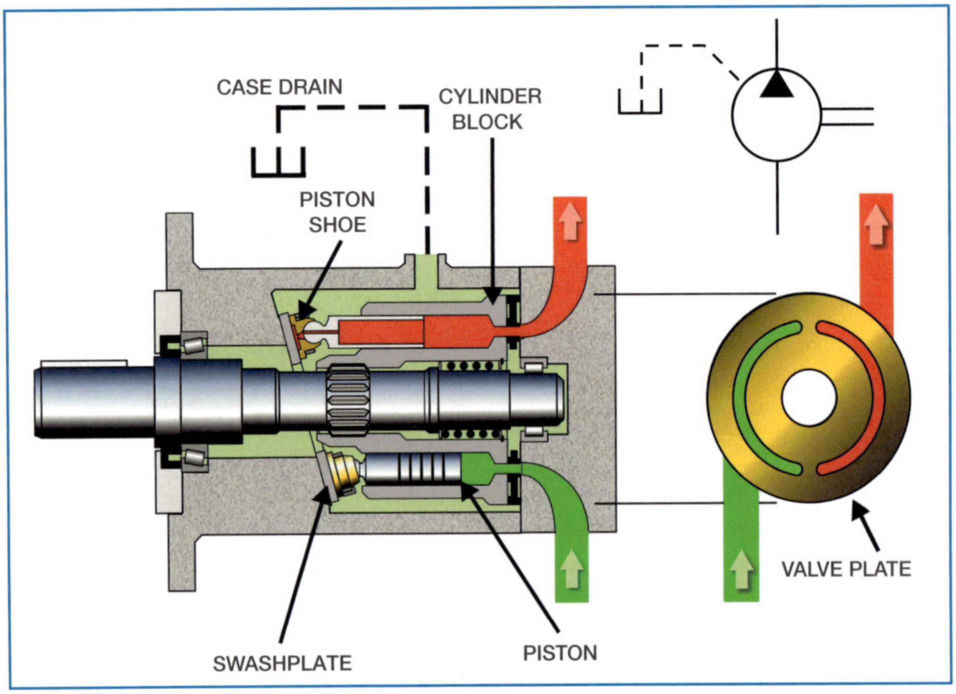

Fig. 2.13 Axial piston pump (Image courtesy of Eaton Corp.)

As the cylinder block and pistons are rotated by the drive shaft, the pistons are reciprocated within their bores, moving in and out once per revolution. As each piston is being retracted from its bore, the connection in the cylinder block is passing the inlet port of the valve plate, thus drawing in fluid to fill the bore behind the piston. During the second half of the cycle, the pistons are being pushed back into their bores by the angle of the swashplate, and at this time they are connected to the outlet via the kidney port of the valve plate, thus creating flow from the pump.

Unlike gear and vane pumps, the internal leakage of a piston pump is drained into the pump case and then back to the reservoir via a separate **case drain connection**. (The internal leakage in gear and vane pumps is simply directed back to the pump inlet port.) The fluid in the casing of a piston pump helps to lubricate all the moving parts, so it is important that when a new pump is installed the case is filled with clean fluid before the pump is first started. It is also important to ensure the pump drain lines are large enough so that excessive pressure is not created in the pump case.

WARNING

Always fill up the cases of piston pumps with clean fluid before they are started for the first time. Internal leakage will then keep the case topped up when the pump is running, and provide the necessary lubrication.

A second common type of piston pump is the **bent-axis piston pump** (Fig. 2.14). This pump works on a similar principle to the one described previously, except that the complete cylinder block is tilted at an angle to the drive shaft. The spherical end on each piston connecting rod is attached to a flange on the drive shaft so that, as the drive shaft and cylinder block are rotated, the pistons are alternately pulled out and pushed into their bores to create the pumping action.

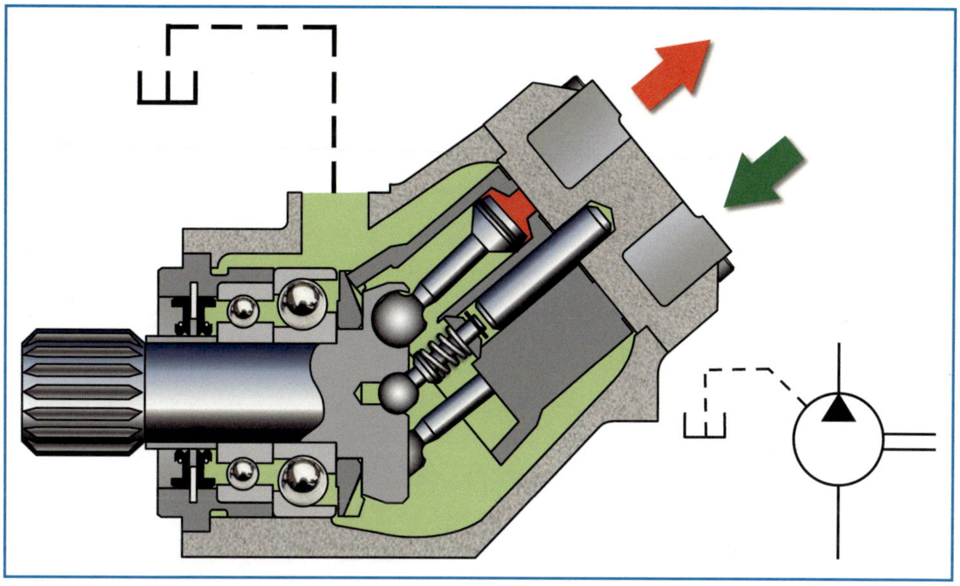

Fig. 2.14 Bent-axis piston pump

A significant advantage of the bent-axis design is that there is no sliding action of the shoes or slippers against a swashplate as in the axial pump. This provides a very compact and robust design capable of generating high pressures.

Summarising the characteristics of piston pumps, therefore:

- they are suitable for high-pressure and high-flow operation
- they are more tolerant of high-water-based fire-resistant fluids than other types of pumps
- they have a long life expectancy provided that operating conditions are well maintained (fluid condition, etc.)
- they can be serviced (but not quite as simply as cartridge-type vane pumps)
- the bent-axis design has no through-drive capability (for multiple pump applications)
- they are expensive.

As mentioned, piston pumps are commonly used when the system pressure requirements are higher than those achievable with gear or vane pumps (typically 250 bar (3600 psi) and higher). However, the good efficiency and durability of piston pumps also make them suitable for applications at lower pressures where continuous operation of the hydraulic system is required.

In many applications the use of a variable-displacement (as opposed to fixed-displacement) pump is beneficial in terms of system efficiency. This is discussed later in this chapter.

POINT OF INTEREST

Bent-axis piston pumps provide a large amount of power from a very small component. They are said to have a high **power density**.

ELECTRIC MOTORS

In most applications the choice of prime mover to drive the hydraulic pump will be either an electric motor or a diesel engine. As mentioned previously, electric drives are usually used for industrial applications, while diesel engines provide the drive for most mobile machinery.

DEFINITION

AC stands for **alternating current**.

The frequency of an AC supply is measured in **hertz (Hz)** (equivalent to cycles per second).

Until relatively recently, electric drive motors (Fig. 2.15) for hydraulic pumps were mostly single-speed three-phase induction motors. The rated speed of the motor is determined by the number of poles (internal windings) and the frequency of the AC electrical supply. Common speeds in Europe (where the electricity supply frequency is 50 Hz) are 1000, 1500 and 3000 rpm, with 1500 rpm being the most common for hydraulic pump drives. In areas where the supply frequency is 60 Hz (e.g. North America), the corresponding speeds are 1200, 1800 and 3600 rpm. Electric motors are characterised by having high starting torques, although starting off-load may be advantageous in terms of the maximum current draw at start up (and hence the size of the operator's electricity bill).

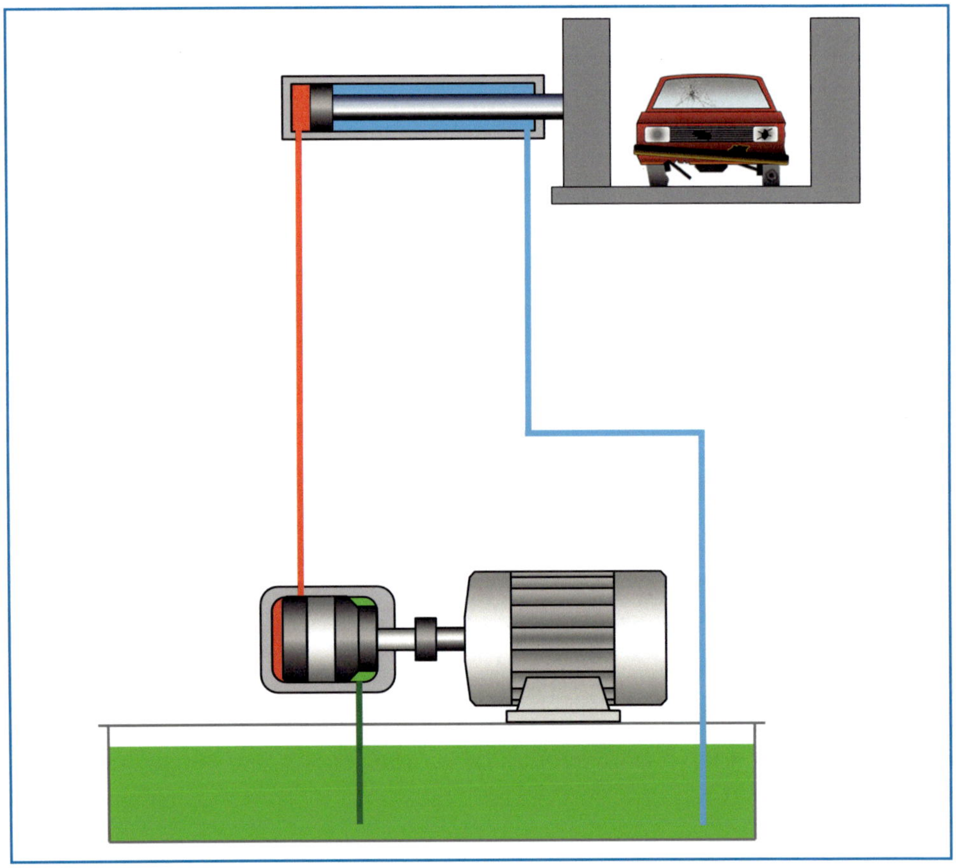

Fig. 2.15 Electric motor and reservoir

Recently, variable-speed drives have become increasingly common as the cost of the control electronics required has reduced. Variable-speed electric motors for simpler applications have basically the same construction as fixed-speed versions, the speed variation being achieved by varying the frequency of the electrical supply by means of a controller (Fig. 2.16). Controllers of varying degrees of sophistication are available for different applications, depending on the requirements.

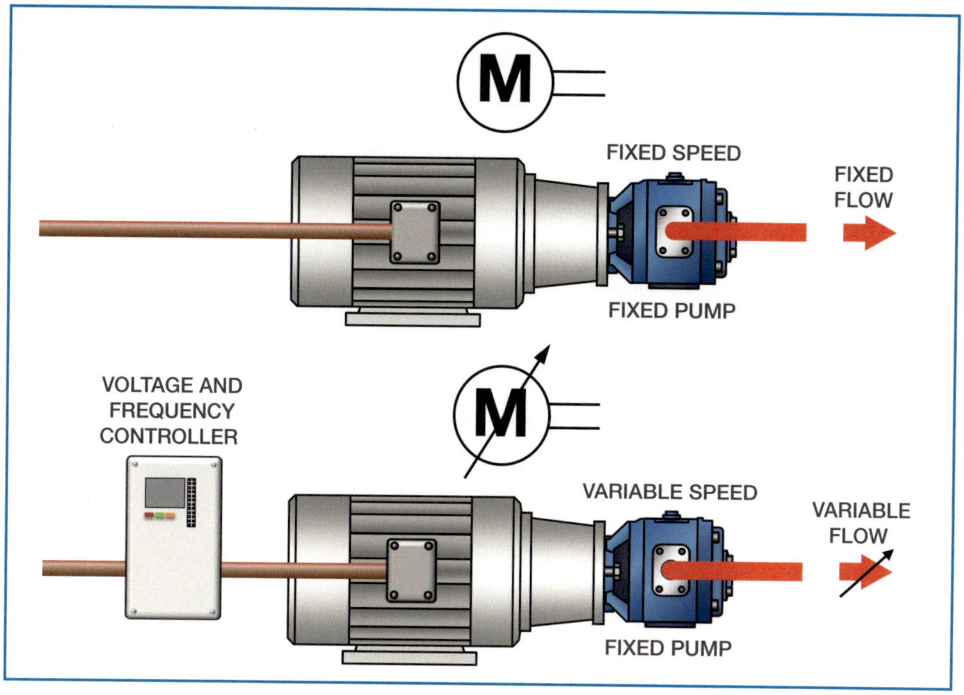

Fig. 2.16 Electric motors

The benefit of a variable-speed drive is that the pump speed and hence flow rate can easily be varied to match the load requirements. This provides a more efficient system and may be an alternative to using a variable-displacement pump.

Where control requirements are more demanding (e.g. when drive is required over a very wide speed range, including down to zero speed), **permanent magnet AC drive motors** operating via a closed-loop electronic controller can be used.

DIESEL ENGINES

The pump drive from diesel engines can be either direct, where the engine drives only the hydraulic pump, or via a **power take-off (PTO)**, where the pump drive is shared with other vehicle functions (e.g. a mechanical transmission) (Fig. 2.17).

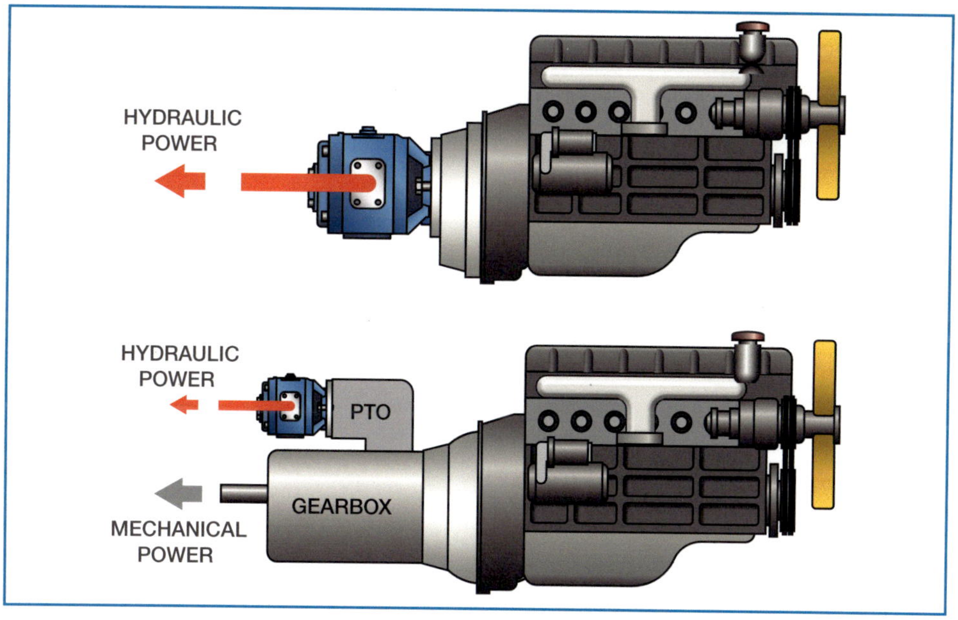

Fig. 2.17 Diesel engine pump drive options

A PTO is basically a gearbox, so the pump drive speed can be either less than, equal to or greater than the engine speed, depending on the gear ratio chosen. A PTO can also be permanently engaged or include a clutch mechanism for disengaging the pump drive when the hydraulic systems are not required.

Unlike electric motors, diesel engines have a relatively low starting torque, so being able to unload the pump at start-up is a useful procedure, especially when operating in very cold environments. Unloading a pump may involve reducing its output flow or pressure, or both, in order to reduce the drive power required. The drive speed of a diesel engine can be varied simply by opening or closing the throttle. However, both torque and power output will vary with the engine speed, so most systems will be designed to operate over a relatively narrow speed range either side of the maximum power or torque speed.

RESERVOIRS

The next component of the hydraulic system to consider is the fluid reservoir or tank. The size and construction of the reservoir will vary between industrial and mobile systems (for practical reasons). The following description applies primarily to an industrial system (Fig. 2.18).

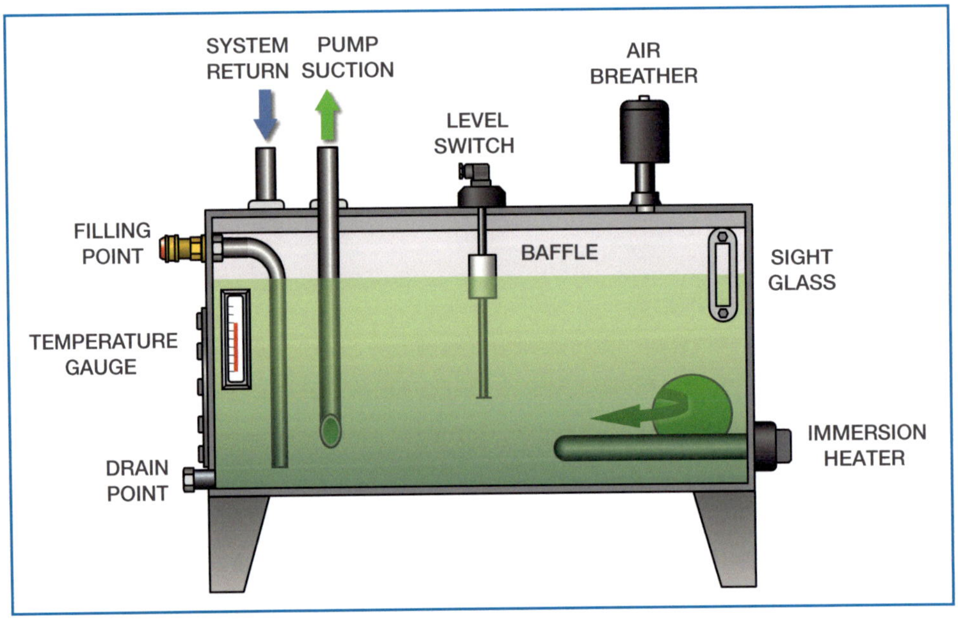

Fig. 2.18 Industrial hydraulic reservoir

The purpose of the reservoir in a hydraulic system is to:

- store fluid that is not circulating around the system
- provide volume and storage for fluid displaced from the cylinders or accumulators
- provide a degree of cooling for the system fluid
- enable heating of the system fluid when necessary
- allow air to separate from the fluid.

Typically, for industrial systems the size of the reservoir is calculated on a rule-of-thumb basis: reservoir size = 3–5 times the pumped flow per minute. For example, a system using a 100 L/min (26 gpm) pump will require a reservoir with a fluid capacity of 300–500 L (80–130 gallons). However, other factors may determine the reservoir size, such as the maximum draw-off required or the amount of heat that has to be dissipated.

Lifting the reservoir off the floor will promote better air circulation and thus increase the amount of heat dissipation. Access doors are required for periodic cleaning out of the reservoir. Shaping the bottom of the tank into a shallow 'V' will ensure that any water or heavy contamination will collect at the bottom of the 'V', where a drain plug can be fitted. Baffles inside the tank will ensure that the return fluid has to follow a certain path inside the tank before it is drawn into the pump suction again, which also aids efficient cooling. Thermostatically controlled immersion heaters can be fitted to heat the fluid to a minimum temperature, and basic instrumentation will include sight glasses, level switches and temperature gauges.

All pipes entering the tank should be sealed by appropriate gaskets to prevent contaminants entering from the atmosphere. Ideally, the filling point for the reservoir should be via a filter or quick-release coupling to ensure that contaminated fluid cannot be simply poured in. As the level of fluid rises and falls, with the movement of cylinders for example, air must be allowed into and out of the reservoir, but this should be only via an air breather, again to prevent contaminants from entering the system from the environment. In some cases the **air breather** will also incorporate a drying medium to limit the amount of moisture drawn into the system.

If air entrainment is a particular problem, **diffusers** can be fitted to the end of the return lines entering the tank in order to dissipate the return flow more evenly. Angled mesh screens can also be incorporated to prevent air bubbles from reaching the pump suction line.

RELIEF VALVES

Whichever type of pump is used in a hydraulic system it will invariably be a 'positive displacement' pump, which means that the inlet and outlet ports are effectively sealed from one another (unlike in centrifugal pumps, which are typically used on central heating systems, for example). This means that if the pump shaft is rotating, then fluid must be being pushed out of the pump (Fig. 2.19). If that fluid has nowhere

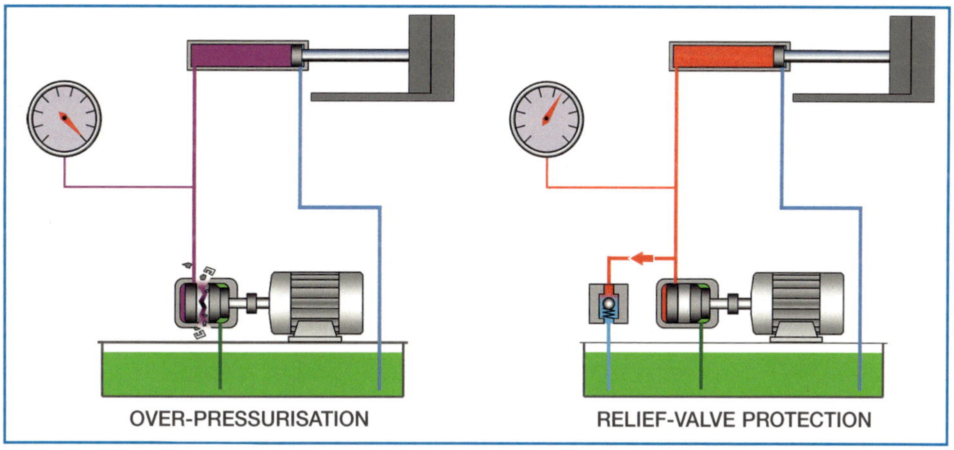

Fig. 2.19 Pressure-relief valve function

POINT OF INTEREST

Due to space limitations, reservoirs on mobile machinery tend to be smaller than those on industrial systems. Special care has to be taken, therefore, to ensure that heat generation and air entrapment in the fluid are minimised.

WARNING

Where regular maintenance activities are required in a hydraulic system (e.g. topping up the reservoir fluid), the design of the system should be such that the correct way of carrying out the activity is also the easiest way.

TOP TIP

The main system relief valve should be fitted in such a way that the pump outlet has direct access to it at all times.

to go, for example when a cylinder piston has reached the end of its stroke, the pressure will build up in the system until either something breaks or the drive motor stalls. Either occurrence is obviously not desirable, so to prevent an excessive build-up of pressure, a pressure-relief valve is nearly always a basic requirement in a hydraulic system.

In its simplest form, a pressure-relief valve consists of a seat, poppet and spring. Usually there is an adjusting device for the spring so that the valve can be set at different pressures as required. This is referred to as a **direct-acting relief valve** (Fig. 2.20), because the pressure acts directly on the main poppet of the valve.

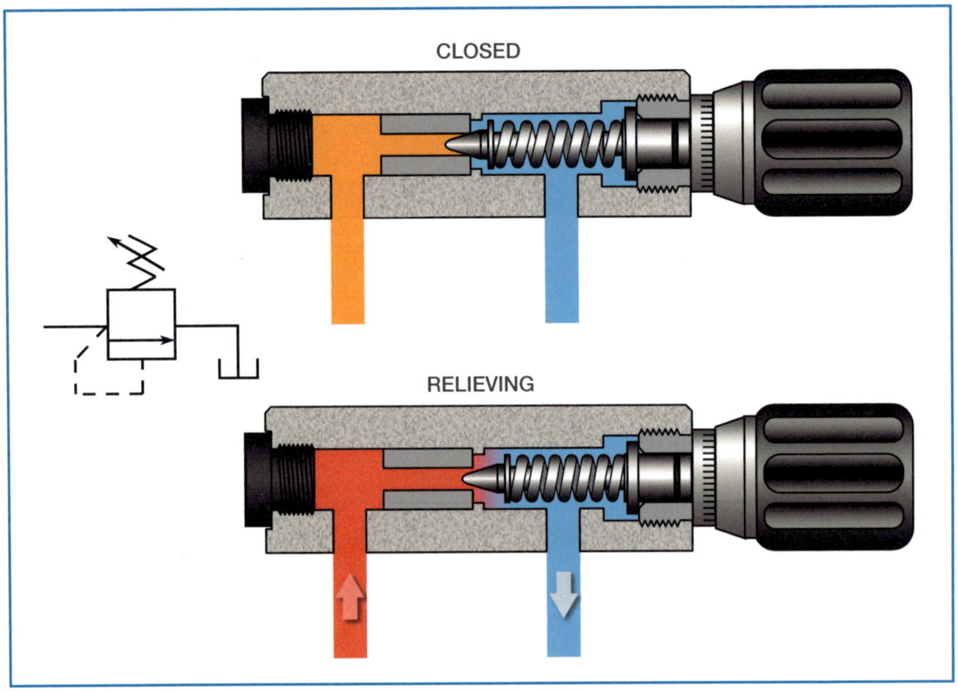

Fig. 2.20 Direct-acting relief valve (Image courtesy of Eaton Corp.)

As long as the pressure on the inlet port of the valve is low, or less than the valve setting, the spring holds the poppet onto its seat and prevents flow through the valve. Eventually the pressure will rise high enough to overcome the spring force and push the poppet off its seat, allowing flow to pass through the valve and back to the reservoir. The valve will, therefore, limit the maximum pressure on its inlet port (and thus in the pressure line of the system) to the setting of the spring, protecting the system from over-pressurisation and potential damage. With very few exceptions, just about every hydraulic system using a positive-displacement pump will include at least one relief valve to limit the maximum system pressure. However, depending on the application, a direct-acting valve may not always be the most suitable choice.

In order to protect the whole system, the size of the poppet and seat of a direct-acting relief valve must be large enough to pass the full pump flow without undue restriction. The area of the poppet exposed to the inlet pressure will be the same as the seat area, so with large flow rates and high pressures a very stiff spring will be required to hold the valve closed up to the required pressure setting.

If the spring is very stiff there will be a significant difference in pressure between that required to start lifting the poppet off its seat (**cracking pressure**) and the pressure required to open the valve fully (**full-flow pressure**). This pressure difference is known as the **pressure override** of the valve. However, having set up the valve to the required maximum system pressure with the valve passing the full pump flow, the valve could start to open and pass a small flow at the lower (cracking) pressure, thus creating heat and robbing the system of useful flow.

To overcome the pressure override problem, most main system relief valves use the **pilot-operated (two-stage) principle**. In this case the direct-acting valve acts as a pilot stage to control a main poppet. The main poppet incorporates a small orifice connecting the top and bottom, and is held onto its seat by a light spring (typically 2 bar (30 psi)). When the system pressure is below the setting of the pilot-stage valve, its poppet remains seated and no flow takes place through the main poppet orifice. The pressures on the top and the bottom of the main poppet will, therefore, be equal, but the light spring will hold the poppet onto its seat and the relief valve is closed (Fig. 2.21).

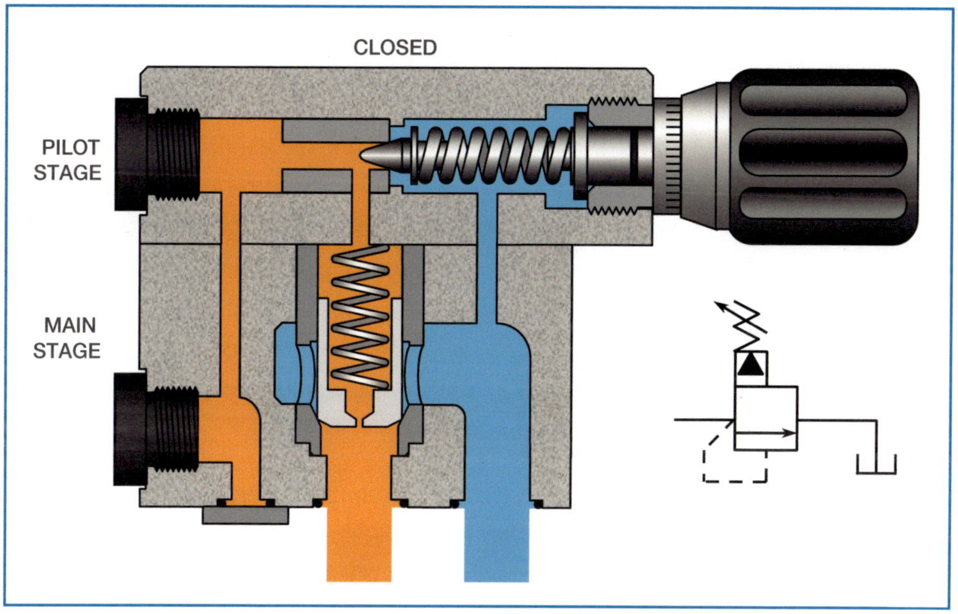

Fig. 2.21 Two-stage (pilot-operated) relief valve – closed
(Image courtesy of Eaton Corp.)

When the system pressure reaches the setting of the pilot valve (Fig. 2.22), the poppet lifts off its seat and allows a small flow across it into the main tank port of the valve. This small pilot flow has to pass through the orifice in the main poppet, which therefore creates a pressure difference across it. When the pilot flow creates enough pressure difference to overcome the light spring (approximately 2 bar (30 psi)) the main poppet will also lift, thus relieving the full pump flow to tank. As opening the main poppet only involves compressing a very light spring, the pressure override will be much less than that of a direct-acting valve.

If the system pressure then drops, the pilot poppet will re-seat, blocking off the flow through the main poppet orifice. This will equalise the pressures again on the top and the bottom of the main poppet, and the light spring will push it back onto its seat and close off the main flow path through the valve.

Although pilot-operated relief valves are the most common type of relief valve used for the main system, some characteristics of the direct-acting valve may be

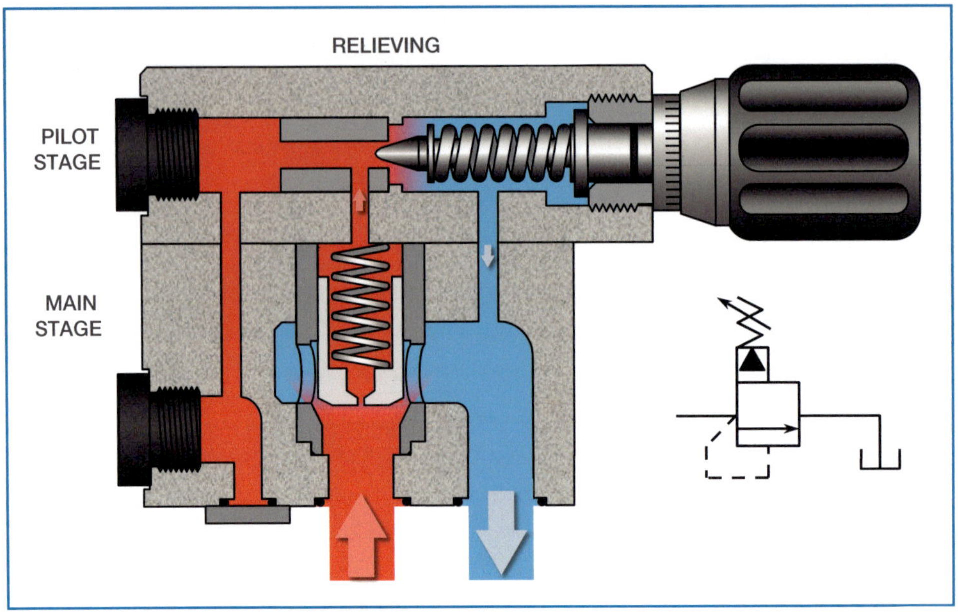

Fig. 2.22 Two-stage (pilot-operated) relief valve – relieving (Image courtesy of Eaton Corp.)

advantageous in other parts of the system. For example, direct-acting valves are very fast acting and so may be used where a relief valve is required to open very quickly to relieve a sudden peak pressure. Direct-acting valves are also very simple valves with very little that can go wrong with them. Two-stage relief valves, on the other hand, are more prone to fail in an open position due to, for example, an orifice blockage in the main poppet. In general, a relief valve failing open is safer than it failing closed, unless the valve is used to control a runaway load on an actuator, for example, in which case it would be undesirable.

WARNING

Heat is generated whenever flow passes across a relief valve at pressure. A blowing relief valve is, therefore, likely to be very hot.

Although relief valves provide very necessary protection in hydraulic systems, it must be appreciated that whenever flow passes across a relief valve at pressure, heat is generated. This is because the hydraulic power entering the valve (flow × high pressure) is greater than that leaving the valve (flow × low pressure), and the difference in power input and output is accounted for in the form of heat.

During machine idling periods, therefore, it would be very wasteful and require a lot of cooling capacity simply to allow a pump to pass its full flow across a relief valve at full pressure. To avoid this, many two-stage relief valves incorporate an extra port known as the **vent** connection, which is blocked, or plugged, for normal operation of the relief valve (Fig. 2.23). Opening the vent port to tank effectively collapses the pressure setting of the valve to a point where the inlet pressure needs only to overcome the light spring on the main poppet to open the valve. This then allows the pump flow to pass freely back to tank at very low pressure, thus avoiding significant heat generation when operation of the machine is not required.

POINT OF INTEREST

Solenoid-vented relief valves are normally arranged such that in the event of an electrical power failure the valve is vented to low pressure ('fail-safe').

In practice, this unloading function of a two-stage relief valve is often achieved by a small solenoid valve mounted on top of the relief-valve pilot stage. Normally the relief valve will be 'vented' (pump unloaded) when the solenoid is de-energised, and operate normally (full pressure setting) when the solenoid valve is energised (Fig. 2.24). This means that if there is an electrical failure of the solenoid valve, the system will revert to low pressure ('fail-safe').

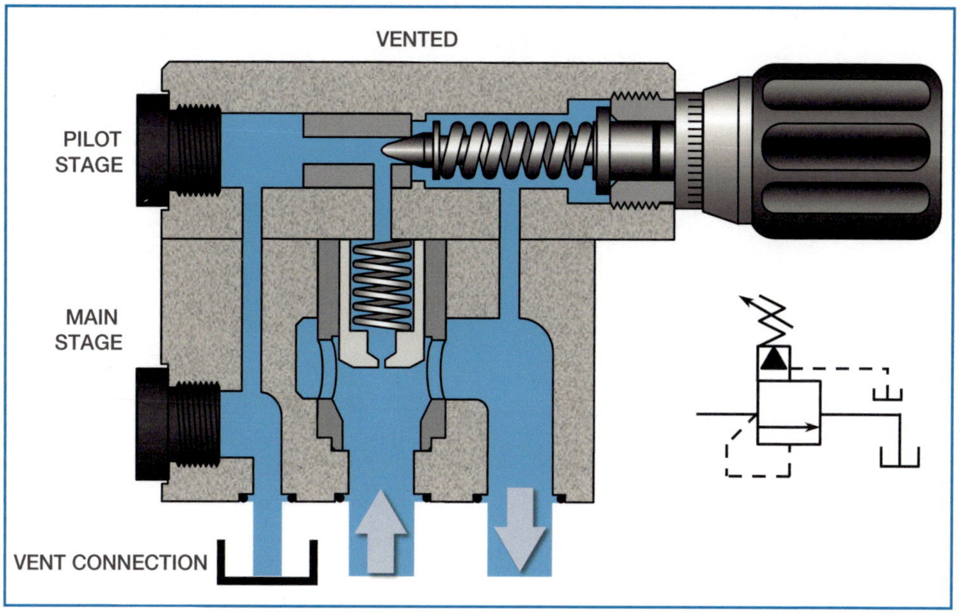

Fig. 2.23 Two-stage relief-valve vent port (Image courtesy of Eaton Corp.)

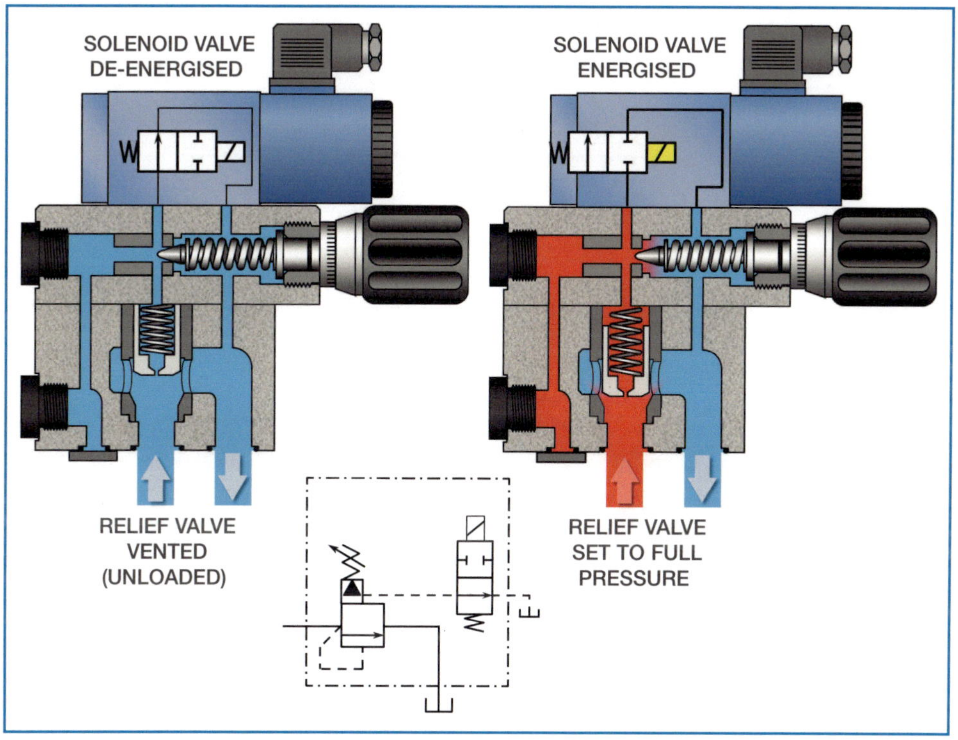

Fig. 2.24 Solenoid-vented relief valve (Image courtesy of Eaton Corp.)

DIRECTIONAL VALVES

Having protected the system from over-pressurisation, it is now necessary to consider how the cylinder will be moved backwards and forwards. This requires some form of directional valve, which can be thought of as a 'hydraulic switch'. In one position the valve will direct pump flow to the full-bore side of the cylinder, while exhaust flow from the annulus side of the cylinder will be directed to tank, thus extending the cylinder rod. Switching the valve across will reverse this action, and the cylinder rod will retract (Fig. 2.25). In some cases it may be possible to have a third (central) position that can be used, for example, to stop the cylinder mid-stroke.

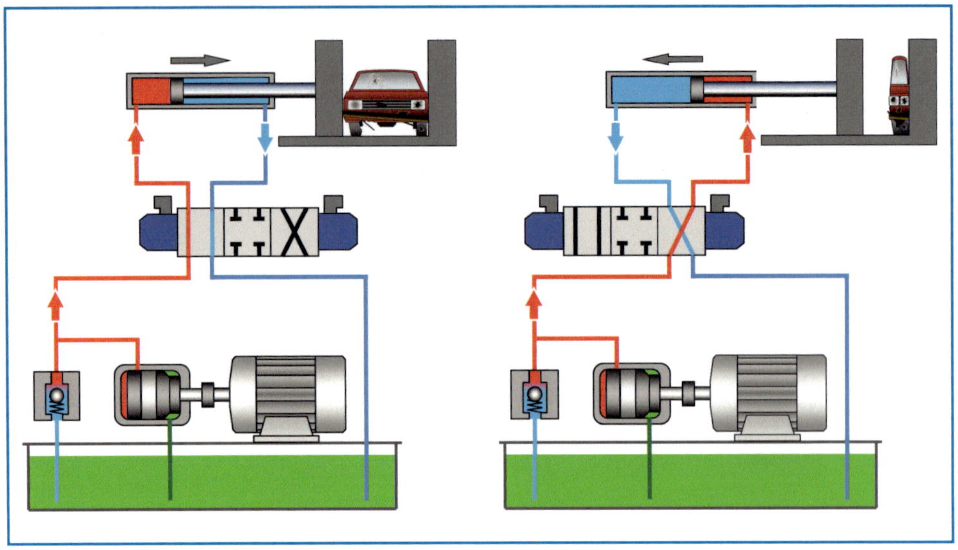

Fig. 2.25 Directional valve function

For low flow rates a **manually operated rotary directional valve** can be used, as shown in Fig. 2.26. In this case an optically flat rotary spool with pressure-loaded seats ensures either zero or near-zero leakage between ports. Two- or three-position valves are available and the flow can be throttled by adjusting to intermediate positions (i.e. partially opening the flow path).

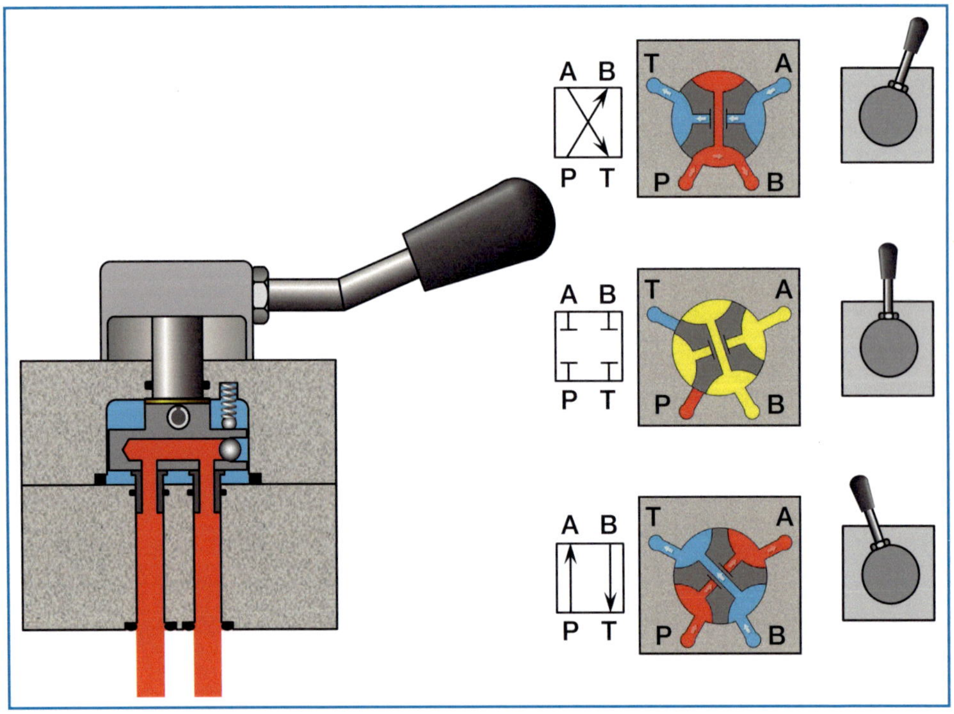

Fig. 2.26 Rotary directional valve

When electrical operation is required, the most commonly used directional valve is a **solenoid-operated sliding-spool valve**. In fact this is probably the most common of all hydraulic components. The valve consists of a body incorporating four main flow ports: a pressure port (P), a tank port (T) and two service ports (A and B) (Fig. 2.27). A bore in the valve body houses a closely fitting spool, which is moved by means of electrical solenoids on one or both ends. Springs can also be fitted to return the spool to a certain position when the solenoid is de-energised.

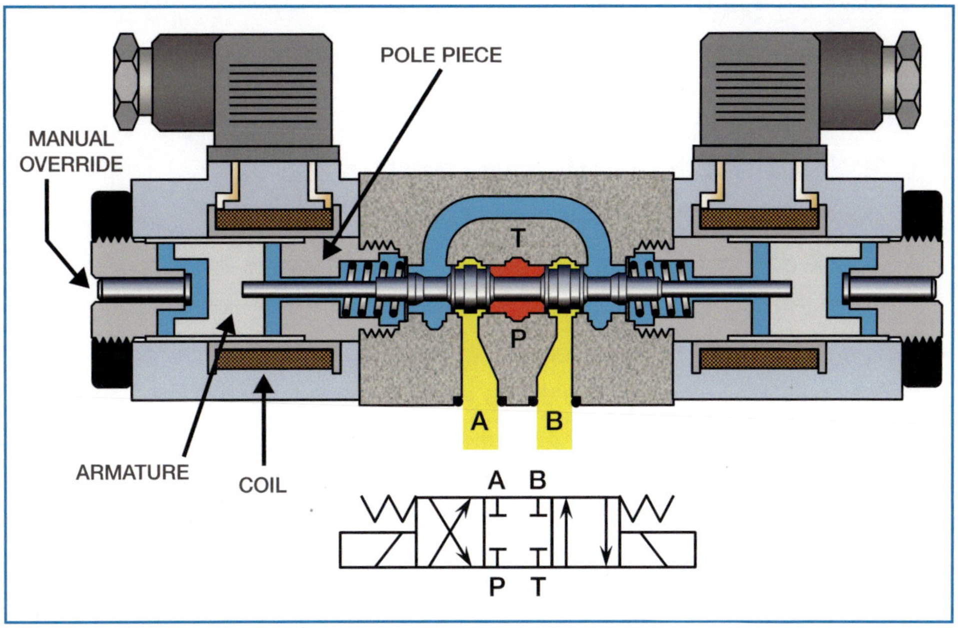

Fig. 2.27 Direct-acting solenoid directional valve

The solenoids consist of a **wire coil** surrounding a **core tube**, inside of which is a sliding **armature** connected to the spool via a **push pin** (Fig. 2.27). When an electrical current is passed through the coil it creates an **electromagnetic field**, which attracts the armature towards the pole piece. As the armature moves it pushes the spool across within the valve body to open flow paths between the four ports, depending on the configuration of the spool (Fig. 2.28). The magnetic field collapses when the solenoid coil is de-energised, allowing a spring on the opposite end of the spool to push it back to its initial position.

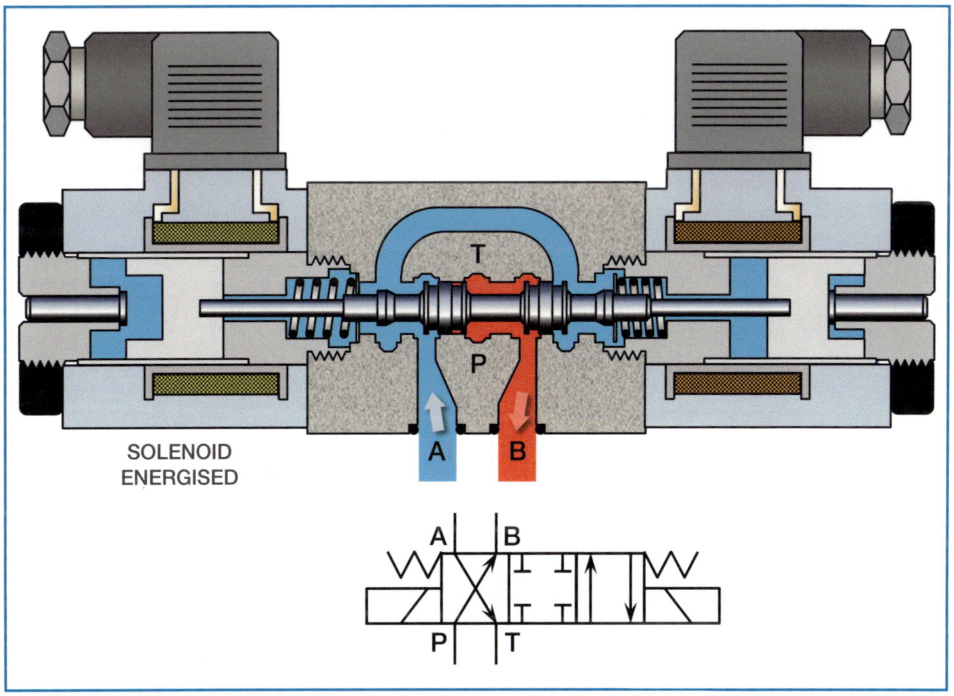

Fig. 2.28 Direct-acting solenoid directional valve (energised)

Apart from the basic valve size, many different configurations and options for solenoid valves are available. These include the following:

- Valve configuration:
  - single solenoid, two position, **spring offset**
  - double solenoid, three position, **spring centred**
  - double solenoid, two position, **detented** (i.e. the valve stays in position when the solenoid is de-energised until the opposite solenoid is energised).
- Flow path options – many options are available, in particular in the centre condition of three-position valves. The four options illustrated in Fig. 2.29 are probably the most common, but there are several more.
- Solenoid supply voltage (and frequency for AC voltage solenoids).
- Type of electrical connector used.
- **Soft-shift** options – these slow down the speed of spool movement (to reduce shock in the system). This is normally achieved by orifice plugs in the solenoid itself.
- **Manual override button** options – these are typically recessed buttons in the ends of the solenoids which enable the valve spool to be moved manually in the event of an electrical power failure or during machine commissioning.
- Indicator lights or LEDs – these are used to signal if a solenoid is energised or de-energised.
- In-built switches: to detect the actual position of the valve spool (often required in safety circuits).
- On-board electrical power switching – this enables the valve to be connected directly to a programmable logic controller (PLC).
- **Proportional solenoids** – these are used instead of simple on/off solenoids to provide additional control (see Chapter 5 for further discussion of these).

WARNING

Be very careful when using manual overrides to ensure that dangerous operation of the machine cannot occur when the valve is operated.

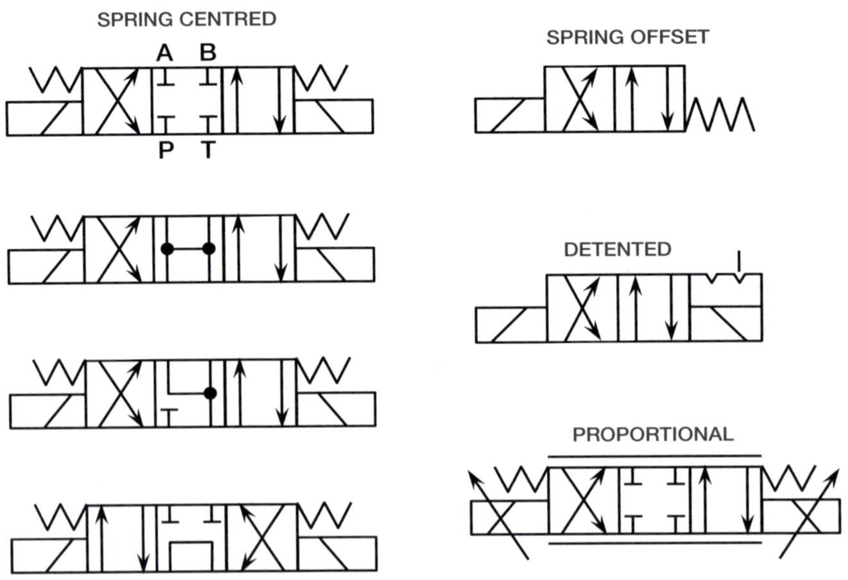

Fig. 2.29 Directional valve symbols

In some cases the centre condition of the valve could be used to unload the pump when the machine is idling (Fig. 2.30). If the centre condition connects the P port to the T port, the pump flow can flow freely back to tank at low pressure. This avoids excessive generation of heat from having to pass the flow across the relief valve at full pressure.

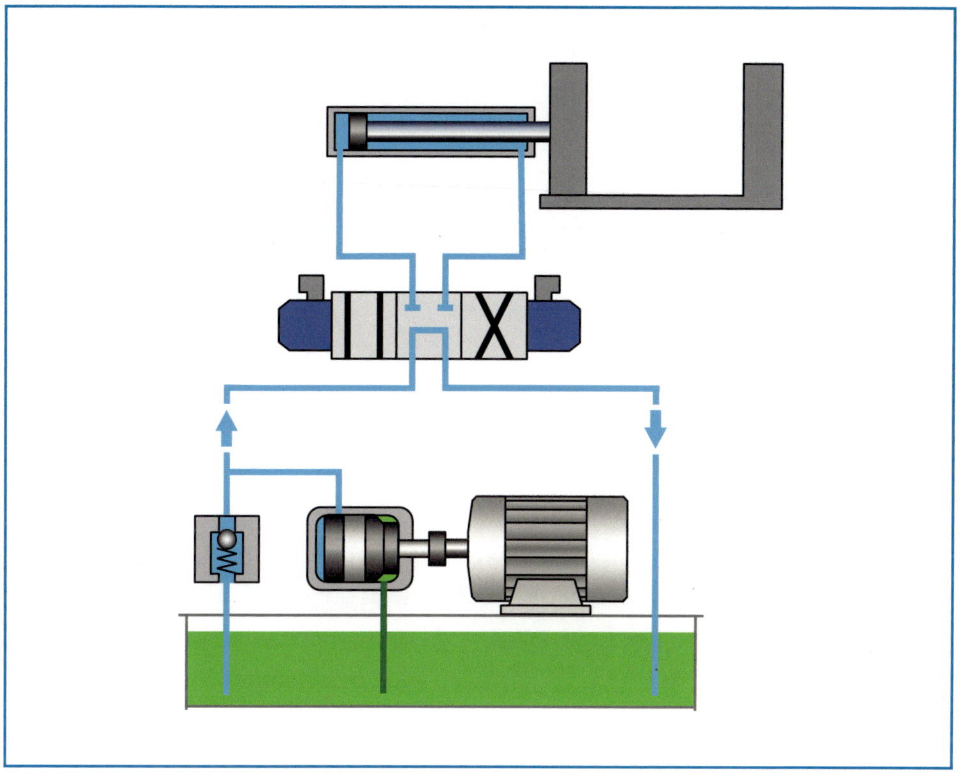

Fig. 2.30 Pump unloading

Solenoid directional valves intended to be mounted on a **subplate** or **manifold block** normally have a standardised interface design. The mounting specification was originally defined by **CETOP** (Comité Européen des Transmissions Oléohydrauliques et Pneumatiques), the European umbrella organisation for national trade associations in the hydraulics industry. Such valves are therefore often referred to as 'CETOP valves'. Subsequently, the International Organization for Standardization (**ISO**) took over responsibility for the specification of direct-acting valves (ISO 4401), and there are three common sizes: ISO 02, 03 and 05. The standard simply defines the position and size of the ports, holding down bolt holes and dowel pins (where fitted), so that one manufacturer's valve should be directly interchangeable with another's. It does not, however, define the overall size of the valve or its flow rating.

When the required flow rate through the directional valve is higher than can be handled easily by an ISO 05 valve (approximately 120 L/min (30 gpm)), it becomes advantageous to use a two-stage valve (Fig. 2.31). In this case the main spool is pushed across in the valve body by means of hydraulic pilot pressure acting on one end of the spool. In turn, this pilot pressure is controlled by a direct-acting solenoid valve mounted on top of the main valve body. The valve is thus still electrically controlled but the main spool is moved hydraulically, hence the description **electrically controlled pilot-operated valve** or, more simply, two-stage or **piggy-back valve**.

POINT OF INTEREST

In the USA, the National Fluid Power Association (**NFPA**) standard defines directional-valve interfaces (denoted D03, D05, etc.). It is equivalent to the ISO standard.

POINT OF INTEREST

Standards are updated from time to time, and the latest version should always be followed. These can be found on the relevant organisation's (ISO, DIN, etc.) website.

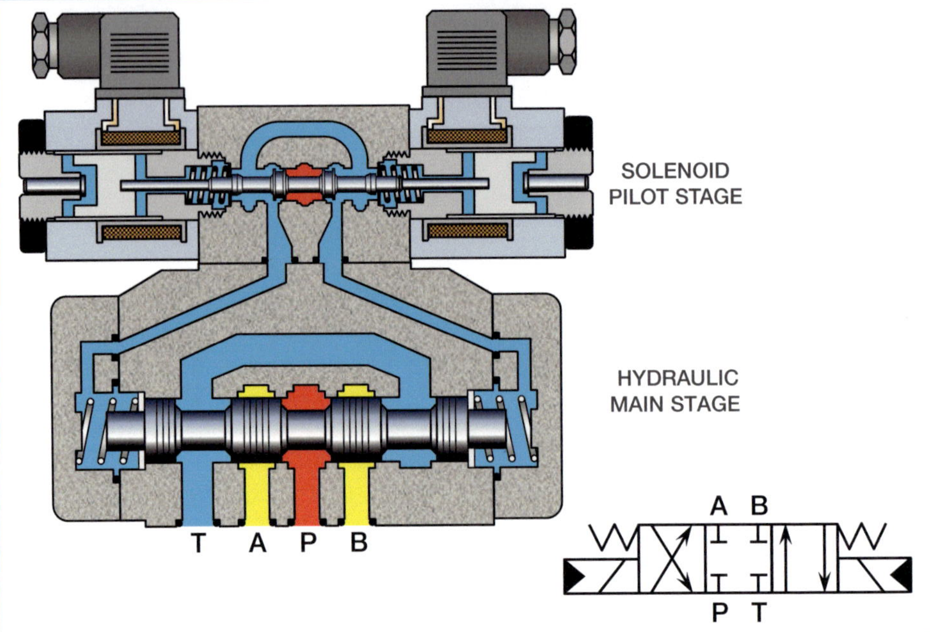

Fig. 2.31 Two-stage (pilot-operated) directional valve

Pilot operation is indicated by the addition of solid black triangles to the solenoid symbol (see Fig. 2.31). The symbol shown in Fig. 2.31 illustrates the centre condition of the main spool only. The centre condition of the pilot spool may well be different (normally the P port is blocked with the A and B ports connected to T in the centre). Two-stage valves also have standardised ISO interfaces in sizes ISO 05, 07, 08 and 10, and sometimes larger.

WARNING

The pilot stage of spring-centred two-stage directional valves has the P port blocked and A and B connected to T when in the spool-centre condition, irrespective of the main spool configuration.

FLOW-CONTROL VALVES

If the speed of movement of an actuator in a hydraulic system is to be controlled, it is necessary to control the flow rate either into or out of the actuator. The simplest way to achieve this is to use some form of **flow-control valve**, which in some cases may be no more complicated than a **needle valve**.

DEFINITION

Needle valves are also known as **throttle valves**.

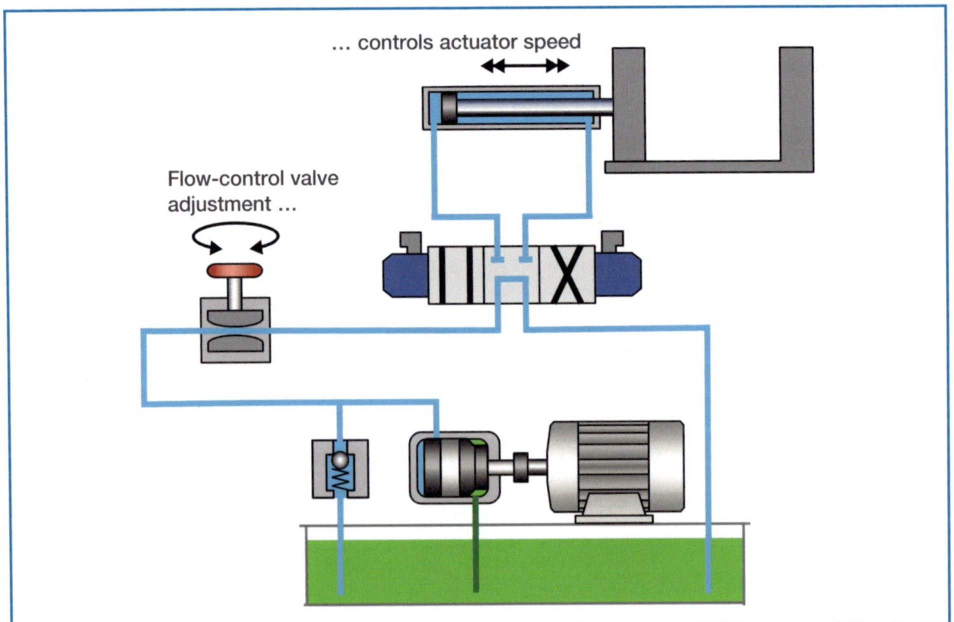

Fig. 2.32 Speed control

A needle valve operates in much the same way as a kitchen tap, which can vary the flow of water through it (i.e. it is not just simply open or closed). As illustrated in Fig. 2.33, a tapered 'needle' can be manually adjusted to vary the opening between it and the valve seat, thus varying the size of the opening through which the flow has to pass.

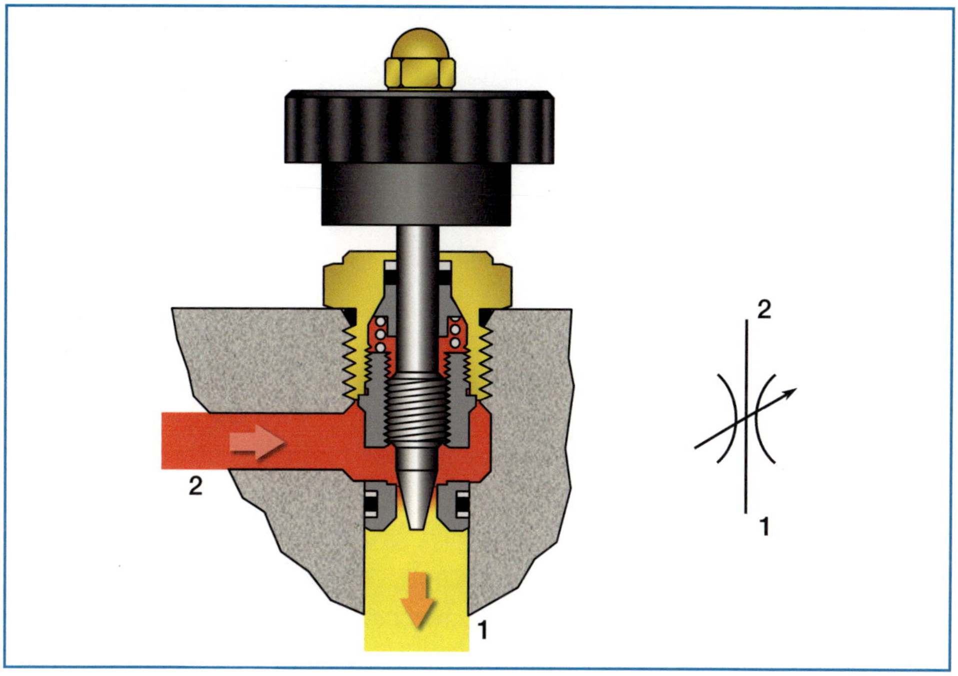

Fig. 2.33 Needle valve (Image courtesy of Eaton Corp.)

In many applications this type of valve is adequate for the degree of control required. However, it does have the limitation that the flow through the valve depends not only on the amount by which the valve is opened but also on the pressure difference across the valve (as well as, to some extent, the viscosity or temperature of the fluid). This means that, having adjusted the valve to provide a certain actuator speed, if the supply pressure varies or the load (and load pressure) changes, then the flow rate through the valve will also change, and the actuator will either slow down or speed up. If the supply pressure remains the same the actuator will tend to slow down with a heavy load and speed up with a light load. If, however, the variation in pressure is small or the variation in speed is unimportant to the operation of the machine, this simple type of flow-control valve can be (and often is) used satisfactorily.

When it is important to maintain a constant flow rate (speed) irrespective of load or supply pressure, a more sophisticated valve is required. A **pressure-compensated flow-control valve** (Fig. 2.34) consists of an adjustable valve, similar to the needle valve just described, in series with a compensating **hydrostat**.

The hydrostat in this case is a spring-biased sliding spool that senses the pressure either side of the adjustable needle valve. When this pressure is sufficient to compress the bias spring, the hydrostat spool is pushed across to further restrict the flow through the valve. In effect, therefore, the flow through the valve is restricted twice. First, the adjustable restriction sets the required flow rate, then a second restriction automatically opens and closes with pressure variations, maintaining a constant flow through the valve (within its limits of accuracy). In some cases the two parts of the

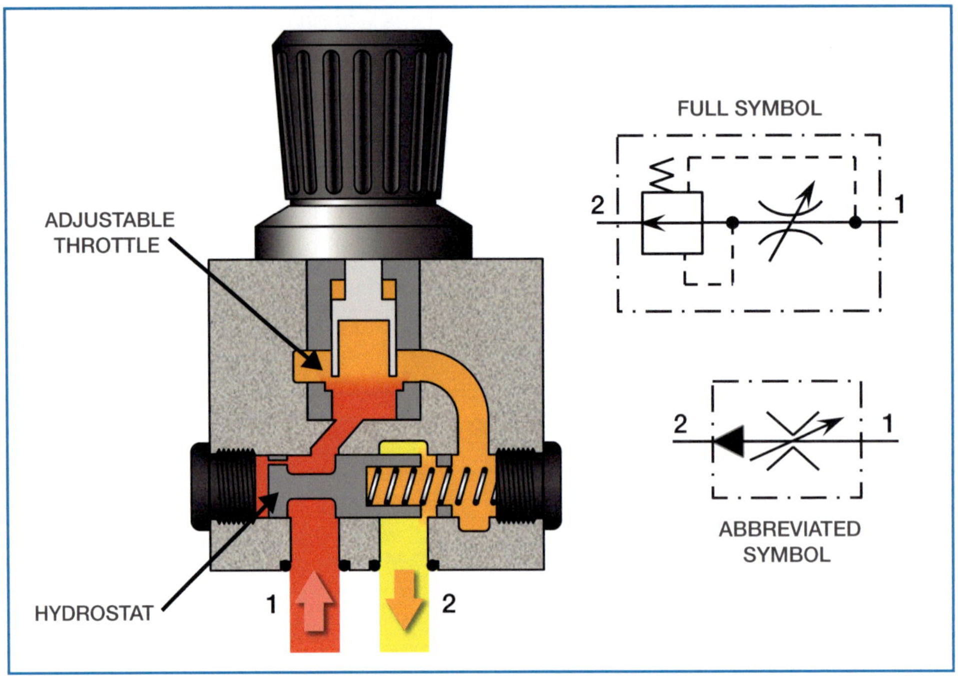

Fig. 2.34 Pressure-compensated flow-control valve (Image courtesy of Eaton Corp.)

valve may be separate components (i.e. a simple needle valve can be converted to a pressure-compensated valve by adding a hydrostat component in series).

> **WARNING**
>
> Whenever flow is restricted in a hydraulic system there is always the potential to create heat.

If a flow-control valve is used in conjunction with a fixed-displacement pump consideration must be given to what will happen to that portion of the pump flow that does not pass through the flow-control valve. In many cases this excess flow will have to pass across the relief valve, which means that the full relief-valve pressure has to be generated at the pump outlet (irrespective of the load on the actuator) and the flow passing across the relief valve will create heat (Fig. 2.35).

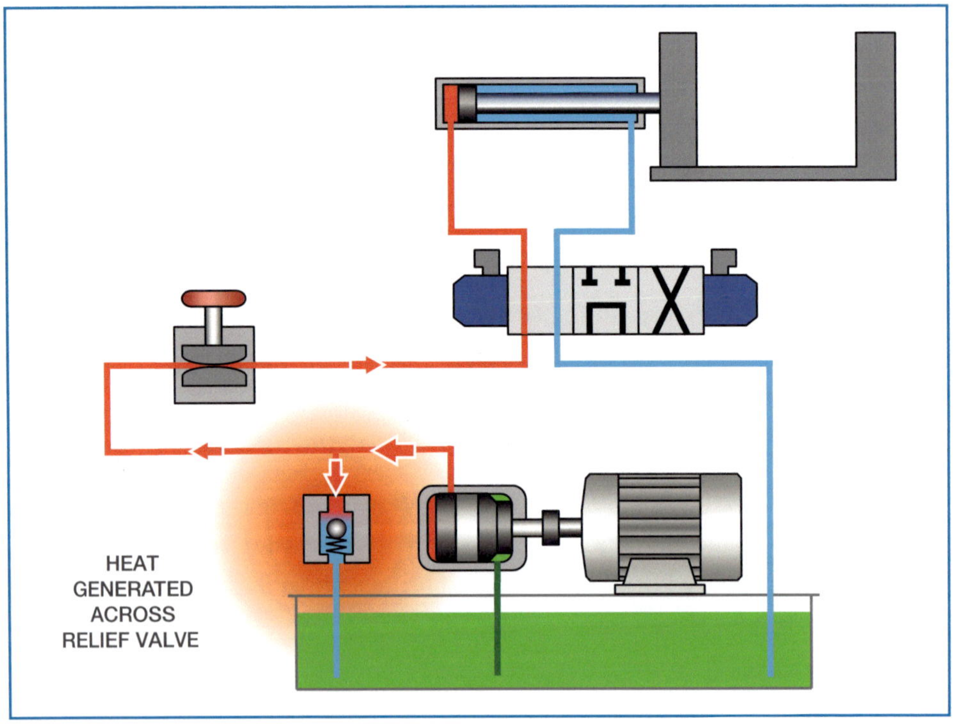

Fig. 2.35 Heat generation

In such situations, therefore, it may be advantageous to use a **variable-displacement pump**, which will only create the amount of flow actually needed to move the actuator at the required speed. Although the pump itself will be more complicated, and therefore more expensive, in most cases it will be a cost-effective solution by virtue of the lower energy consumption and the reduced heat generation.

VARIABLE-DISPLACEMENT PUMPS

With the exception of external and internal gear pumps, all the fixed-displacement pumps already described have variable-displacement versions.

Variable-displacement vane pumps

Rather than having the elliptical cam ring as in the fixed-displacement pump, the variable-displacement vane pump has a fully circular ring that can be moved from side to side relative to the rotor and vanes. With the cam ring pushed fully over to the right-hand side, as shown in Fig. 2.36, fluid is drawn in from the bottom of the pump, carried around the right-hand side in the clearance between ring and rotor, and then squeezed out of the outlet port on the top. The virtually zero ring-to-rotor clearance on the left-hand side means that fluid cannot pass back from the outlet to the inlet.

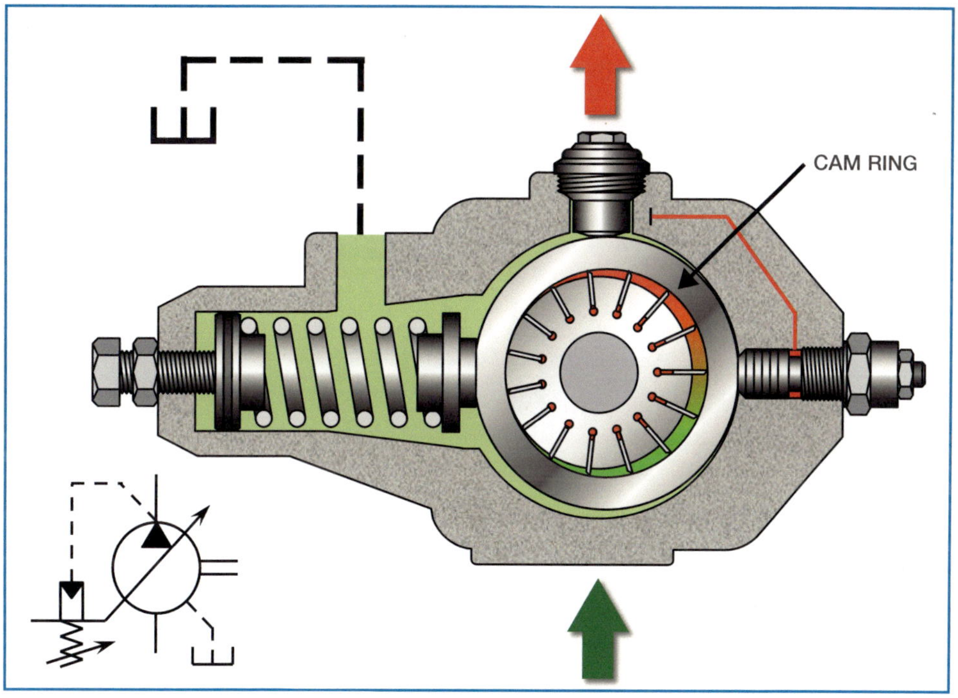

Fig. 2.36 Variable-displacement vane pump (Image courtesy of Eaton Corp.)

If, however, the cam ring is moved across towards the left, the eccentricity between the ring and the rotor is reduced, which means that less fluid will be carried around the right-hand side and some fluid will pass back from the outlet to the inlet on the left-hand side. The net effect will therefore be a reduced flow from the outlet port. Once the ring has moved to a concentric position relative to the rotor and vanes, the pump flow has been reduced to zero. The flow from the pump is therefore variable between zero and maximum, depending on the position of the cam ring.

There are several ways in which the cam-ring position, and hence pump flow, can be varied, but the simplest method is shown in Fig. 2.36. In this case, a piston on

> **DEFINITION**
>
> A variable-displacement pump that reduces its flow at a predetermined pressure is known as a **pressure-compensated pump**.

the right-hand side is fed with pressure from the pump outlet port and tends to push the ring towards the zero-flow position. This, however, is resisted by an adjustable spring on the left-hand side. Therefore, the pump is held at maximum flow until the outlet pressure is high enough to compress the spring, allowing the control piston to start reducing the pump flow rate. The flow rate will then reduce to a level that is sufficient to maintain the outlet pressure at the setting of the spring.

The use of a circular cam ring means that, unlike the fixed-displacement vane pump, the variable-displacement version is no longer balanced from the point of view of internal pressure forces. In practice this limits the maximum pressure capability of the pump, typically up to approximately 210 bar (3000 psi). However, the design does provide a relatively low-cost, low-noise variable-displacement pump that is ideally suited to applications such as machine tools and food machinery.

Variable-displacement piston pumps

A typical variable-displacement axial piston pump is illustrated in Fig. 2.37. The pump works on the same principle as the fixed-displacement version except that the swashplate can now be tilted to different angles, which varies the stroke length of each piston as it rotates, and hence the output flow of the pump. When the swashplate is perpendicular to the drive shaft the flow will be zero, but when tilted to its maximum angle (typically 18–20°) the flow will be at a maximum.

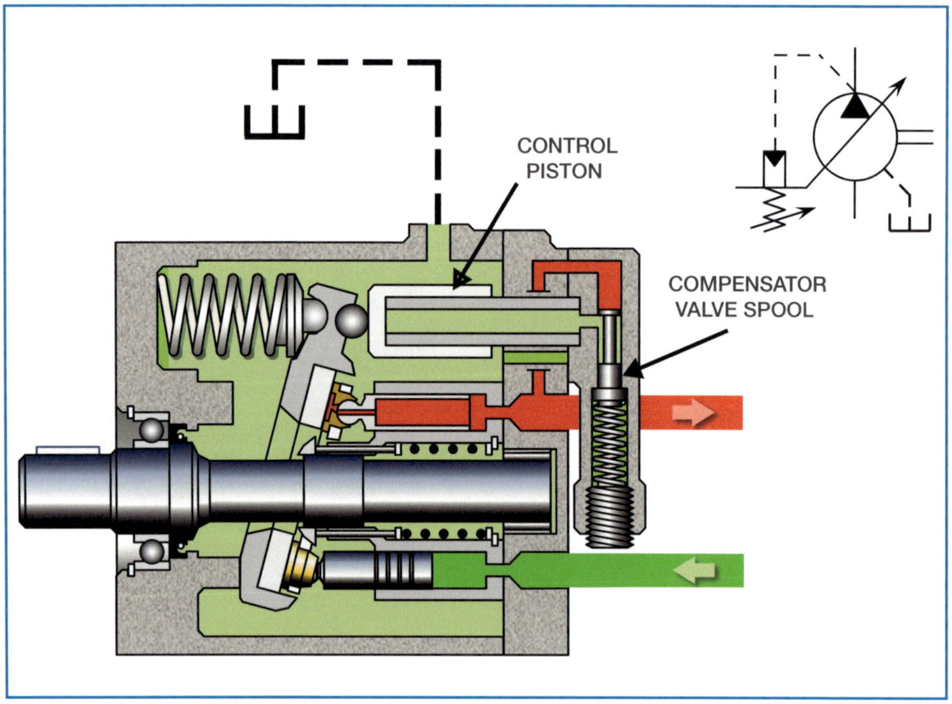

Fig. 2.37 Variable-displacement axial piston pump (Image courtesy of Eaton Corp.)

Normally a control piston at the top of the pump case is used to tilt the swashplate to the required angle, although in some cases the swashplate can be moved mechanically or electrically. A spring, together with the internal forces acting on the swashplate, tends to move the swashplate to the maximum angle (maximum flow) position so that, when pressurised, the control piston will extend and act to reduce the flow of the pump. The fluid pressure inside the control piston is, in turn, regulated by an adjustable **pressure-compensator valve** mounted on the rear of the pump.

When the pump outlet pressure is less than the setting of the spring in the compensator valve, the spool remains pushed upwards to connect the fluid in the control piston to tank via the pump case. The swashplate then moves to the maximum angle and the pump provides maximum flow. When the outlet pressure rises to the spring setting, however, the spool is pushed down, and fluid under high pressure enters the control piston to start to de-stroke the pump.

As with the variable-displacement vane pump, the pump flow will reduce to a level where the outlet pressure is maintained at the setting of the spring in the compensator valve. If the pump outlet is blocked (e.g. by a closed port directional valve), the pump flow reduces to virtually zero – just sufficient to maintain leakage in the system. If the pump outlet is restricted (but not blocked completely), the pump produces whatever flow the restriction can pass at the pressure setting of the compensator valve.

The same general principle applies to variable-displacement bent-axis piston pumps (Fig 2.38). In this case the angle of the cylinder block can again be varied to change the piston stroke and hence the pump flow. However, the control piston this time is mounted on the rear of the pump and tilts the cylinder block by means of a link rod.

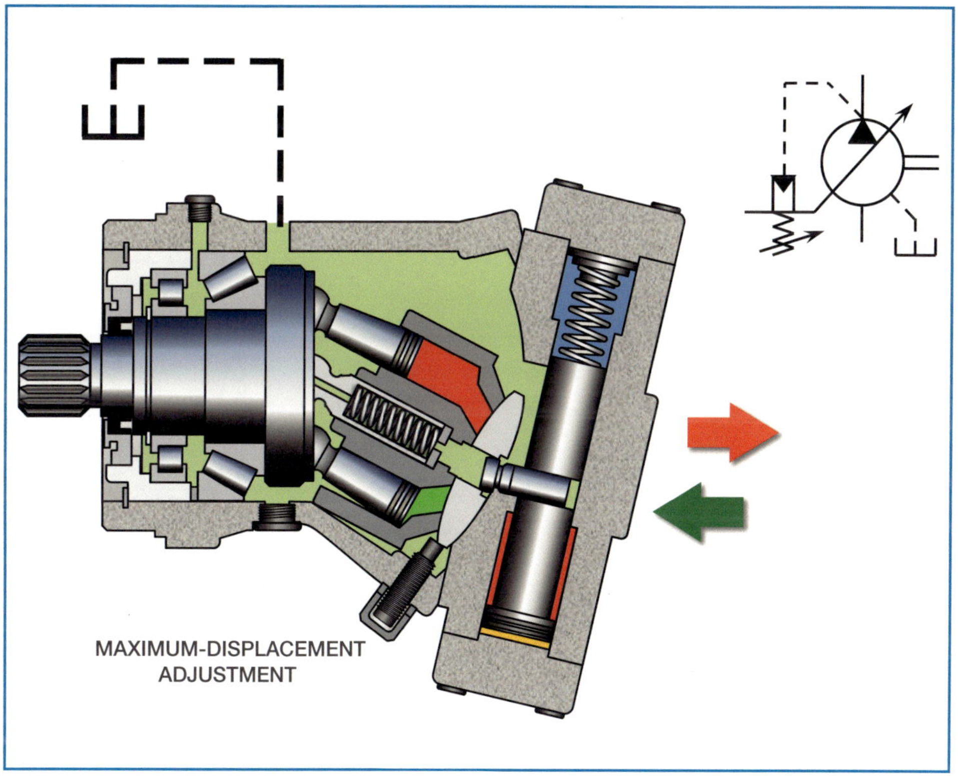

Fig. 2.38 Variable-displacement bent-axis piston pump

With variable-displacement pumps there is normally an option to limit the maximum flow of the pump by means of fixed or adjustable stops, which prevent the cam ring, swashplate or cylinder block moving to its maximum flow position. So, by using a variable-displacement pump in conjunction with a flow-control valve, the system is now more efficient in that the pump will only produce the required flow rate as set by the flow control, and no flow will be wasted over the relief valve (Fig. 2.39).

WARNING

A pump relief valve is often required in conjunction with a pressure-compensated variable-displacement pump in order to relieve peak pressures and also to provide protection should the control piston or its control valve jam. However, the pressure-relief valve should normally be set at least 20 bar (300 psi) higher than the compensator setting.

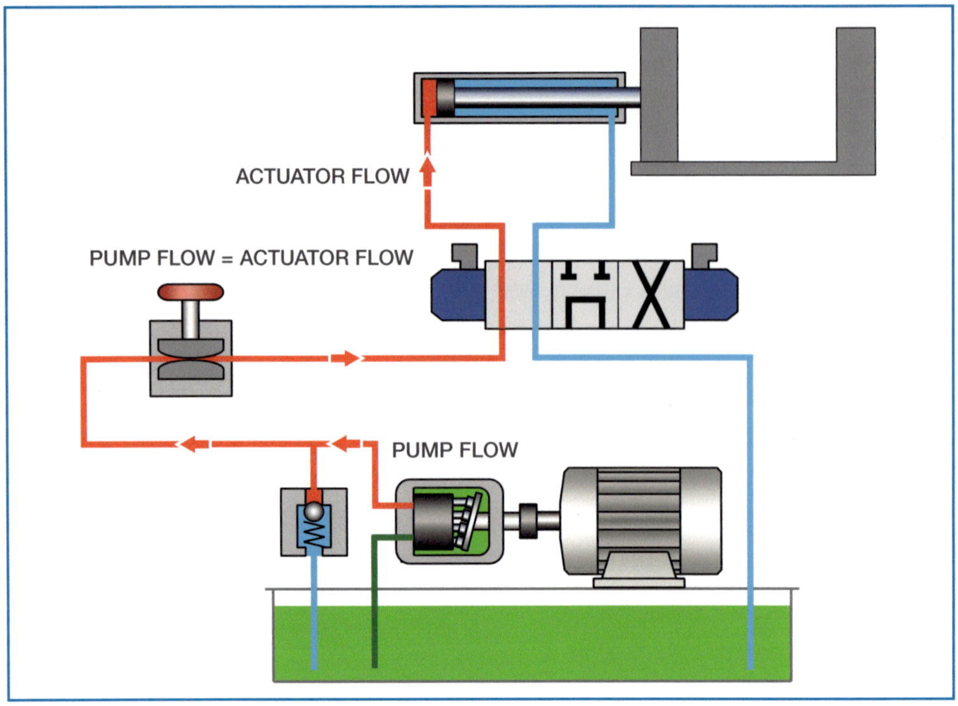

Fig. 2.39 Variable-displacement pump

FILTER

The final component required in the simple system considered here is a filter of some description to ensure that the system fluid is maintained to the correct level of cleanliness. Inevitably some contamination may remain in the system from its manufacture and installation, and as components wear and maintenance work is carried out further contamination may enter the system. Fluid cleanliness is discussed in more detail in Chapter 4, but for now it can be assumed that a filter is required in the main pressure line of the system (Fig. 2.40).

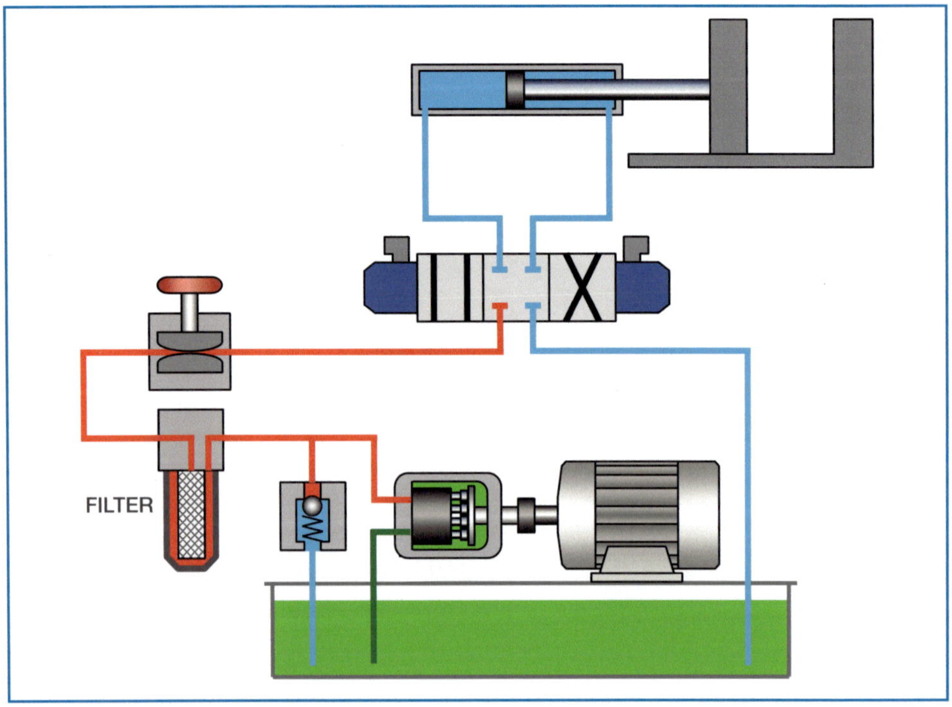

Fig. 2.40 Filter

There are many different types of filter available for hydraulic systems but they all have certain features in common (Fig. 2.41). All filters have an element through which the fluid passes and which is designed to trap and retain particles of dirt. Different materials are used for the filter element depending on the fineness of filtration required, ranging from coarse wire mesh strainers to extremely fine glass-fibre type materials.

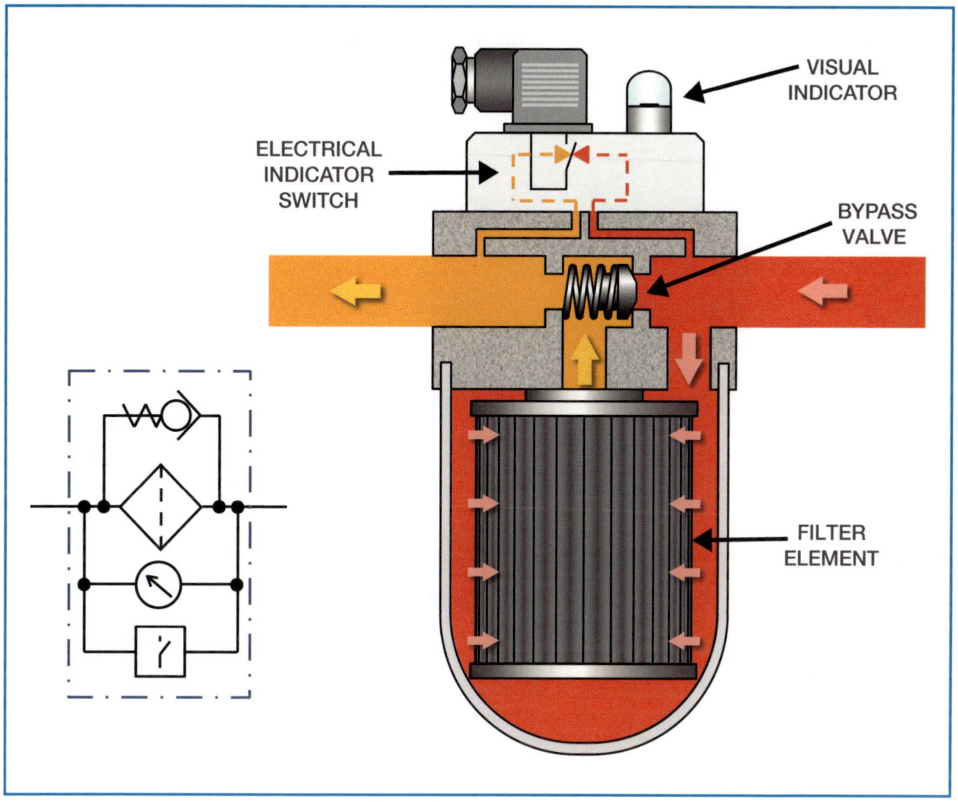

Fig. 2.41 Filter construction

Most industrial systems use filters fitted with indicators, either visual or electrical (or both), which sense the pressure difference across the filter element. When this difference reaches a predetermined level, indicating that the element is close to being fully clogged with contamination, the indicator is triggered and, in the case of electrical indicators, the appropriate control action can be taken (normally some form of alarm). In most cases the filter will also incorporate a bypass or low-pressure relief valve which will open to protect the element from physical damage if no action is taken when the filter element becomes clogged. The pressure difference required to open the bypass valve is slightly higher than that required to trigger the indicator.

ACCUMULATOR

Hydraulic accumulators are devices that can store fluid under pressure. They act in a similar way to a rechargeable battery in an electrical circuit. In the early days of hydraulic machinery, accumulators consisted of a weight-loaded piston within a cylinder, but modern accumulators are normally gas-charged devices. Because fluids are only slightly compressible, storing a volume of fluid under pressure means that only a small amount of the fluid can be used before the pressure drops to zero. A gas, however, is much more 'elastic' than a fluid, and this property enables a useful volume of fluid to be stored under pressure.

WARNING

To avoid the risk of an explosion **never use compressed oxygen or compressed air in an accumulator.**

An accumulator consists of a rubber bag or bladder inside a metal shell (Fig. 2.42). A valve on the top allows the inside of the bladder to be filled with gas to a predetermined pressure known as the **pre-charge pressure**. The poppet valve at the bottom of the accumulator prevents the bladder from being pushed down the port and becoming damaged when the bag is charged with gas. When dealing with flammable hydraulic fluids, the gas used must be an inert gas. Nitrogen is the gas that is generally used due to its low cost.

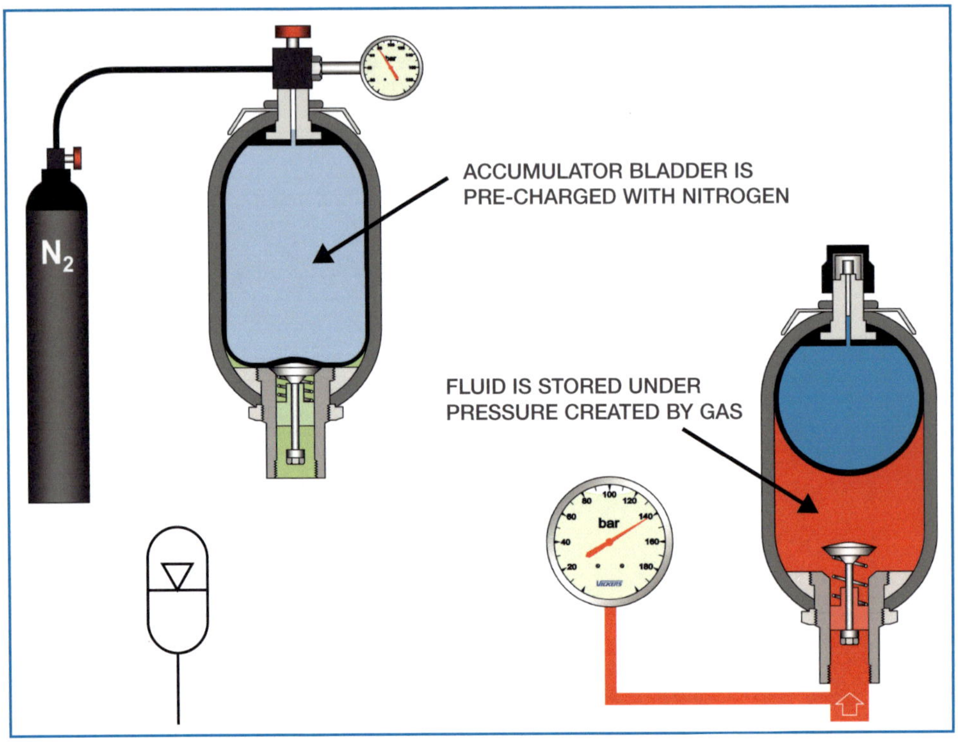

Fig. 2.42 Accumulator

When the accumulator has been charged with gas (from a nitrogen bottle), it is ready for use. However, fluid will not start to fill the accumulator until the fluid pressure reaches or exceeds the gas pre-charge pressure. Once this happens, fluid enters the accumulator shell, thus compressing the gas further. The more fluid that is pushed in the higher the pressure rises. Normally the accumulator is only filled to a maximum of approximately two-thirds of its total volume with fluid so that the bladder is not damaged by too much compression. When fluid is drawn from the accumulator, the pressure will gradually drop from its maximum down to the gas pre-charge pressure. Provided the pre-charge pressure is higher than the minimum required system pressure, all the stored volume will be available to do useful work.

Accumulators are therefore often used to supplement the pump flow when actuators have to move rapidly, or to provide an emergency source of energy in the event of a power failure. They are, however, potentially dangerous devices, because there may be a considerable amount of stored energy within an accumulator when it is fully charged. Accumulators must, therefore, be fitted with a **drain valve** (automatic or manual) to drain the fluid from the accumulator before any maintenance work is carried out on the system. A permanently mounted pressure gauge enables maintenance personnel to ensure that the pressure has been released, and a protective relief valve on the fluid port protects the system against over-pressurisation in the event of a fire.

SYMBOLS AND CIRCUIT DIAGRAMS

Hydraulic components are represented by a set of graphical symbols which represent the function of a particular component. These are specified in the international standard ISO 1219-1. While the symbols can convey a lot of information about the component, they do not generally provide information about its construction. For example, the symbol for a fixed-displacement vane pump is the same as that for a gear pump.

The ISO symbols for the components discussed so far in this chapter are shown in the corresponding figures. It would be impossible for ISO 1219-1 to cover every possible hydraulic component, so it gives rules for constructing specialised symbols as required.

In order to convey how the individual components are interconnected to create a system, they are laid out on a **circuit diagram**, which gives further information on the interconnection components (hoses, pipework, etc.), component settings, identification numbers, etc. The layout of circuit diagrams is covered in the second part of the standard (ISO 1219-2). A highly simplified circuit diagram for the car crusher system considered in this chapter is shown in Fig. 2.43.

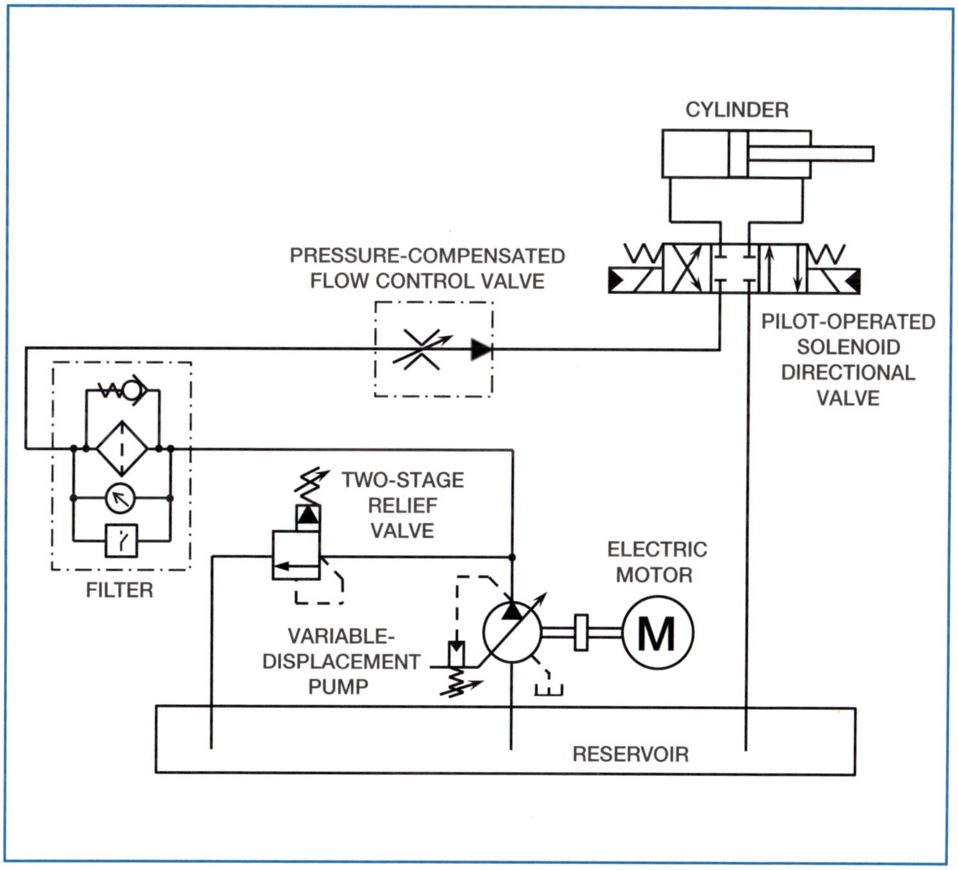

Fig. 2.43 Circuit diagram

As will be described later, circuit diagrams are vital pieces of information for people maintaining or troubleshooting hydraulic systems. With a little practice and experience, the symbols are reasonably self-explanatory. To fully understand the operation of a complete hydraulic system, however, it is usually necessary to know the function and the requirements of the machine itself.

POINT OF INTEREST

Symbols shown on a circuit diagram are drawn in their 'at-rest' position (i.e. with solenoids de-energised and no pressure acting on valves, etc.).

FURTHER READING

For a chart of typical hydraulic graphical symbols, see:

www.webtec.com/education

A BASIC MOBILE HYDRAULIC SYSTEM

POINT OF INTEREST

Mobile systems often have to be designed for more variable operating conditions (loads, speeds, temperatures, etc.) compared with industrial systems.

The basic principles of hydraulic systems are the same whatever type of machine or vehicle the system is controlling. However, there will often be some practical differences in the types of components used between industrial and mobile systems.

Consider the fork-lift truck that is used to transport the car to the crusher, in particular the operation for raising and lowering the forks (Fig. 2.44). The basic components used in this system would be similar to those already discussed (Fig. 2.45).

Fig. 2.44 Fork-lift truck

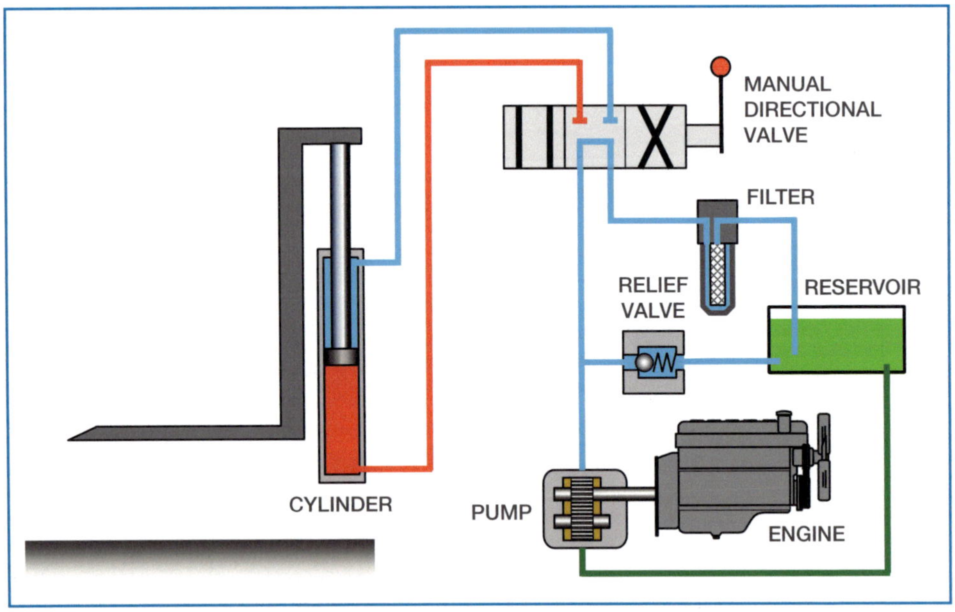

Fig. 2.45 Fork-lift truck system

The choice of pumps for use in mobile systems is essentially the same as for industrial systems (i.e. gear, vane and piston pumps, either fixed or variable displacement). However, whereas an industrial system such as a plastic injection moulding machine may operate continuously for two shifts a day and five days a week, many mobile machines are operated relatively infrequently (e.g. a crop sprayer may be in use for only two or three weeks a year). So, for many mobile applications long life and high-pressure operation are not significant factors, and the rugged, low-cost gear pump is a popular choice. Although gear pumps are fixed-displacement pumps, it is still possible to vary the pump flow by varying the drive speed. Where the pump is driven by a diesel engine, as is the case in many mobile applications, the pump flow can be varied simply by increasing or decreasing the engine speed.

A relief valve is still required to limit the maximum system pressure. The valve, which will often be a two-stage (pilot-operated) type valve, operates on the same principle as already described for an industrial system. However, because the size and weight of components used in mobile systems are more of a consideration, the valve will tend to be designed to the smallest possible dimensions, and may even be incorporated in the body of a directional valve. It may also be tailored to a specific application (because production numbers are generally greater for mobile machines) and be non-adjustable.

MOBILE DIRECTIONAL VALVES

Whereas it is commonplace for industrial systems carrying out defined, repetitive cycles to be controlled by electronic controllers (typically **programmable logic controllers (PLCs)**), many mobile machines are still controlled, in part at least, by human operators. Industrial systems therefore tend to use valves controlled electrically or electronically, but manually operated valves are used extensively in mobile applications.

A typical example of a manually operated valve is shown in Fig. 2.46. In this case the hand lever acts directly on the valve spool to move it within the valve body and open or close the appropriate flow paths. As with solenoid valves, manual valves can be spring offset or spring centred, and sometimes are detented.

DEFINITION

A **detented** valve is one that stays in position when released, as opposed to spring-offset or spring-centred valves that return to an offset or central position when released.

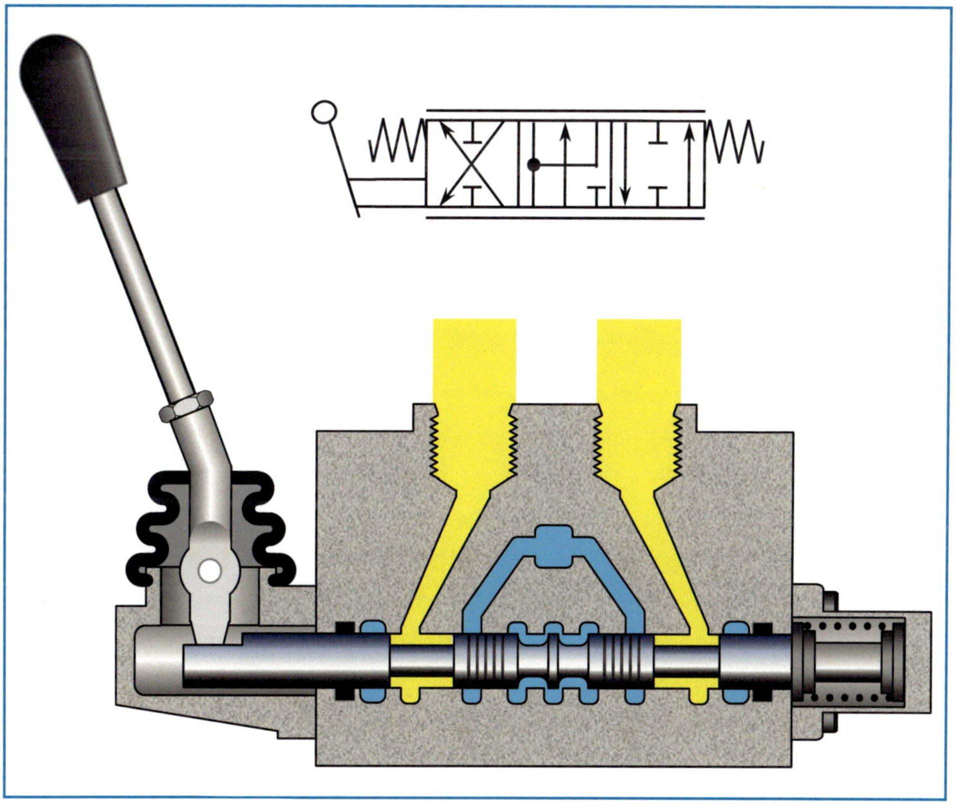

Fig. 2.46 Manual directional valve (open centre)

As well as acting as simple directional valves, however, manual valves can often regulate the flow passing through the valve. This is determined by the amount of spool opening, which in turn is proportional to the amount of handle movement. In effect, therefore, the manual valve acts both as a directional valve and as a flow-control valve.

Although in its simplest form the flow through the valve will also vary with pressure difference or load pressure, any variation in load speed that this creates can be corrected by the operator adjusting the valve opening. In effect the operator is acting as a human compensator. More sophisticated valves include hydrostats, similar to those used in pressure-compensated flow-control valves, to perform this task automatically.

DEFINITION

An **open-centre directional valve** will unload the pump flow back to tank when the valve spool is centred. It is therefore normally used with a fixed-displacement pump.

When used with a fixed-displacement pump the valve will normally use an **open-centre** configuration, where fluid entering the valve's P port can pass freely through the valve when centred and out through the T port to tank, thus unloading the pump. As the lever and spool are moved away from the central position, this unloading flow path is gradually blocked off, allowing pressure to build up and move the load (Fig. 2.47).

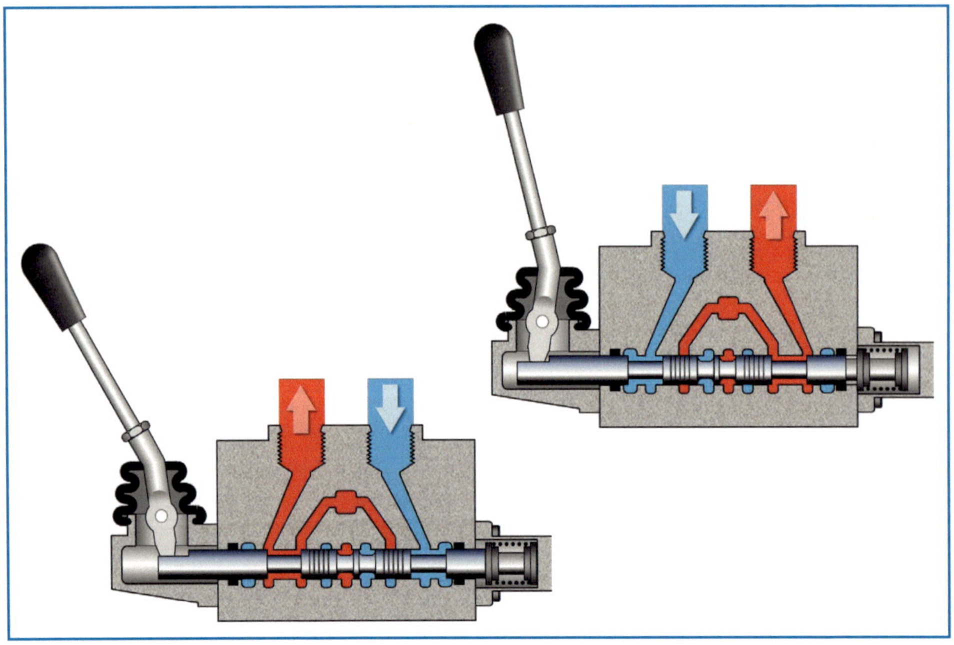

Fig. 2.47 Manual directional valve operated

The throttling effect of the valve spool, together with the gradual build-up of pressure as the valve is moved away from the mid-position, thus provides the operator with a good degree of control over the load movement. In many applications the operator may have several actuators to control, some simultaneously. On an earth-moving machine, for example, this could be the boom, arm and bucket. Mobile valves are therefore often mounted in sections or 'slices' (Fig. 2.48), where several valves are stacked together, thus enabling the operator to control two or more valves at the same time.

In such cases the bypass connections of all valves are connected in series so that when all the valves are centred the pump flow has a free passageway back to tank and the pump is unloaded. As soon as any one valve spool is moved away from centre, however, the bypass passage closes off, allowing pressure to build up in the system.

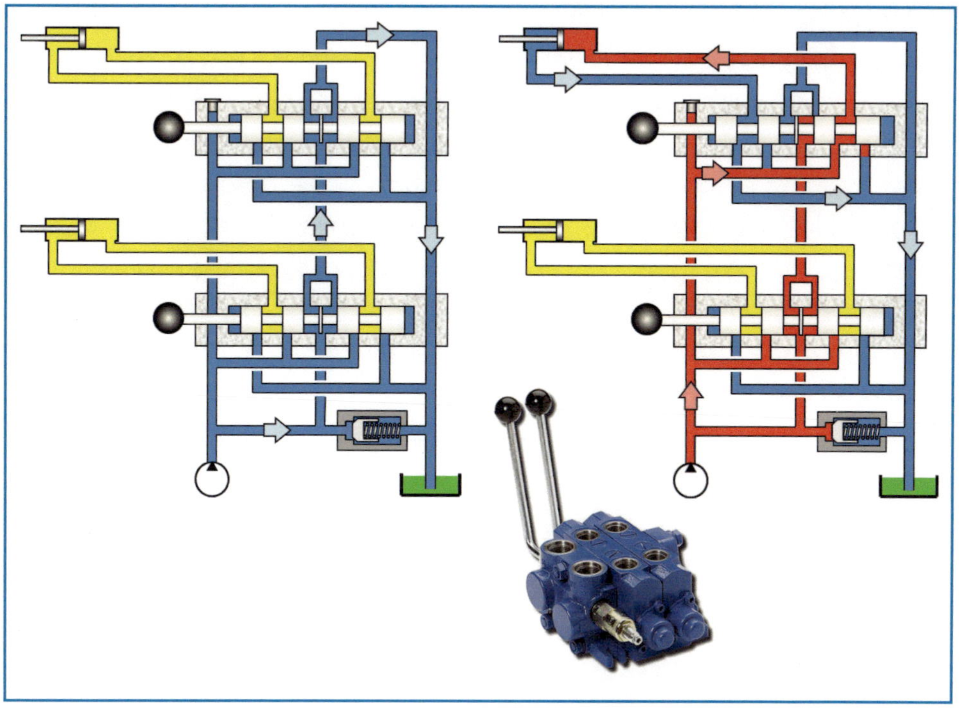

Fig. 2.48 Two-section valve (Image courtesy of Eaton Corp.)

For safety reasons, control valves often have to be mounted outside of a vehicle cab, which therefore requires some means of operating the valves remotely. This can be achieved by using pilot pressure to move the valve spools, in which case a low-pressure hydraulic joystick is mounted within the vehicle cab. Alternatively, the valve spool may be moved by means of a proportional solenoid, which simply requires an electrical joystick located at the operator's position.

Where a simple on/off function is required (i.e. no flow control), solenoid valves can also be mounted as slices. A typical two-stage valve is shown in Fig. 2.49.

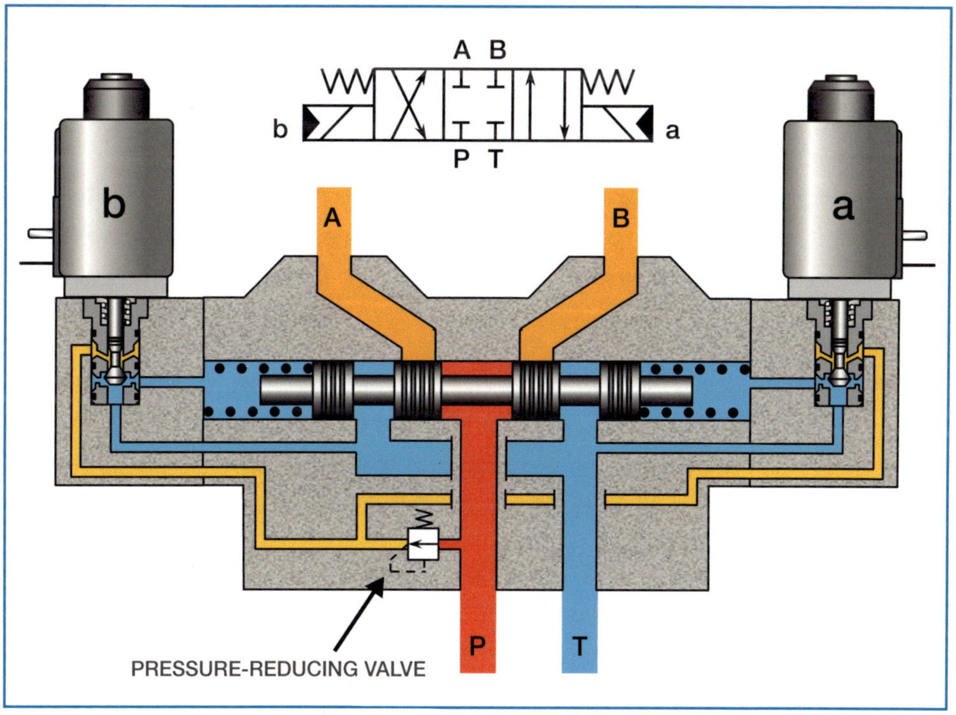

Fig. 2.49 Mobile two-stage solenoid directional valve

POINT OF INTEREST

The pilot supply of a two-stage valve often operates at a reduced pressure to avoid excessive shock or possible damage during movement of the main spool. A small pressure-reducing valve built into the main valve body can be used to achieve this (see Fig. 2.49).

LOAD CONTROLS

Another control function that needs to be considered in the fork-lift application is when the load is being lowered and the cylinder is subject to a negative (runaway) load (Fig. 2.50). In this situation gravity is acting on the load in the same direction as the hydraulic action. Although careful opening of the directional valve by the operator may be sufficient to control the downward speed of the load by restricting the exhaust flow from the cylinder, the possibility still exists for the load to lower at an excessive speed if the directional valve is opened too far. This would then create a negative pressure (vacuum) on the inlet side of the cylinder, as the piston moves at a rate faster than fluid is entering.

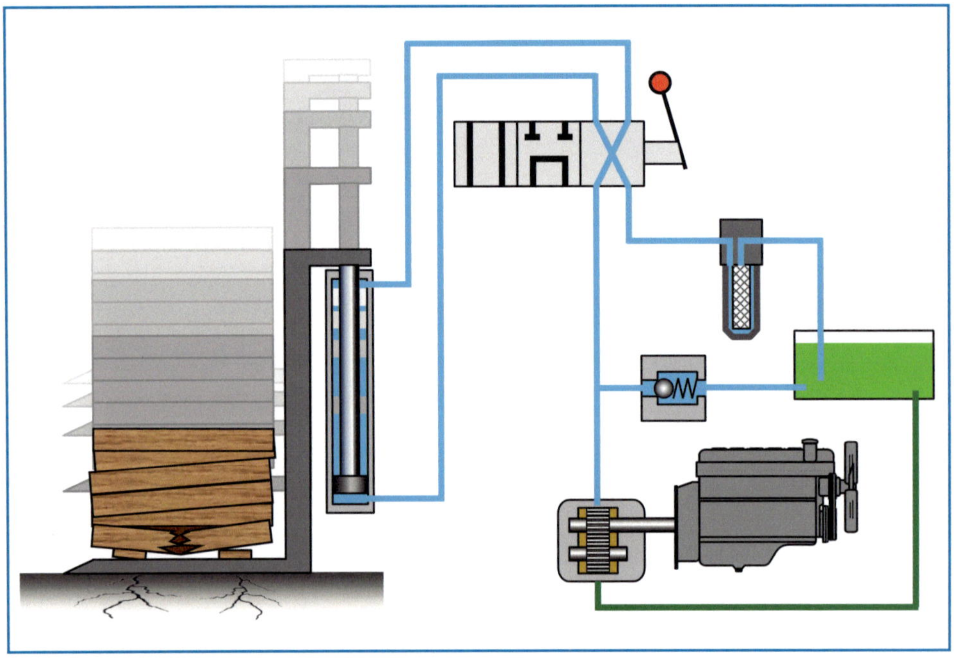

Fig. 2.50 Negative (runaway) load

To prevent this from occurring, a 'braking' valve can be mounted in the exhaust line from the cylinder. This valve will remain closed unless there is sufficient pressure on its pilot port to overcome the spring acting on the valve. Taking the pilot pressure from the inlet port of the cylinder means that if the load begins to run away the pressure on the inlet side of the cylinder will drop, causing the braking valve to close off sufficiently to slow the load down again. This means that the piston and load will only descend at the rate at which it is pushed down by the controlled amount of flow entering the cylinder.

The type of valve used in this application is normally referred to as a **counterbalance valve** (Fig. 2.51) because its function is basically to counterbalance the effect of the load. To raise the piston and load back up it is necessary to reverse the flow through the counterbalance valve, and so a free-flow check valve poppet is normally incorporated for this purpose (Fig. 2.52).

As shown in Fig. 2.53 the counterbalance valve normally has an internal pilot connection as well as the external connection. This enables the counterbalance valve to act also as a relief valve to limit any shock or intensified pressures on the annulus side of the cylinder caused, for example, by a sudden stopping of the load.

WARNING

Counterbalance valves have to be selected and adjusted very carefully to ensure that they operate as required over all working conditions of the machine.

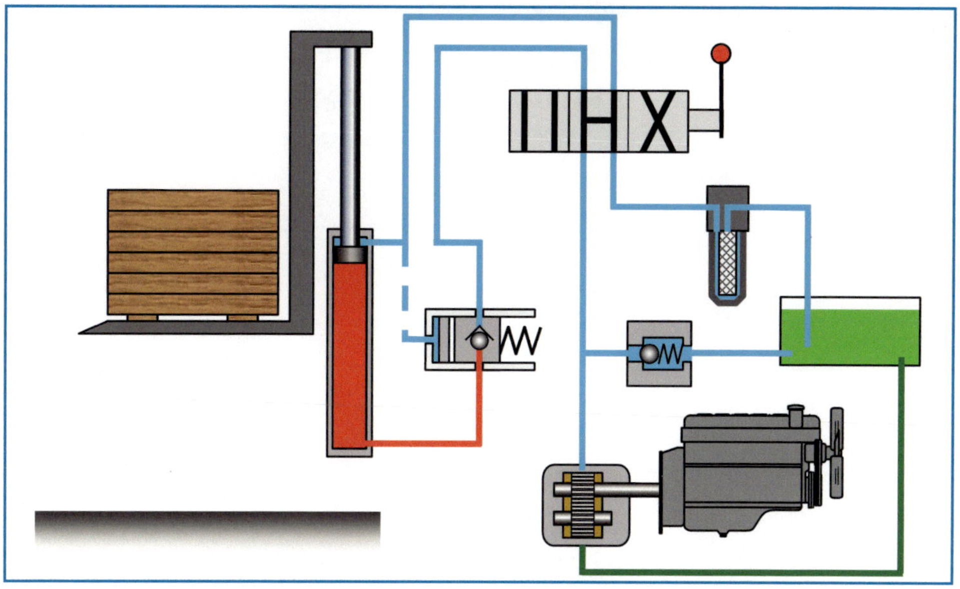

Fig. 2.51 Counterbalance valve

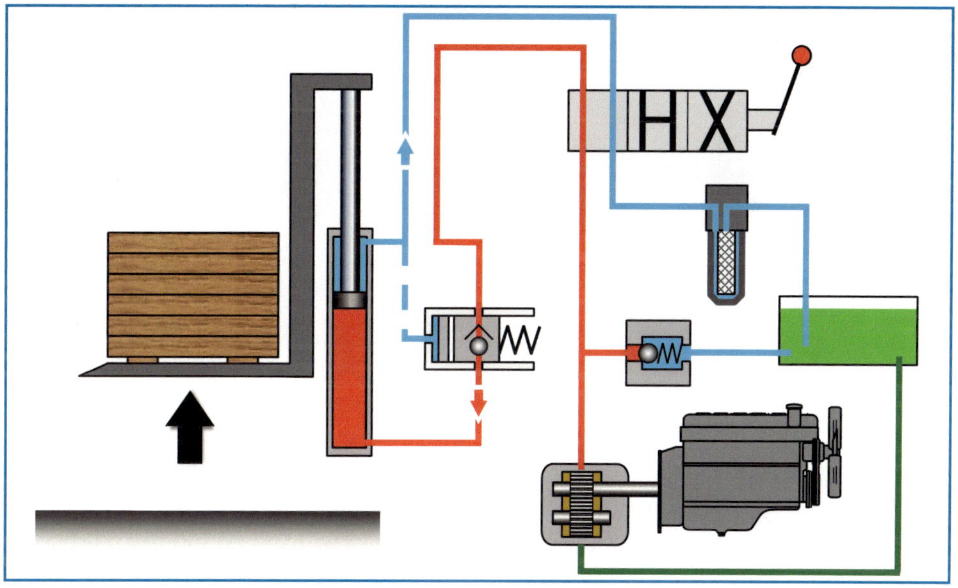

Fig. 2.52 Counterbalance valve – free reverse flow

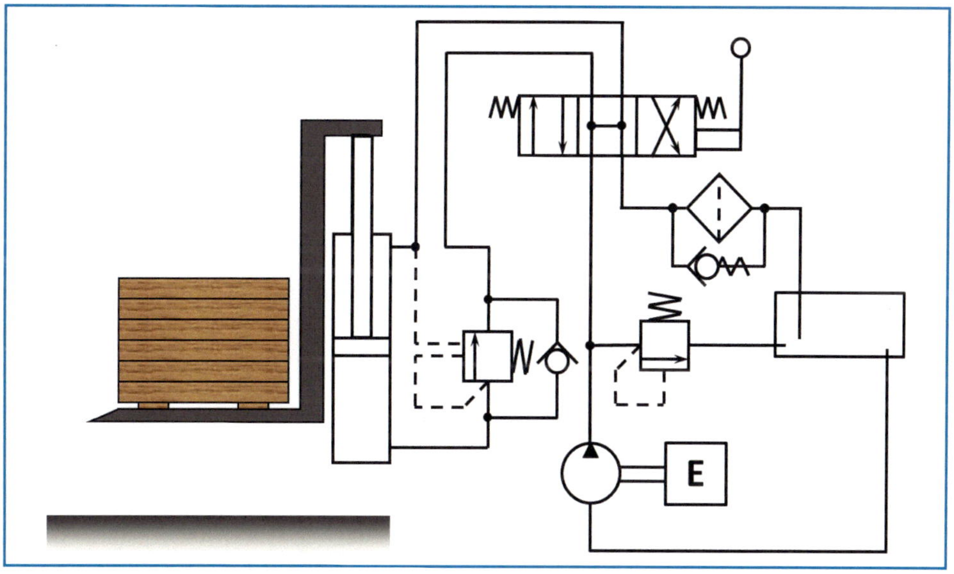

Fig. 2.53 Fork-lift circuit

Figure 2.54 illustrates a typical poppet-type counterbalance valve. The counterbalance function is provided when flow is from port 1 to port 2, either by having sufficient pressure at port 1 or by piloting the poppet open via port 3. The free-flow direction is from port 2 to port 1 when the check valve collar can be pushed away from the poppet against a light spring.

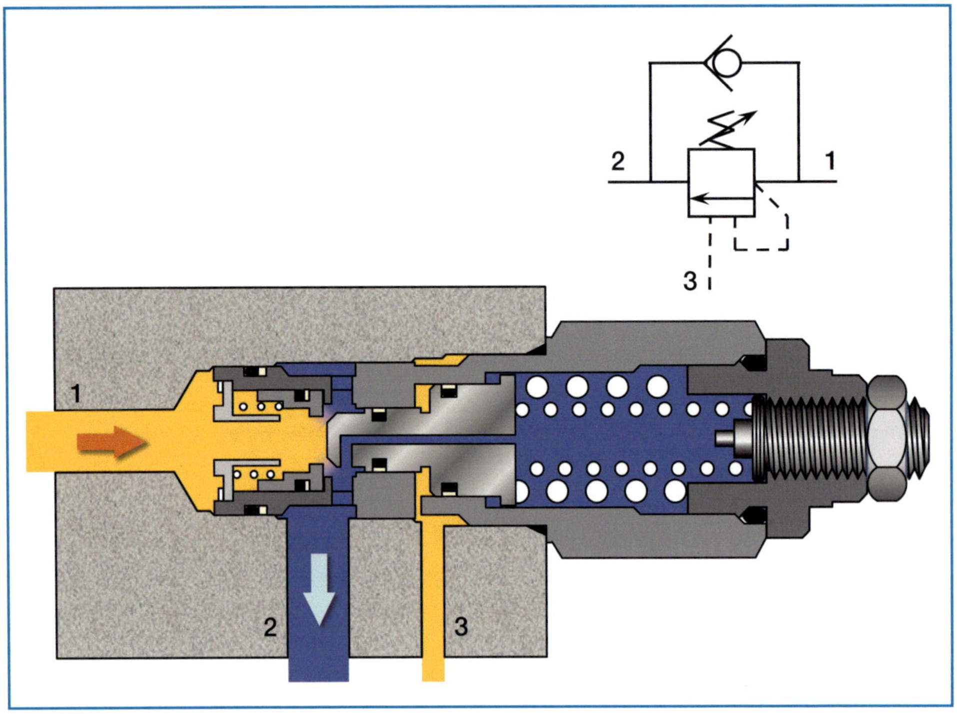

Fig. 2.54 Counterbalance valve construction

FLOW CONTROLS

In many applications the pump flow has to be shared between several different functions on the machine. If the functions are sequential (i.e. only one function operates at a time), then all the pump flow is available to each function. If two or more actuators can operate simultaneously, however, the pump flow has to be shared between them, and some functions may have to take priority over others (Fig. 2.55). Simultaneous operation of functions, such as braking and/or steering, is typical on vehicles.

To avoid having to use a separate pump for the priority functions, a **priority-flow divider valve** (Fig. 2.56) can be used to ensure that all the pump flow is directed to the priority outlet first, with the excess flow being directed to the secondary outlet. The valve also provides a **pressure-compensated** flow to the priority port, which means that, once set, the flow will remain virtually the same regardless of the pressure difference between the valve's inlet and priority outlet ports. This provides for a consistent operation of the function powered by the priority port, whatever the load happens to be.

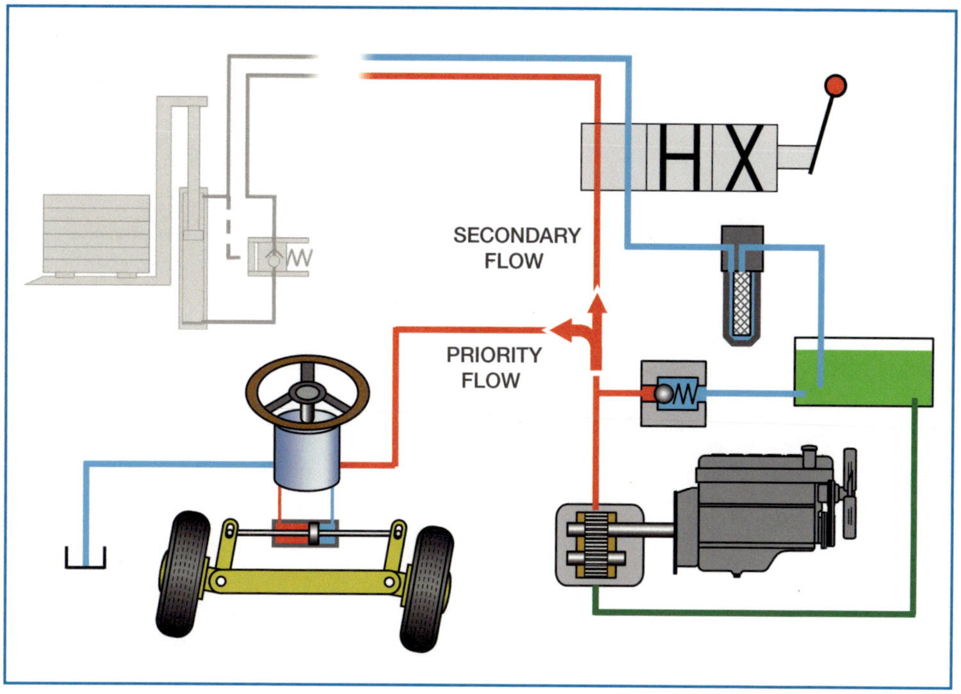

Fig. 2.55 Priority functions

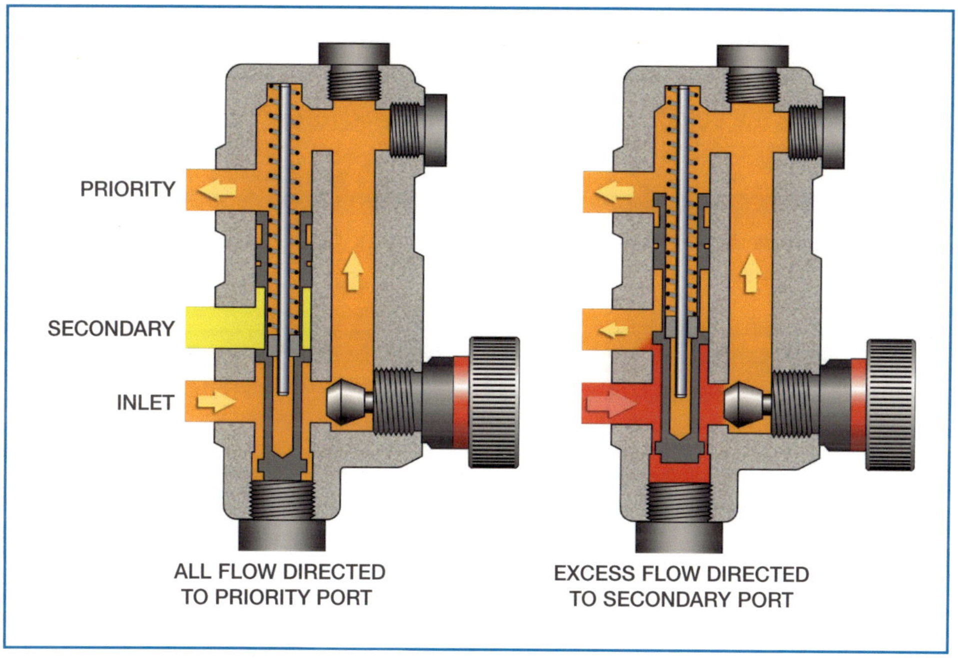

Fig. 2.56 Priority-flow divider valve

A BASIC HYDROSTATIC DRIVE SYSTEM

Hydraulic power is often used for the drive train or transmission of a vehicle, especially for off-road, utility and material-handling vehicles. In this case a rotary output is required from the hydraulic system in order to drive the vehicle wheels or tracks (Fig. 2.57).

Fig. 2.57 Vehicle transmission

DEFINITION

Hydrostatic – power transmission using the pressure of a fluid.

Hydrodynamic – power transmission using the momentum of a fluid.

A common configuration is a **closed-circuit system**, where flow from the pump is used to drive a hydraulic motor and the exhaust flow from the motor is connected straight back to the pump inlet rather than to the fluid reservoir (Fig. 2.58). By using a variable-displacement (variable-flow) pump, the speed of rotation of the hydraulic motor can be controlled directly by varying the pump flow. In addition, by using a pump that can create flow in either direction (an 'over-centre' pump, see below), the direction of rotation of the motor can also be controlled directly by the pump.

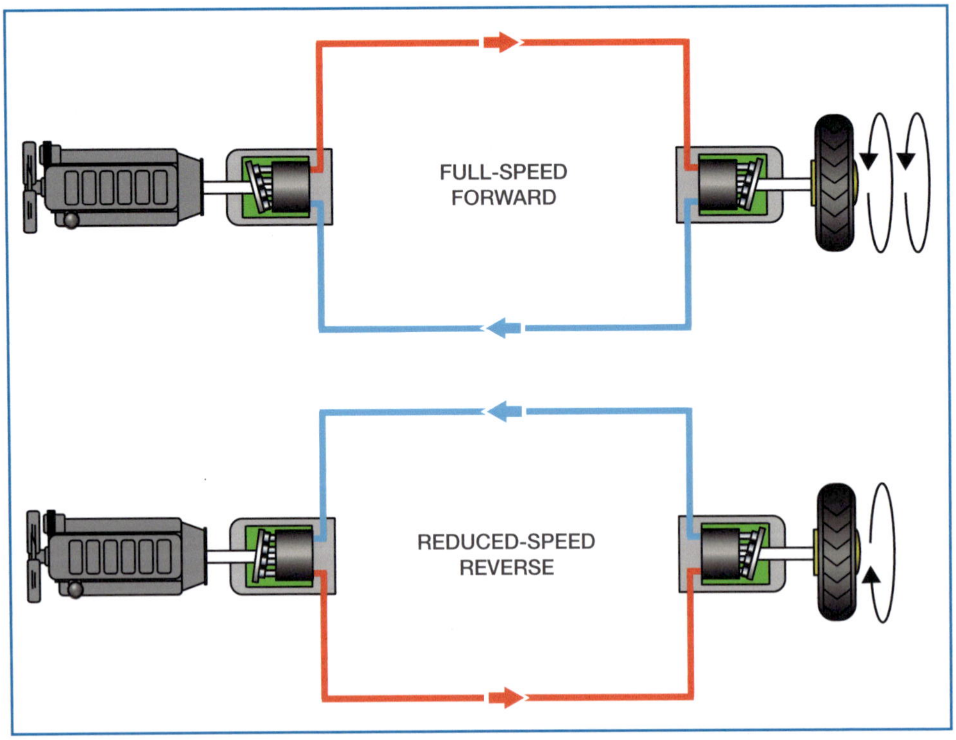

Fig. 2.58 Closed-circuit transmission

This arrangement provides an efficient system, because the system flow is not restricted by flow-control or directional-control valves and the pump always produces the required amount of flow at the pressure required to move the load at any one time. It is often referred to as a closed-circuit hydrostatic transmission to differentiate it from a hydrodynamic transmission, which uses a torque converter in conjunction with an automatic mechanical gearbox.

The type of pump normally used for closed-circuit applications is the variable-displacement axial piston pump described earlier. However, the pump is modified to allow the swashplate to tilt in either direction from the zero flow position, and is therefore known as an **over-centre pump** (Fig. 2.59). This means that, although the shaft is always driven in the same direction (and often at a fixed speed), the output flow can not only be varied but also reversed, depending on which side of centre the swashplate is moved.

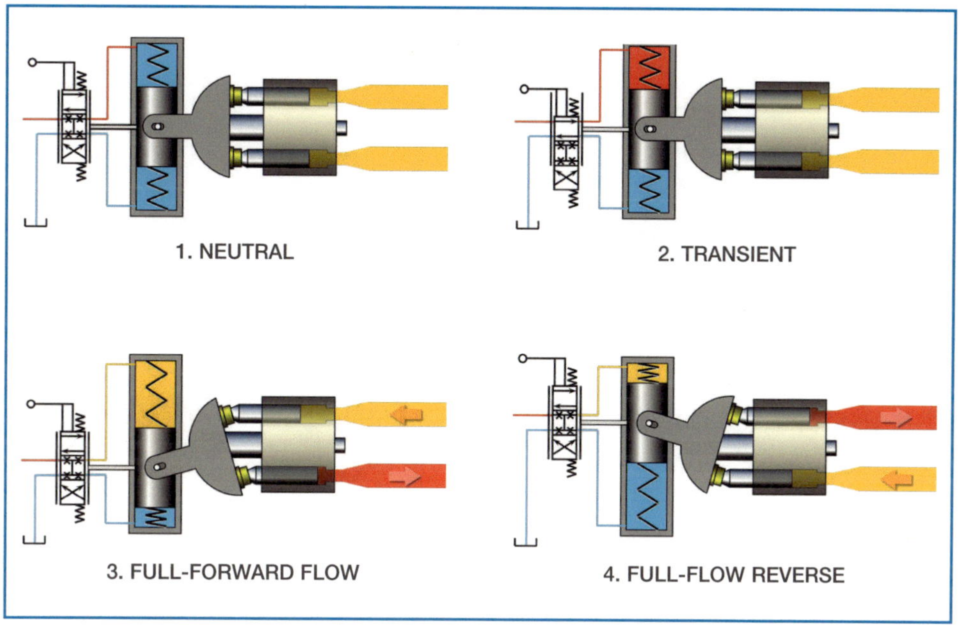

Fig. 2.59 Over-centre transmission pump (Image courtesy of Eaton Corp.)

In smaller pumps the swashplate can be moved directly by a lever connected to the operator's position by linkages or cables. Larger pumps use a power-assisted servo arrangement similar to that used for vehicle power steering. Control may still be achieved by a mechanical lever or linkage, but this now operates a directional valve spool within the pump mechanism (Fig. 2.59). As the spool is offset by an input movement, hydraulic fluid is directed to one end or the other of an internal cylinder, the piston of which is connected to the pump swashplate. The piston and swashplate then move to change the pump displacement and flow, but in doing so also move the spool valve sleeve (effectively the body of the directional valve).

When this movement corresponds to the initial input movement, the spool and sleeve are centred relative to each other, and the piston and swashplate come to rest in a position and direction directly proportional to the direction and amount of input movement. The mechanical input of the operator that controls the pump flow is therefore **servo-assisted** (or power-boosted), and the large control forces required by high-power pumps can easily be achieved.

DEFINITION

Servo – a device that accurately amplifies the power or force of an input control signal.

As mentioned above, the rotary output required from a hydrostatic transmission drive requires some form of hydraulic motor. Where the output requirement is a relatively high speed, the type of motor used will be very similar in construction to some of the pumps already discussed. Typically these are axial or bent-axis piston units, either fixed or variable displacement. The use of a variable-displacement motor in conjunction with a variable-displacement pump will increase the practicable speed range (i.e. the effective ratio between the pump input speed and the motor output speed). Where a large ratio is not required, however, a fixed-displacement motor is a common (and less expensive) choice.

WARNING

As with piston pumps, the cases of piston motors should be filled with clean fluid before they are started for the first time.

For slower-speed drives (often where the hydraulic motor is connected directly to the vehicle wheel or track), radial-piston or gerotor motors are often used. **Radial-piston motors** use multiple pistons acting either inwards on an **eccentric cam** (Fig. 2.60a) or outwards on a wave-type cam profile. In a **cam-lobe** motor (Fig. 2.60b) the pistons are often retractable. This provides a 'free-wheel' function, which may be useful if a vehicle is being towed, for example.

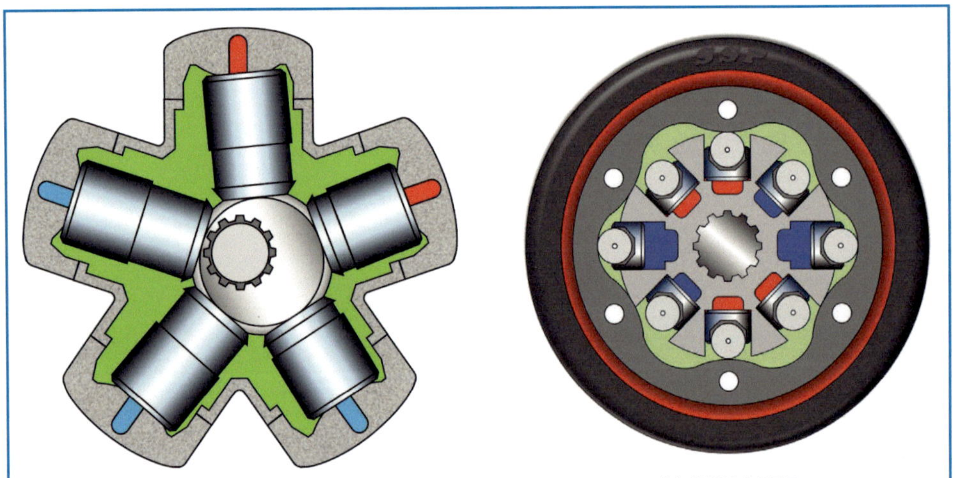

Fig. 2.60 Radial-piston motors

The **gerotor motor** (Fig. 2.61) uses a star-shaped rotor that rotates in an orbital manner, again to provide a relatively slow-speed drive but with a high torque capability.

Both piston and gerotor motors use spool- or plate-type distributor valves to direct the incoming high-pressure fluid to the appropriate pistons or cavities and thus generate the torque on the motor shaft.

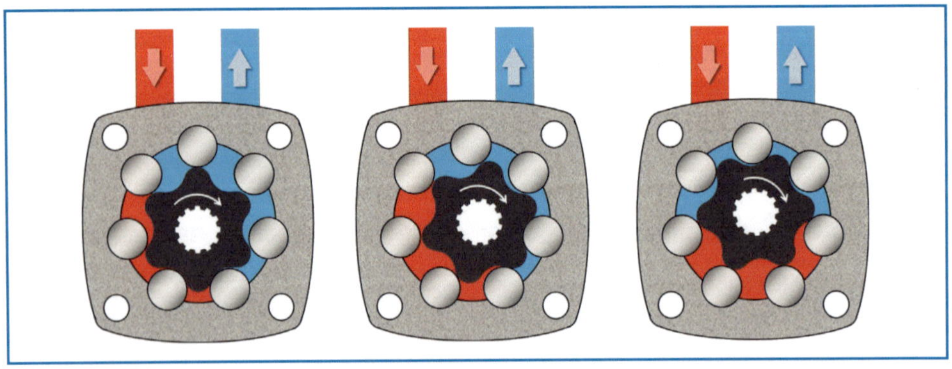

Fig. 2.61 Gerotor motor (Image courtesy of Eaton Corp.)

Although the basic circuit shown in Fig. 2.58 has only two components and no reservoir, in practice other components will be required. The internal leakage from both the pump and motor has to be drained back to a reservoir, although the reservoir used will often be much smaller than in an equivalent open-circuit system. In order to replace the fluid being drained from the closed loop, a **charge (or boost) pump** is required (Fig. 2.62). This draws fluid from the reservoir and feeds it back into the loop on the low-pressure return side. Two non-return valves are required to enable this. The charge pump itself will normally be a low-pressure, fixed-displacement pump (typically a gear pump), and so will require a simple relief valve to limit its maximum pressure.

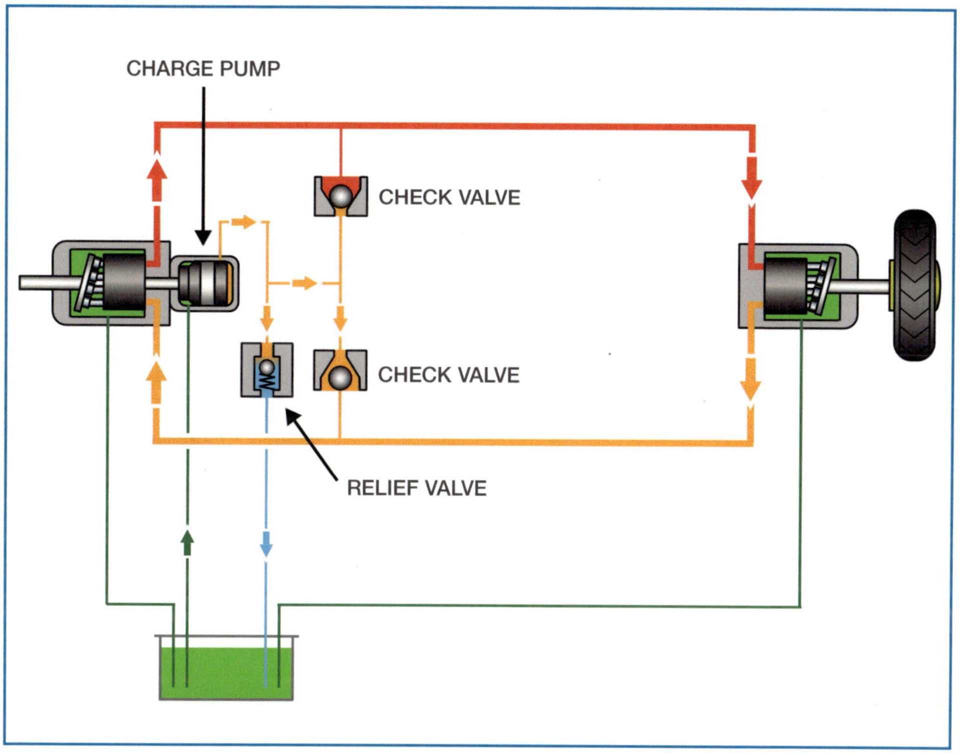

Fig. 2.62 Charge pump relief valve and check valves

The maximum pressure within the loop has to be limited to protect the system components in the event of the motor output being stalled. This can be achieved by the use of relief valves (as described earlier), but to take account of the bi-directional operation two relief valves are often used (one for each direction of rotation), connected across the two main sides of the loop (Fig. 2.63).

The final components required are fitted to enable a certain amount of fluid to be removed from the loop and directed back to the reservoir, where it can be cleaned and cooled (if necessary) before being replaced by the charge pump. This is often referred to as a **hot-oil shuttle valve** (Fig. 2.64), and consists of a simple pilot-operated directional valve spool and relief valve.

The directional valve spool is pushed across by whichever is the high-pressure side of the loop in order to bleed off flow from the low-pressure side of the loop. The relief valve ensures that a minimum pressure is always maintained on the low-pressure side (by the charge pump flow passing across it) and thus prevents possible cavitation of the main system pump.

WARNING

As mentioned previously, a relief valve passing flow at high pressure will create heat. In a closed-circuit hydrostatic system (with a relatively small reservoir) the temperature of the fluid can rise very rapidly in such situations.

DEFINITION

Cavitation damage is caused primarily by the violent collapse of fluid vapour bubbles at the outlet (high-pressure) port of a pump. The main cause of cavitation is a restricted inlet flow to the pump.

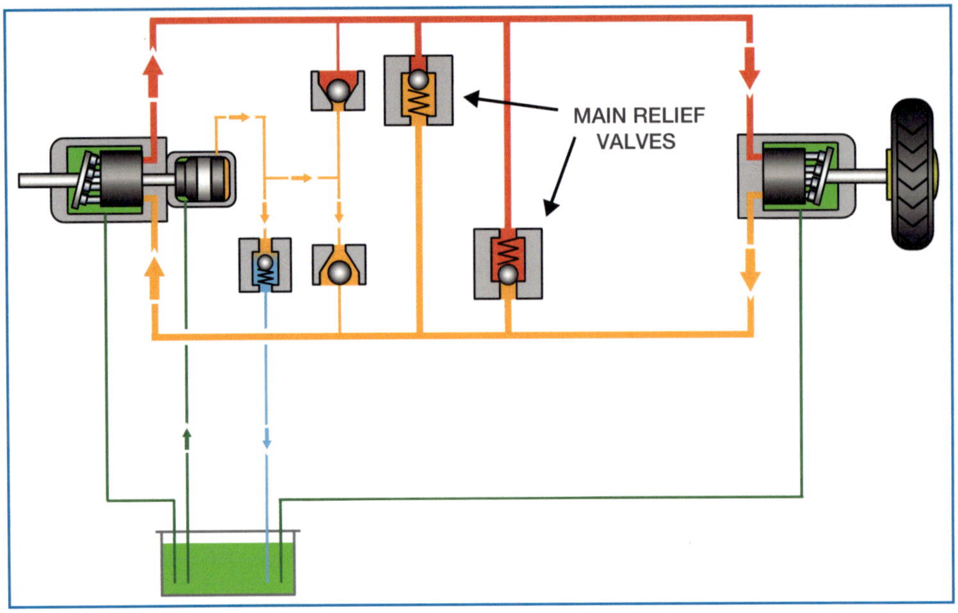

Fig. 2.63 Main loop relief valves

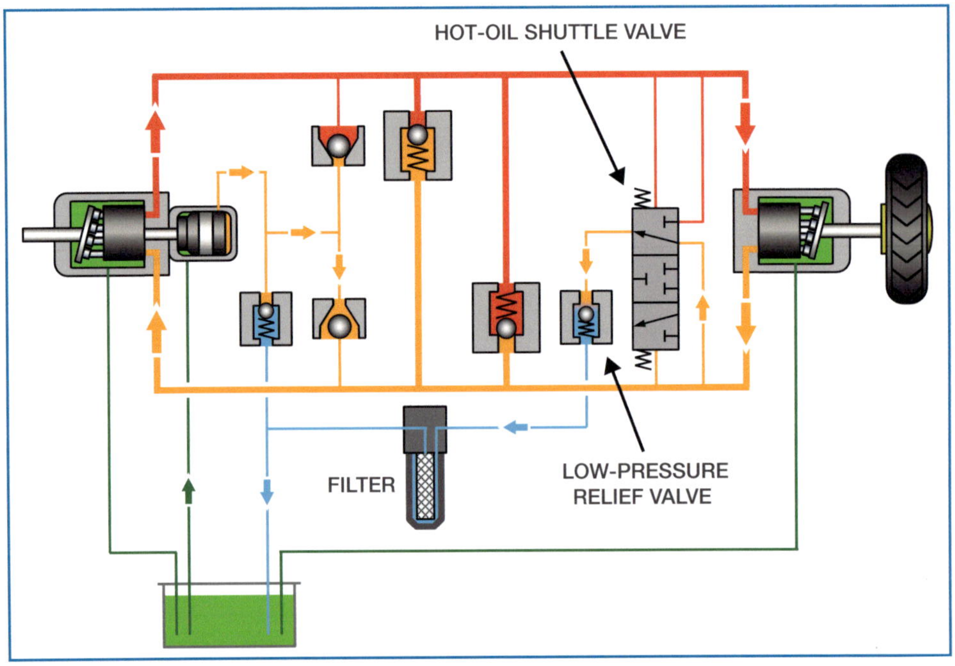

Fig. 2.64 Hot-oil flushing valve and filter

Many of the components required in a closed-circuit system can be built into either the main pump or, sometimes, the motor. In Fig. 2.65, for example, the main system pump incorporates the charge pump and relief valve, the check valves and the main system relief valves, and the hydraulic motor includes an inbuilt hot-oil shuttle valve. Such a system is shown schematically in Fig. 2.66.

Although it has only been possible in this chapter to describe a few of the components found in hydraulic systems, the basic building blocks of the components will be common to most of the others also. Hydraulic components are typically constructed from pistons, spools, poppets, springs, solenoids, etc., all of which are manufactured to tight tolerances and assembled with small clearances. To obtain an acceptable operational life from the components, therefore, it must be ensured that they are operated within the manufacturer's recommendations for such specifications as:

TOP TIP

A good understanding of how hydraulic components work is essential when maintaining or troubleshooting hydraulic systems.

This chapter has considered only a very small number of the hydraulic components in use today, so always consult the manufacturer's data sheets for operation and application information.

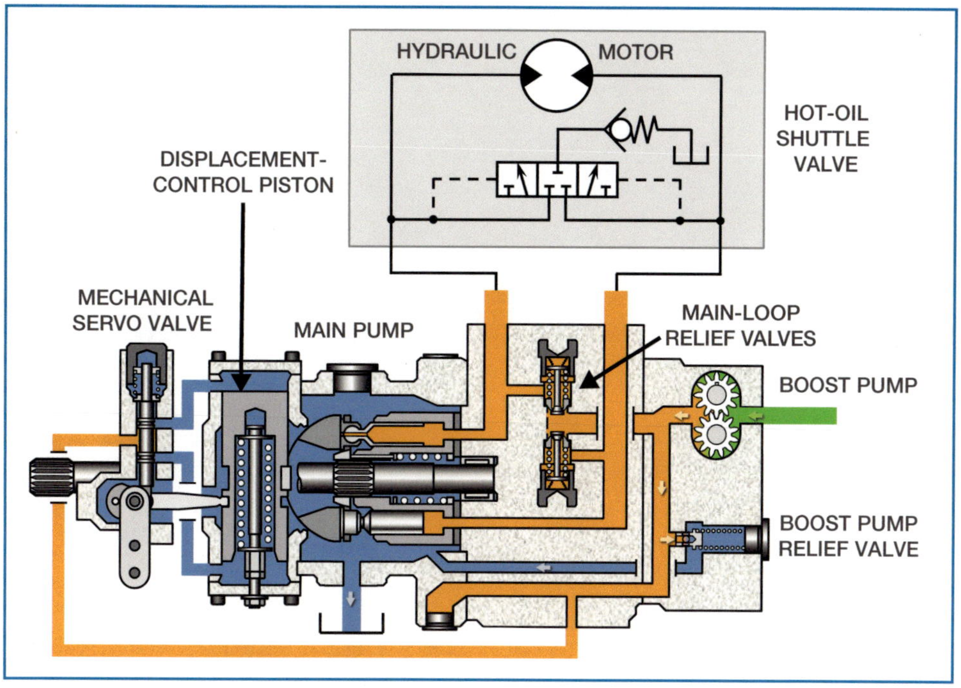

Fig. 2.65 Closed-circuit transmission system (Image courtesy of Eaton Corp.)

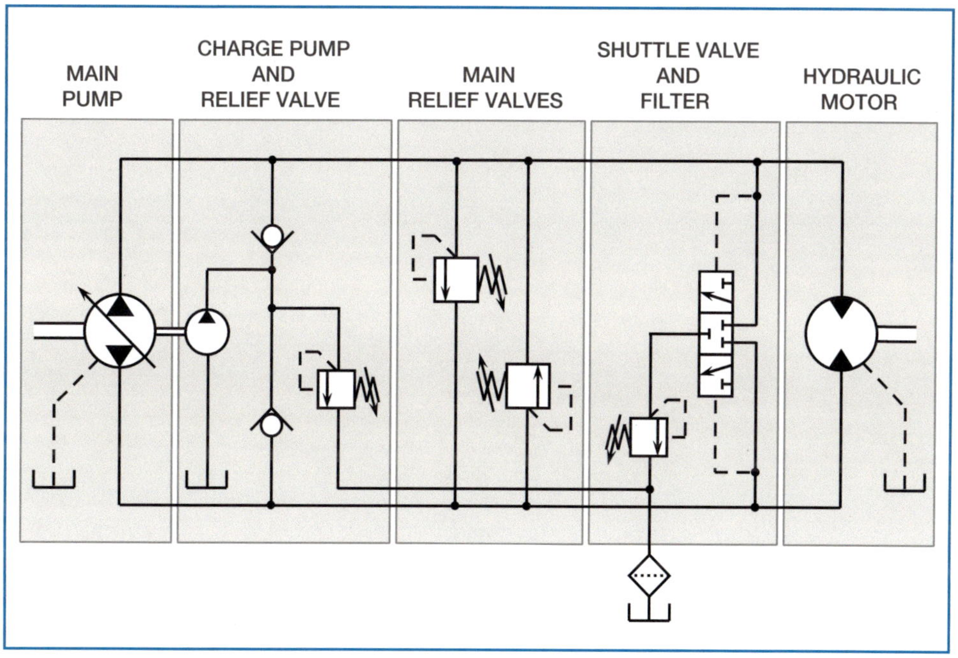

Fig. 2.66 Closed-circuit hydrostatic transmission circuit (Image courtesy of Eaton Corp.)

- operating pressure (inlet and outlet)
- operating drive speed
- supply voltage
- operating temperature (minimum and maximum)
- external shock and vibration
- fluid compatibility
- fluid type and condition
- fluid cleanliness.

FURTHER READING

For a chart of hydraulic symbols and white papers on how to select flow and directional control valves, see:

www.webtec.com/education

CHAPTER

3 HYDRAULIC FLUIDS

INTRODUCTION

The main purpose of the fluid in a hydraulic system is to transmit the power through the system from the pump at one end to the actuator at the other. In order to do this the fluid needs to have two main properties: it needs to be able to flow freely and be incompressible. As mentioned in Chapter 1, all liquids can be compressed to a degree but, unlike gasses, the amount of compressibility is very small. The ability of a liquid to flow freely is measured by a property known as its **viscosity**. A very thick fluid like treacle (molasses) is said to have a high viscosity, whereas a thin fluid such as water has a low viscosity. The viscosity of a fluid is dependent on its temperature, and the viscosity of most hydraulic fluids tends to reduce (i.e. they become thinner) when they are hot. When specifying the required viscosity of a fluid, therefore, it is necessary also to specify the temperature at which the viscosity is to be measured.

Over the years, many different means of defining a fluid's viscosity have been used, such as SAE numbers, SUS figures and so on. Currently, the generally accepted method for hydraulic fluids is the International Organization for Standardization (ISO) **viscosity grade (VG)** number, which defines a fluid's approximate viscosity in **centistokes (cSt)** at a temperature of 40°C (100°F). Commonly used hydraulic fluids have a VG between ISO VG15 (low viscosity) and ISO VG100 (high viscosity). It should be remembered that increasing pressure will also increase a fluid's viscosity, in some cases doubling the viscosity over a pressure range of 350 bar (5000 psi).

The viscosity of some fluids will vary more than others over a given temperature range. The measure of this property is known as the **viscosity index (VI)**. A fluid with a high VI will have a smaller change in viscosity with temperature compared with a low-VI fluid. Use of a high-VI fluid is often more important in mobile hydraulic systems, which may have to operate over large temperature ranges from start-up on a cold winter's morning to fully operational on a hot summer's afternoon, rather than the relatively stable temperature environment of a factory.

The viscosity of a fluid will affect the internal frictional losses within a system. The higher the viscosity the harder it is to push the fluid along pipes, hoses, etc., so in this respect a low-viscosity fluid is beneficial. However, the secondary purpose of the fluid is to lubricate the components of the system, in particular pumps and motors, which are subject to high mechanical loads and speeds. A fluid that is too thin (viscosity too low) may not be able to prevent metal-to-metal contact between components, which will lead to rapid wear. The correct choice of fluid viscosity is, therefore, often a compromise between sufficiently low viscosity that the fluid flows easily with low pressure losses and sufficiently high viscosity to provide the required

DEFINITION

The **centistoke (cSt)** is a measure of a fluid's **kinematic viscosity** in the SI system and is defined as its resistance to deformation by shear stress. For comparison, water has a viscosity of approximately 1 cSt, whereas the viscosity of a typical hydraulic oil is 50–100 cSt at room temperature.

POINT OF INTEREST

When it was first devised, the VI was an arbitrary scale ranging up to 100, which was the highest value obtainable at the time. However, modern hydraulic fluids are available with VI values much higher than 100.

level of lubrication. It will also be beneficial for the viscosity to remain relatively stable over the working temperature range of the system (i.e. a high VI).

In the early days of hydraulic systems, the fluid used was almost exclusively water. Water is inexpensive and the operating speeds of the components used were slow (most systems were powered by hand pumps), so its poor lubrication properties were not too much of a disadvantage. At the turn of the 20th century, however, when pumps started to be driven by electric motors, mineral (petroleum) oil became the fluid of choice, and it has remained so ever since. Despite the requirement for specialist fluids in certain applications, the majority of hydraulic systems in use today still use a mineral oil based fluid. However, mineral oil is a flammable fluid, so in applications where a fluid spillage or leak will create a significant risk of fire, a fire-resistant fluid is normally used instead. Also, mineral oil can pollute the environment if a spillage occurs in a sensitive area (such as forests, lakes or rivers). In such situations the use of a less environmentally damaging fluid, known generally as a biodegradable fluid, may be required.

Apart from the main properties of incompressibility and viscosity already discussed, and the possible requirements for **fire resistance** and **environmental friendliness**, there are several other properties that an ideal fluid should have, including:

- it should not react chemically with any of the materials (metals, rubbers, plastics, etc.) commonly used in hydraulic components
- it should not vaporise easily, in order to minimise the likelihood of cavitation
- it should have a low density (light weight), in order to further reduce pressure drops around the system
- it should be capable of separating out air bubbles and water quickly in the reservoir, in order to avoid them circulating around the system
- it should be a non-irritant if it comes into contact with skin or eyes
- it should remain chemically stable over long periods
- it should be as low cost as possible, both to purchase initially and to dispose of at the end of its useful life.

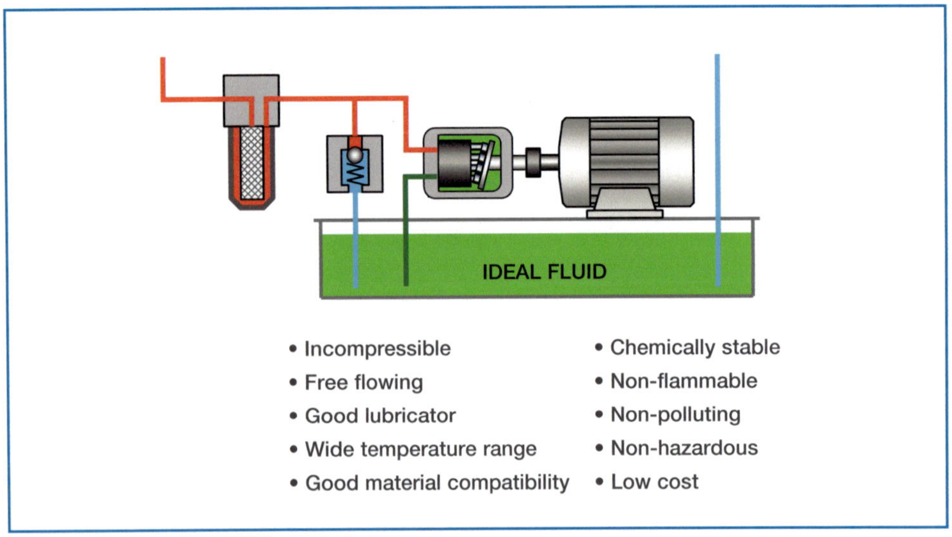

Fig. 3.1 The ideal hydraulic fluid

Unfortunately, no one has yet devised the ideal fluid that is perfect for all applications, so in practice different fluids are used depending on the priorities of the application. Three main types of fluid are in common use: mineral oil based fluids, fire-resistant fluids and biodegradable fluids.

MINERAL OIL BASED FLUIDS

Mineral oil is a good fluid in terms of its lubrication properties and that it can be made as free-flowing as necessary; however, it can be improved further by incorporating additives. Additives are chemicals dissolved or suspended in the base oil in order to enhance its properties specifically for use in hydraulic systems. For example, additives may be incorporated in the base fluid in order to:

- make the fluid more easy-flowing at low temperatures (**pour-point depressants**)
- increase the VI of the fluid (**VI improvers**)
- reduce the effects of high temperatures on the fluid (**anti-oxidants**)
- further increase the lubrication properties of the fluid (**extreme-pressure additives**)
- reduce the corrosive effects of water vapour (**rust inhibitors**)
- increase the rate at which water and air can separate from the fluid and thus reduce 'foaming' (**demulsifiers** and **anti-foam agents**).

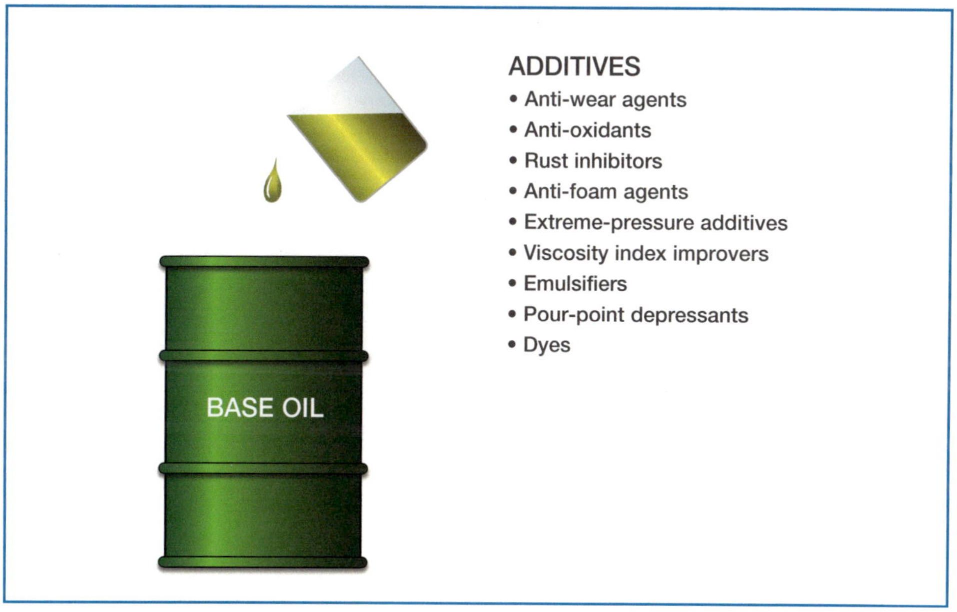

Fig. 3.2 Fluid additives

Typically, additives comprise 2–10% of the total fluid volume, but it is important to realise that the additives in a hydraulic fluid do not last forever. Some will become less effective in the presence of water or contamination in the fluid, some form coatings on the surface of metal components and are therefore 'sacrificial', and others will lose effectiveness with use (such as the long-chain molecules used as VI improvers). The condition of a hydraulic fluid is, therefore, dependent not only on the condition of the base fluid but also on the state of the additives and their ability to perform their specific tasks.

What causes a fluid to 'wear out'?

Heat

In most cases, excessive heat in a hydraulic system will result in a reduction in the fluid's viscosity. This in itself may cause wear in heavily loaded components such as pumps and motors if the fluid is no longer able to perform its lubrication function. Heat will also accelerate the rate at which the fluid oxidises, which is often characterised by a darkening of the fluid's colour (Fig. 3.3) and sometimes a noticeable smell. As the fluid oxidises it becomes more acidic and causes waxes and varnishes to form, which can coat the surface of components or block up small orifices and clearances.

DEFINITION

The acidity of a fluid is normally quantified by determining its **total acid number (TAN)**. The TAN of a hydraulic fluid can be used as a measure of when it has reached the end of its useful life. Typically, for mineral oils the TAN value should not exceed 2.0.

Fig. 3.3 Fluid degradation with excessive heat – darkening due to oxidation (Image courtesy of GPM Hydraulic Consulting Inc.)

The question arises then of how to define 'excessive' heat. There is no hard and fast boundary between an acceptable operating temperature and an unacceptable one, so normally it is necessary to follow the supplier's guidelines. Inevitably there will be variations from one fluid to another, but around 65°C (150°F) is normally regarded as the maximum operating temperature of a mineral oil based fluid. This does not mean that at 66°C the oil will instantly break down, but it does mean that the oxidation rate of the fluid (which often determines its useful life) starts to noticeably accelerate above this temperature.

Certainly by the time fluid temperatures reach around 80°C (180°F) permanent damage to standard seals and hoses is likely to occur, unless the system has been specially designed to operate at higher temperatures. The recommended maximum temperature for a hydraulic fluid should not be regarded as the normal operating temperature, as then there would be no factor of safety should something go wrong. A stall in the machine operation causing a relief valve to blow for even a short period of time could cause a rapid increase in fluid temperature. A blowing main relief valve in a closed-loop hydrostatic drive system, for example, could easily heat up the fluid by 20°C (36°F) per minute or more.

The effect of excessive heat on both the base oil and the additives is often the most important factor in determining the useful life of the fluid in a hydraulic system. Indicators such as colour and smell may provide clues about the condition of the fluid. However, a proper fluid analysis (e.g. to measure the acidity level of the fluid) is really the only way to determine for sure the true condition of the fluid and whether or not it has reached the end of its useful life. Such analyses are generally readily available from fluid or component suppliers and third-party test laboratories.

The way to avoid excessive heat build-up in a hydraulic system is to design the system to be as efficient as possible under all working conditions so that waste heat is not generated in the first place. Sometimes this is easier said than done, and heat exchangers or coolers have to be fitted into the system to remove the excess heat. In stationary applications, water coolers are normally the preferred component. However, on mobile machinery, air-blast radiators will normally be the only practical option. In this latter case either an electric or hydraulic motor drives a fan to blow air through a cooling matrix, with the system normally being thermostatically controlled.

Water

Water in an oil-based hydraulic fluid is probably second only to heat in its destructive effect on the fluid itself. Water not only causes corrosion (rust) of components and reduces the lubrication properties of the fluid but can also destroy the effectiveness of the additives in a hydraulic fluid. Again, it is not always easy to define how much water is too much, but it is of the order of a few hundred parts per million.

DEFINITION

ppm – parts per million

Depending on the temperature and the base oil itself, a certain amount of water can be dissolved in oil naturally. This is not normally visible to the naked eye but may still have detrimental effects on heavily loaded components such as shaft bearings in pumps and motors. The maximum amount of water that can be dissolved in an oil is referred to as the 'saturation level'. Targeting an operational level of no more than approximately 50% of the saturation level provides a factor of safety and will have a positive effect on bearing life in particular. Levels of water above the saturation level start to become visible as cloudiness or a milky appearance of the fluid, and at this level of water damage to components and the fluid is almost certain to occur (Fig. 3.4).

Fig. 3.4 Water-contaminated mineral oil

If a mineral oil fluid is cloudy or milky in appearance, there is definitely too much water present, but it does not follow that if there is no such visible cloudiness the water content is acceptable. Removing water from a hydraulic system fluid is not always an easy task, but the options include:

- Draining water from the tank – a shallow V-shaped profile at the bottom of the tank helps to collect any water, which can then be removed periodically via a drain connection situated at the bottom of the 'V'.
- Water-removal filters – these combine the tiny droplets of water into larger drops, which then fall to the bottom of the filter bowl and can then be drained off via a drain connection.
- **Centrifuges** – these can be used to separate large volumes of water from the hydraulic oil. They are normally used on a temporary off-line basis to recover the fluid from a catastrophic failure such as a burst cooler.
- **Vacuum dehydration** – this effectively 'sucks' the water out of the fluid. It can be used either online or off-line.

Preventing water from getting into a hydraulic system is usually much easier than trying to remove it. Entry points for water into a hydraulic system are through ineffective sealing of the system, via the air breather or from a leaking water cooler (Fig. 3.5). All connections into a hydraulic reservoir or components mounted on it must be effectively sealed to prevent the ingress not only of dirt but also of water from rain, splashing or machine wash-down. When the level of fluid drops in a reservoir, due to the movement of cylinders or charging of accumulators, air is drawn into the

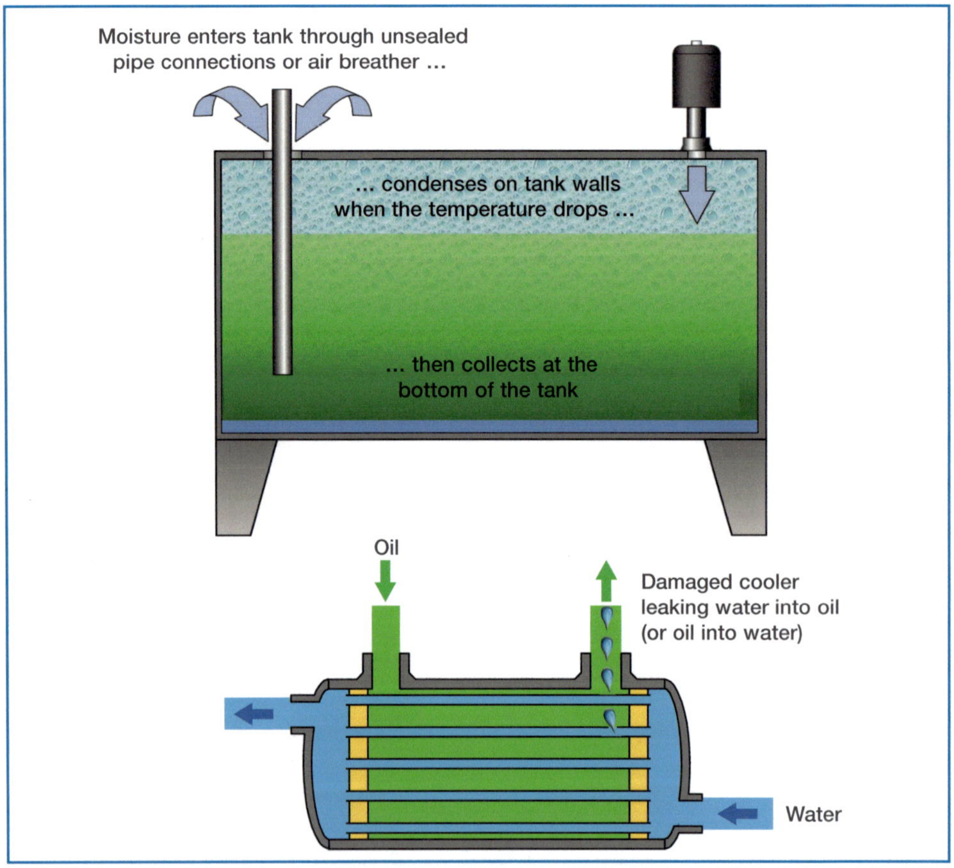

Fig. 3.5 Water contamination

reservoir to replace the fluid volume. In warm humid environments the air can contain a significant amount of water vapour, which may condense on the walls of the tank when the temperature drops (e.g. overnight or when the machine is switched off). Over a period of time this can build up to a significant amount of water collecting at the bottom of the reservoir (as water has a higher density than oil).

To reduce the amount of water vapour entering a reservoir, a **desiccant** material or water-impervious membrane can be included in the air breather to absorb or prevent moisture being drawn into the reservoir. A burst or leaking water cooler, however, has the potential to dump a large volume of water into the system in a short space of time, so effort should be made to protect the cooler from pressure peaks, etc., which could cause such damage.

DEFINITION

A desiccant is a material that absorbs water. An everyday example is the small sachets of silica gel often packaged with electronic items to reduce humidity and possible corrosion.

For critical applications, online sensors can be used to monitor the water content of a fluid and thus provide a warning of when action needs to be taken. Such sensors are now available at moderate cost.

Dirt

As will be discussed in Chapter 4, dirt can have a very damaging mechanical effect on hydraulic components, causing wear, erosion, jamming of components, etc. What may not always be appreciated, however, is that dirt can also have a detrimental effect on the fluid itself. Fine particles of solid contamination can interfere with water-separation additives and can act as a catalyst for the oxidation process, thus reducing the useful life of the fluid.

Ageing

As mentioned previously, some of the additives used in hydraulic fluids are sacrificial (i.e. they are 'used up' during the natural operation of the system). Anti-wear additives, for example, which tend to form a low-friction coating on metal surfaces, have a finite life. VI improvers, which are composed of long-chain molecules that 'tangle' together to increase a fluid's viscosity, tend to get 'chopped up' by the mechanical action of some hydraulic components, a phenomenon known as 'shearing down'.

Prolonging the life of a hydraulic fluid

While some of the above processes may be obvious from a simple visual inspection of the fluid or comparison with a sample of new fluid, others will not be. A good maintenance procedure should therefore include a fluid analysis that is carried out at regular intervals depending on the type of fluid and the duty cycle of the system. In short, however, the key to prolonging the life of a hydraulic fluid can be summarised as:

Keep it cool – Keep it clean – Keep it dry

FIRE-RESISTANT FLUIDS

In applications such as steel making, die casting and furnaces, a leak or spray of flammable hydraulic fluid from a burst hose or fitting has the potential to cause a serious fire. In such cases, therefore, a fire-resistant fluid must be used rather than a flammable mineral oil fluid. Fire-resistant fluids can broadly be divided into two categories: **water-based** and **synthetic** fluids.

Probably the commonest fire-resistant fluid is **water glycol** which, as its name suggests, is a combination of water and polyglycol (a fluid similar to car anti-freeze). The proportions of the two fluids will vary from one manufacturer to another, but are typically in the ratio 40% water to 60% glycol. The water component provides the fire resistance while the glycol thickener provides the lubrication and other properties required by the hydraulic components. As with mineral oils, additives help to improve the lubrication properties of the fluid and help with corrosion resistance, air release, etc.

Inevitably, however, the lubrication properties of any water-based fluid will never be as good as those of mineral oil, so the pressure (and sometimes speed) rating of hydraulic components may have to be lower than if mineral oil is used. This is especially true for pumps where inlet conditions and outlet pressures are very important to the life expectancy of the pump. In some cases, pumps can be modified (with surface coatings) to further improve their compatibility with water glycol fluids.

DEFINITION

High water content fluid (HWCF) is sometimes referred to as **high water based fluid (HWBF)**.

DEFINITION

An **invert emulsion** is a fluid where the oil content is greater than the water content.

A second main type of water-based fire-resistant fluids are known as **water–oil emulsions**. Unlike water glycol, these are not true solutions but mixtures of very small droplets of either water in oil or oil (or synthetic chemicals) in water. In some cases the water content may be as high as 90–95%, in which case the fluid is referred to as a 'high water content fluid' (HWCF) or an oil-in-water emulsion. In other types the oil/water ratio is similar to that of water glycol (i.e. approximately 60% oil and 40% water) (Fig. 3.6), in which case they are then known as water-in-oil or, more commonly, invert emulsions.

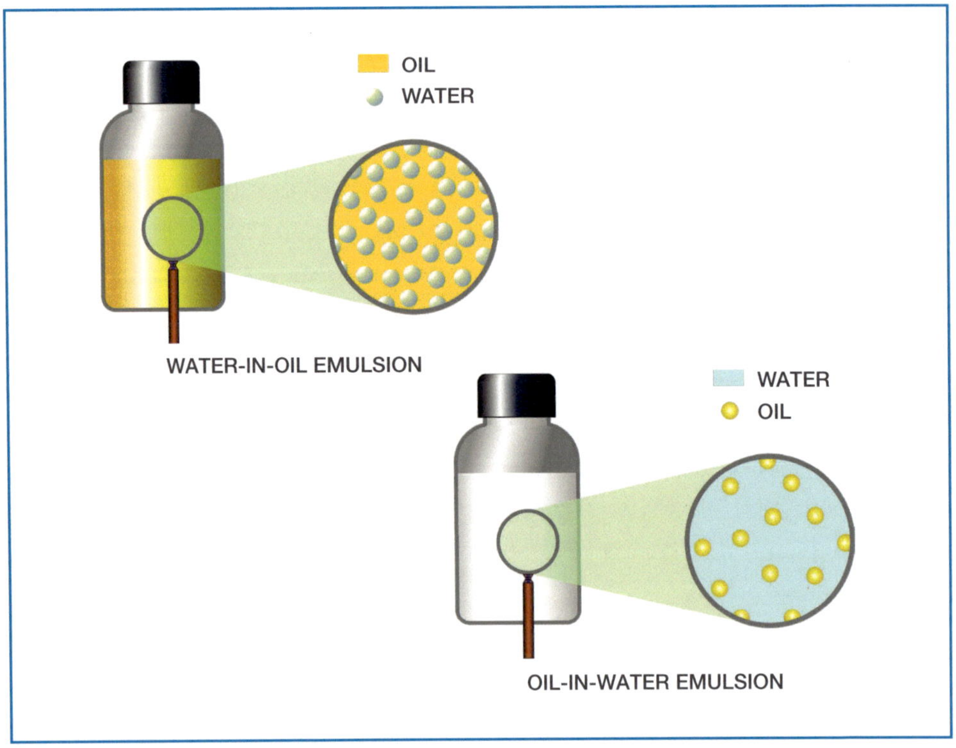

Fig. 3.6 Water and oil emulsions

As with water glycols, the water content of water–oil emulsions provides the fire resistance, and the oil lubricates the components of the system. However, the oil or chemical content of a HWCF fluid is relatively low, so components often have to be de-rated in terms of pressure, speed or both when this type of fluid is used. The use of HWCFs is normally confined to applications where a high level of fire resistance is required and the hydraulic components in the system are not heavily loaded or do not have continuous duty cycles. This type of fluid was originally developed for use in the coal-mining industry.

All water-based fluids will require a relatively high level of maintenance to keep them in good condition. Inevitably the water in the fluid will tend to evaporate over time, resulting in a more viscous fluid with lower fire-resistance properties. In such situations the water content must be restored to its correct level, usually by the addition of distilled or de-ionised water. Bacterial growth can also be a problem with water-based fluids, causing clogging of the system, especially filter elements. Bactericides can be added to the fluid to limit this problem, but compromises may then have to be made with regard to environmental concerns.

Maximum temperatures for water-based fluids will be lower than for mineral oils, with 50°C (120°F) being the normal recommended level to avoid excessive water evaporation. Fluid cleanliness and pump inlet conditions may also require special attention. The relatively high vapour pressure and higher density of water means that pump cavitation can be more of an issue, so positive-head reservoirs are generally recommended. As for mineral oils, a systematic fluid analysis programme is recommended when water-based fluids are used, to ensure that the fluid and additives remain in ideal condition. However, fluid sampling will inevitably have to be carried out more frequently than when mineral oil is used.

Where the demands of the hydraulic system components are beyond the capabilities of a water-based fire-resistant fluid, fully synthetic fire-resistant oils can be specified. These generally have superior lubrication properties. The two common fluids in this category are phosphate esters and polyol esters. The lubrication properties of these synthetic fluids are very good, as is their fire resistance, but they do have some disadvantages, which can be summarised as follows:

Phosphate esters

- are aggressive to standard paints and seal materials, and normally **fluorocarbon** (Viton) or **ethylene propylene diene monomer (EPDM)** seal materials have to be used
- have a high specific gravity (heavier than water), which tends to increase system pressure drops and may affect pump inlet conditions
- are very expensive to purchase and to dispose of
- have a low VI, so system temperatures have to be kept as stable as possible
- are a potential health hazard if they come into contact with skin or eyes, and may produce toxic fumes when heated to high temperatures.

Polyol esters

- are aggressive to some metals and require fluorocarbon seal materials
- have a high specific gravity (but less so than phosphate esters)
- are expensive (but cheaper than phosphate esters).

BIODEGRADABLE FLUIDS

In applications where a spillage of hydraulic fluid could cause an environmental pollution problem, a **biodegradable fluid** may be required. The requirement of biodegradable fluids is that a specified proportion of the fluid must decompose into harmless products in a specified time (e.g. 60% of the fluid must break down in 28 days). This then ensures that a spillage of hydraulic fluid on farmland, in forests, in lakes or in rivers will not unduly harm plant or animal life. The decomposition process is triggered by heat and water, so it is necessary to prevent water from entering the hydraulic system under normal working conditions, and in many cases to limit the maximum working temperature.

Biodegradable fluids are derived from either vegetable oils or synthetic oils, and in some cases may be a combination of the two. Typical vegetable oils used include **sunflower**, **rapeseed (canola)** and **soya**. Although additives may be incorporated in biodegradable fluids (e.g. as viscosity thickeners), it is obviously a requirement that the additives are also environmentally friendly, which tends to rule out traditional anti-wear compounds, for example, which often contain zinc. Synthetic biodegradable fluids are generally more expensive than vegetable-based oils but are able to operate at higher temperatures.

As with fire-resistant fluids, special care needs to be taken with pump inlet conditions, cleanliness levels and reservoir design in order to promote the effective release of air bubbles. In general, therefore, systems using biodegradable or fire-resistant fluids need to be specially designed from the outset.

FLUID CLASSIFICATION

The international standard ISO 6743-4 established a classification system for lubricants and industrial oils, including fluids used in hydraulic systems, which are all coded with the letter H. Figure 3.7 illustrates some of the more common designations for the three groups of fluids discussed in this chapter.

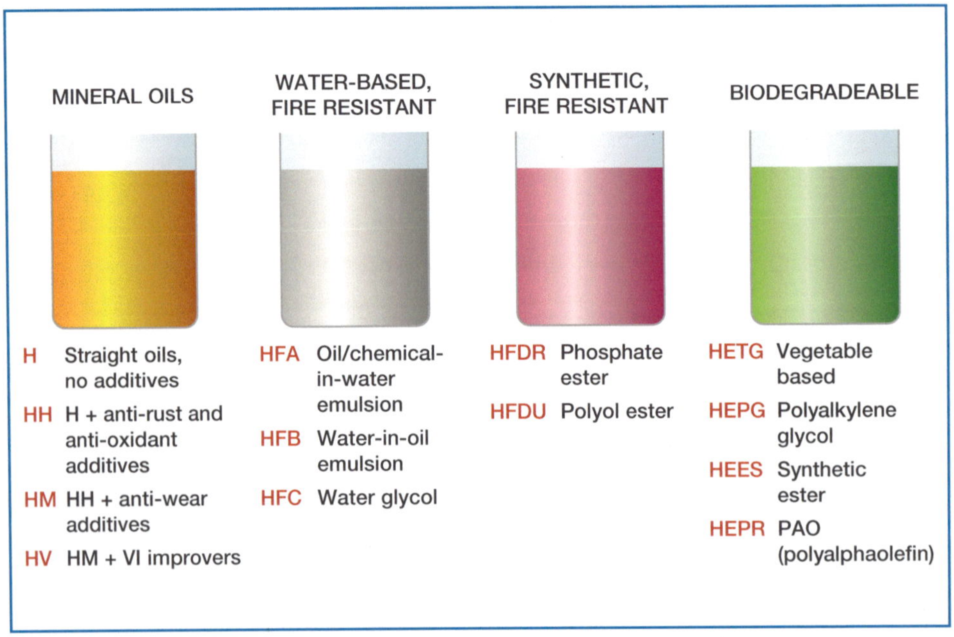

Fig. 3.7 Common hydraulic fluids

The other commonly used standard for mineral oils is the German standard DIN 51524, which uses a similar method of coding. Table 3.1 summarises the designations given in the two standards.

Table 3.1 Classification of fluids used in hydraulic systems – comparison of ISO and DIN standards

Description	ISO 6743-4	DIN 51524
Straight oil with no additives	HH	H
HH plus anti-oxidant and anti-corrosion additives	HL	HL
HL plus anti-wear additives	HM	HLP
HLP plus detergent additives		HLPD
HM plus VI improvers	HV, HR	HVLP
HM plus anti stick–slip additives	HG	

FLUID CHANGEOVER

As mentioned previously, a hydraulic system ideally needs to be designed for the fluid it will use throughout its life. However, situations sometimes arise where it is necessary to change from a mineral oil fluid to either a fire-resistant fluid or biodegradable fluid. In such cases it is important to remove as much of the old fluid as possible, bearing in mind that in many mobile systems there may well be more fluid stored in pipes, hoses and actuators than there is in the reservoir. Even small amounts of mineral oil left in the system can cause problems, so thorough flushing of the system will be required and the flushing fluid will have to be drained and replaced with new fluid as many times as necessary. Whenever it is required to change a fluid in a system, therefore, the fluid supplier should be consulted for guidelines on the process to use and the acceptable levels of cross-contamination.

WARNING

Changing the fluid in a hydraulic system from one type to another may involve more than simply replacing the fluid. Component compatibility and operating parameters must also be checked with the manufacturers.

FURTHER READING

For further practical information on hydraulic fluids see: *Hydraulic Fluids – A Practical Guide* (P104), by David Phillips (available from the British Fluid Power Association (BFPA))

For a web-based app to model the viscosity of hydraulic oil under different temperature and pressure conditions, see: www.webtec.com/education

CHAPTER

4 FLUID CLEANLINESS AND TEMPERATURE CONTROL

INTRODUCTION

As explained in Chapter 3, the key to prolonging the life of a hydraulic fluid can be summarised by the maxim 'Keep it clean, keep it cool and keep it dry'. This chapter explains how this can be achieved in practice and describes the components necessary to maintain the fluid in good condition. In addition to the different types of hydraulic fluid available, there are also many suppliers to choose from, so selecting which fluid to use in a particular system is not always easy. If long fluid life is a prime consideration, then it makes sense to choose a high-quality fluid from a reputable supplier, especially for applications that involve severe duty cycles. In situations where the hydraulic system operates very infrequently or, as in the case of some agricultural equipment for example, for only a few days each year, the requirements of the fluid may be different.

TOP TIP

To prolong the life of a hydraulic fluid:

Keep it clean – Keep it cool – Keep it dry

and check it regularly.

There is an often-quoted statistic within the hydraulics industry that *x*% of breakdowns are caused by poor fluid condition (where *x* seems to vary between 70% and 90%). The hard evidence to support these numbers seems to be fairly thin, but there can be little doubt that lack of fluid cleanliness and poor fluid condition are a significant factor in system failures. Therefore, whether the actual figure is 50%, 70% or 90%, the basic principle of keeping the fluid clean, cool and dry still holds good if unexpected breakdowns are to be avoided.

Why is fluid condition so important in hydraulic systems?

- The fluid is the lubricant for all the hydraulic components, so any loss of lubrication properties will affect heavily loaded components in particular (such as pumps and motors).
- Hydraulic components are precision-made items with very small clearances, and their function is thus susceptible to very small particles of contamination. They can also be damaged by high levels of fluid acidity or a high water content of the fluid.
- Many faults that can occur in a hydraulic system will directly affect the fluid itself. For example, a blowing relief valve will cause excessive heat, a leaking joint may cause aeration and a split cooler will cause water contamination.
- The fluid is the one component of a hydraulic system that connects all the other components together, so any problem with the fluid is rapidly spread around the system.

A micrometre (µm) is 1/1000 of a millimetre. Its former name of micron is still widely used.

POINT OF INTEREST

Other standards are or have been used to define fluid cleanliness, in particular the National Aerospace Standard **NAS 1638**, which is now being gradually superseded by the Society of Automotive Engineers standard **SAE AS4049**. For a comparison of industry standard cleanliness codes visit www.webtec.com/education

MEASUREMENT OF FLUID CLEANLINESS

In order to explain how to achieve fluid cleanliness it is first necessary to explain how cleanliness can be defined and measured. There are several methods of defining cleanliness levels, but the one in widest use is that given in the International Organization for Standardization (ISO) standard **ISO 4406**. In this standard a code number is allocated according to the number of particles per millilitre of fluid larger than 4, 6 and 14 micrometres (µm).

Figure 4.1 illustrates a portion of the coding system. For example, if the number of particles greater than 4, 6 and 14 µm in a 1 mL sample of fluid is 3250, 780 and 290, respectively, the cleanliness level is stated as 19/17/15.

Range code	No. of particles per mL	
	From	To
24	80,000	160,000
23	40,000	80,000
22	20,000	40,000
21	10,000	20,000
20	5,000	10,000
19	2,500	5,000
18	1,300	2,500
17	640	1,300
16	320	640
15	160	320
14	80	160
13	40	80
12	20	40
11	10	20
10	5.0	10
9	2.5	5.0
8	1.3	2.5
7	0.64	1.3
6	0.32	0.64
5	0.16	0.32
4	0.8	0.16
3	0.4	0.8
2	0.2	0.4
1	0.1	0.2

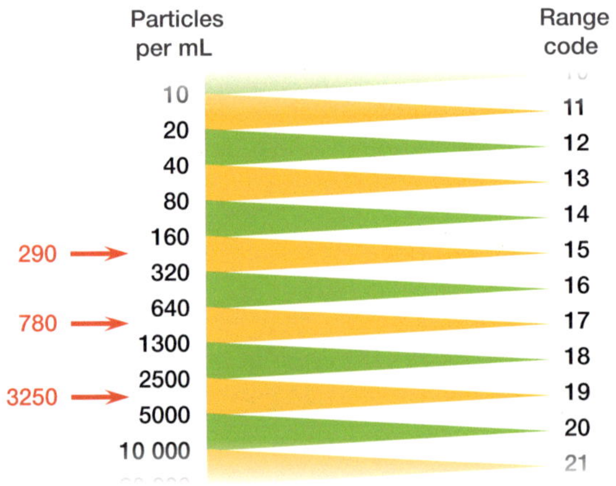

Fig. 4.1 Cleanliness codes given in ISO 4406

Most contamination monitoring is carried out using automatic particle-counting devices, which have to be calibrated using a sample of fluid with a known contamination profile (i.e. the number and size distribution of particles). Originally, the ISO standard considered particle sizes of 5 and 15 µm. Later, as the sensing technology improved, 2 µm particles were incorporated in the standard. Then, in the late 1990s, a forced change in the calibration medium resulted in slightly different results, so the particle sizes were amended to 4, 6 and 14 µm in order to keep the code numbers consistent. However, where particle counting is carried out manually by means of a membrane and microscope, the 5 and 15 µm particle size reference is maintained.

Most equipment manufacturers provide recommended (or sometimes mandatory) cleanliness codes for the fluid used with their components (Fig. 4.2). These guidelines ensure an acceptable life is obtained from components (especially hard-worked items such as pumps and motors) and may also be a requirement of the manufacturer's warranty conditions.

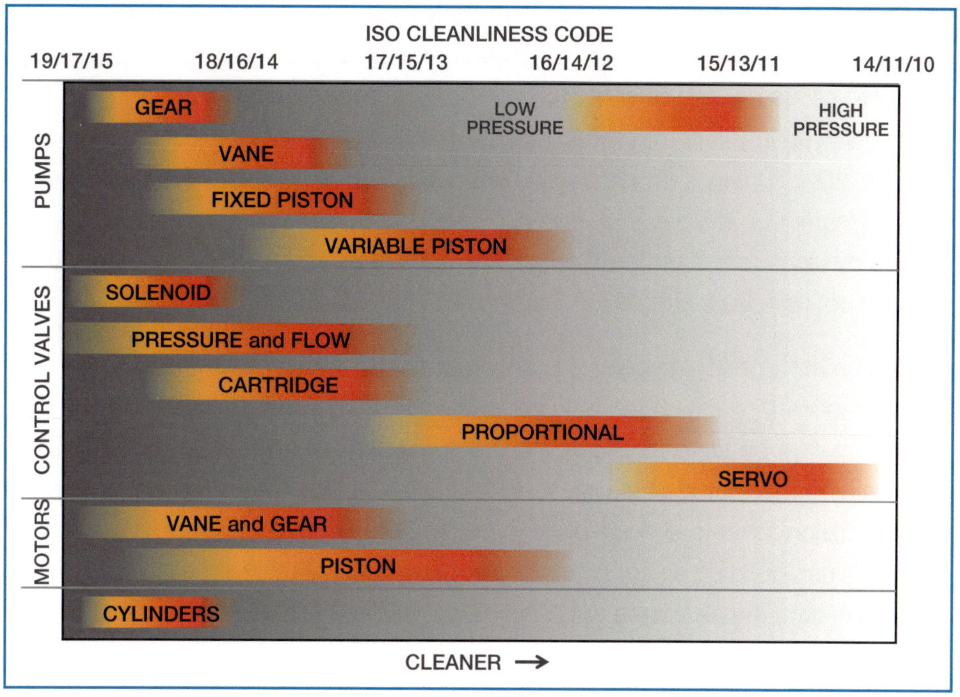

Fig. 4.2 Typical recommended ISO 4406 cleanliness levels

EFFECTS OF CONTAMINATION IN A HYDRAULIC SYSTEM

Particles of contamination in a hydraulic fluid can have a detrimental effect on the components of a system. Large particles, like those created by the breakdown of a seal or inadequate flushing of manufacturing debris, can block up orifices in control components, jam spools in valves or prevent the closure of poppets onto seats (Fig. 4.3).

Smaller particles, of the same order of size as the clearance between spools or pistons and their corresponding bores, tend to get pushed into the clearances by pressure differences. With components such as solenoid-operated spool valves this

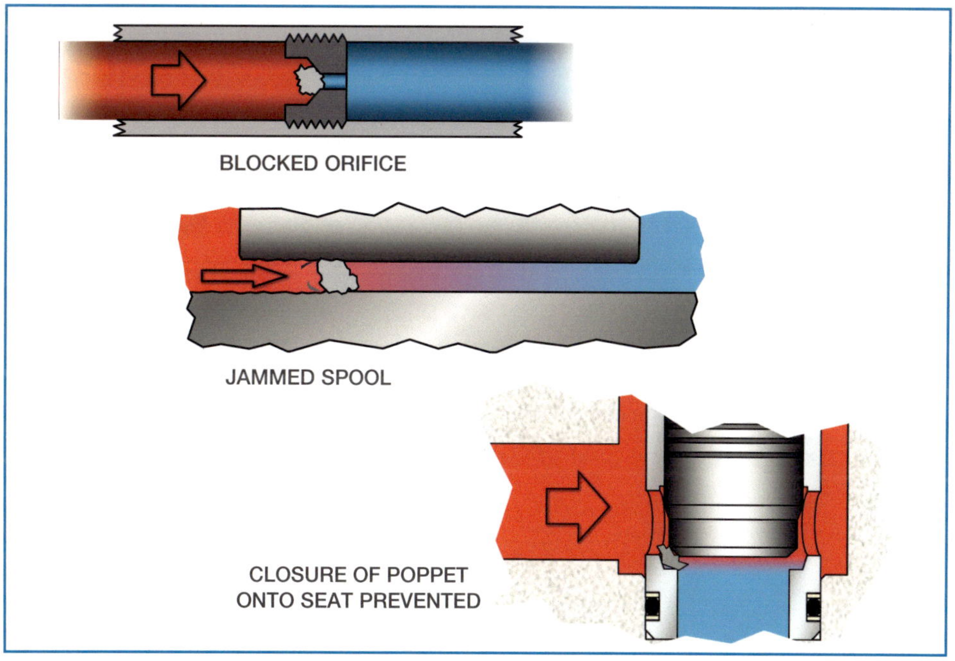

Fig. 4.3 Particle contamination

DEFINITION

Wear caused by two component parts rubbing against each other is referred to as two-body wear. Wear caused by dirt particles trapped in the clearance between two component parts is known as three-body wear.

POINT OF INTEREST

Under ideal conditions and with perfect eyesight, the unaided adult human eye may be able to see particles only as small as 30 µm. Therefore, a fluid sample that looks clean to the eye may not be clean enough for use in a hydraulic system.

can cause an immediate malfunction of the valve as the spool becomes jammed in its bore. In other components, such as pumps and motors, a clearance-sized particle is likely to cause scoring of either or both of the surfaces due to the higher levels of power involved in the component. Typical clearances in hydraulic components vary between 20–25 µm in a simple control valve and 1–2 µm in a servo control valve or under dynamic conditions in a pump. Furthermore, the damage caused to each surface by clearance-sized particles is likely to create more contamination particles which, if not removed, accelerate the degradation process.

It might be assumed that very tiny particles of contamination (i.e. much smaller than the clearances through which they pass) do not create a significant problem. This may not always be the case, however, due to a phenomenon known as silting. This is something that occurs typically in sliding spool valves where the spool remains in one position for long periods of time. In this case, a small logjam of particles in a clearance can build up a very much larger mass of tiny particles behind it, and over a period of time the clearance will become completely blocked and jam the spool in the bore (Fig. 4.4).

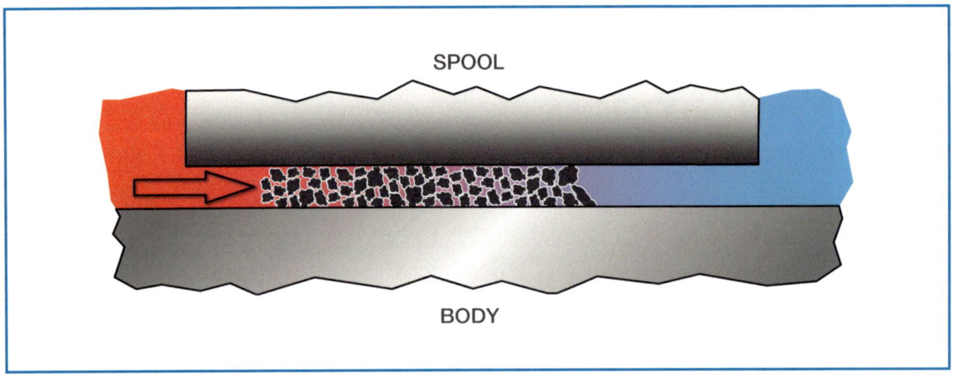

Fig. 4.4 Spool silting

This could be a particular problem in an application where a valve has to operate only on machine shutdown or in the event of a power failure (such as the drain valve on an accumulator). For most of the time the valve will be in an energised state, but on shutdown or a power failure will be required to spring return to a 'safe' condition. If the spool has jammed due to silting, the safe condition will not be achieved.

As it may be very difficult to remove these tiny particles from the fluid (plus the fact that the number of particles tends to multiply over time as larger particles get ground down), a possible solution would be to use a poppet-type valve (which is less prone to silting) or to cycle the valve frequently to prevent the build-up of silt.

As will be explained in Chapter 5, high-performance proportional and servo valves often incorporate a **dither signal** to keep the spool moving very slightly in order to avoid a similar phenomenon.

What is not always appreciated is that as well as having a detrimental effect on the components of a hydraulic system, contamination can also degrade the fluid itself. Small particles of dirt tend to act as a catalyst for oxidation processes, and thus reduce the useful life of the fluid.

As well as contamination caused by solid particles, hydraulic fluids can be contaminated by other fluids, the most likely of which is water. Just as with solid-particle contamination, water in an oil-based fluid can have damaging effects on system components (rusting, loss of lubrication, etc.) but even more so on the fluid itself. As with dirt particles, water tends to accelerate the oxidation process of oils (causing varnish, and sludge to precipitate out), it affects the fluid viscosity and can also reduce the effectiveness of the additives. Therefore, water entering a hydraulic reservoir in the form of vapour needs to be minimised and kept under control, especially in systems that experience a significant variation in temperature.

POINT OF INTEREST

Water contamination of a hydraulic fluid is more likely in mobile systems, which may have to operate outside in rain and snow and be subject to vehicle wash-down. The wide temperature range that mobile equipment may have to operate in will also cause increased condensation within fluid reservoirs.

THE CONTAMINATION PROCESS

In a fully sealed hydraulic system it may be difficult to see how contamination can get in. However, no hydraulic system will be 100% sealed for all of its working life, and the system has to be manufactured in the first place (as will the fluid), so there will always be the possibility of contamination (solid or liquid) being in the system right from the start. An effective flushing process should remove most of the inbuilt contamination, but there is still the possibility of contamination entering or being generated during the life of the system. This can be caused by:

- *Contamination introduced via the fluid*. Generally speaking, new hydraulic fluid, as supplied, will not meet the cleanliness requirements of most systems. Even if it does, its cleanliness will also depend on how the fluid has been stored and transferred from the bulk storage into the hydraulic reservoir. In most cases, therefore, it is recommended to fill (and subsequently top up) the hydraulic system reservoir only with fluid passed through a filter. Figure 4.5 illustrates typical contamination levels of new fluid compared with what may be required for reliable operation.
- *Unsealed openings in the reservoir*. Unsealed openings (e.g. around entry and exit pipework) will allow dirt and moisture to enter the system, especially in dirty or wet environments (Fig. 4.6).

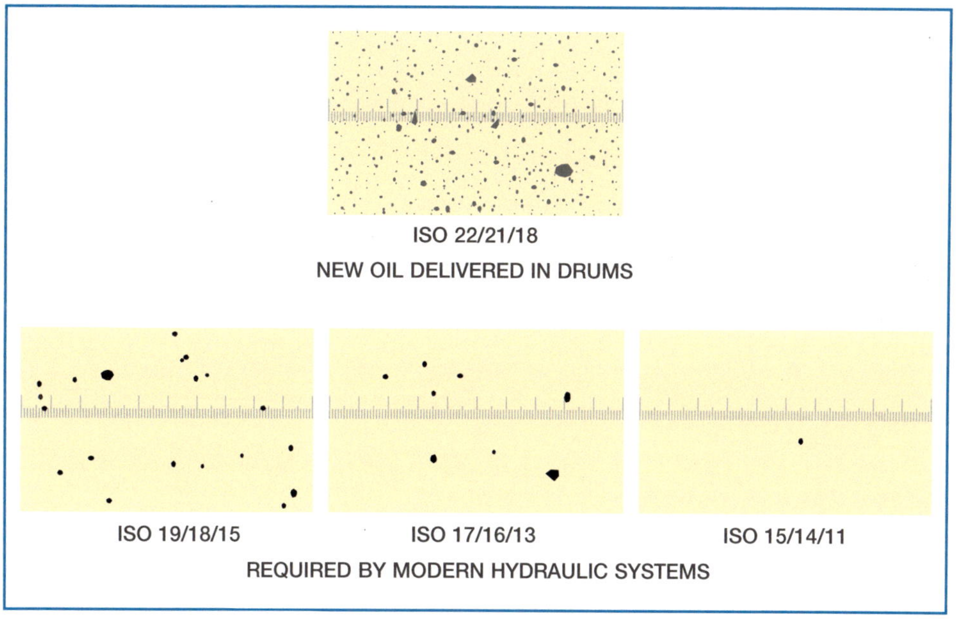

Fig. 4.5 Typical oil samples and ISO cleanliness codes

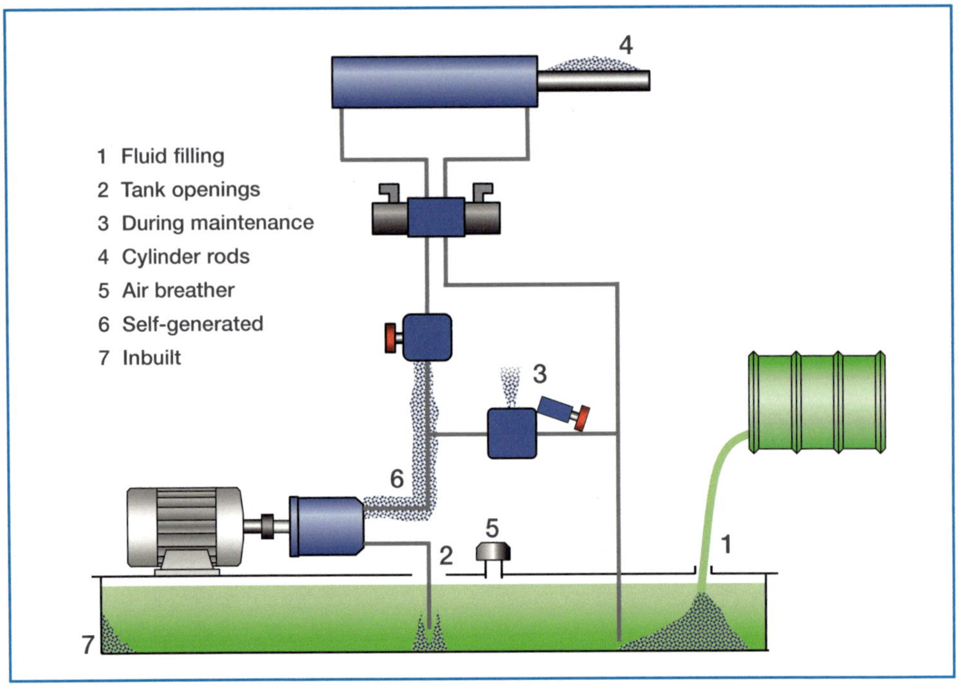

Fig. 4.6 Contamination entry points (Image courtesy of Eaton Corp)

- *Maintenance work carried out on the system*. When pipes, hoses or components have to be temporarily removed from a system, care should be taken to ensure dirt or moisture cannot be introduced.
- *Cylinder rods retracting, pulling dirt into the system*. As described in Chapter 2, cylinders are normally fitted with a **wiper seal** to prevent dirt being drawn into the system when the piston retracts. However, if this seal is worn, missing or ineffective, dirt can enter the system by this means.
- *Ineffective air breather*. An effective, well-maintained air breather should prevent solid contamination entering the system reservoir when the fluid level drops (when cylinders are extended or accumulators charged). However, it will not on its own prevent moisture in the air from entering the reservoir. When the system is shut down and cools, the moisture tends to condense on the reservoir walls and then drop down into the system fluid. However, air breathers are available that block the entrance of moisture or absorb the moisture as the air passes through, and in environments that are particularly wet or humid the use of such devices can be advantageous.
- *Self-generated dirt*. As system components wear (pumps and motors in particular), contamination will be generated internally within the system.

TOP TIP

It is always easier to prevent dirt from entering a system than it is to remove it once it has.

TARGET CLEANLINESS LEVEL

Whenever a hydraulic system is designed, a **target cleanliness level** should be established. This enables a choice of suitable filters, etc., to be made that will achieve and maintain this cleanliness level. The target cleanliness level will depend on a number of factors, including:

- *Type of components used* – some components will be more contamination sensitive than others, so manufacturers' recommendations should be obtained to determine the most sensitive components used in the system.

- *System operating pressure and duty cycle* – generally speaking, the higher the pressure the cleaner the fluid needs to be, as loads on individual component parts increase and clearances tend to reduce.
- *Type of fluid being used* – lower-lubricity fluids (such as water-based fire-resistant fluids) may require a higher level of cleanliness.
- *Operation under temperature extremes* – very hot or very cold conditions can affect the fluid viscosity, clearances within components, etc.
- *Criticality of the system* – where machine operation is critical or breakdown costs are high, system reliability will be improved by operating at higher levels of cleanliness.
- *Expected lifetime of the system* – the longer the design life of the equipment the cleaner the fluid should be maintained. As mentioned previously, the anticipated working life of something like a crop sprayer may be just a few hundred hours, compared with tens of thousands of hours for a process plant in a steelworks.

Having established a target cleanliness level based on the above factors, the system designer can then choose the type, rating and location of the filtration components that will achieve this level of cleanliness.

FILTER CONSTRUCTION AND OPERATION

The simplest type of filter used in hydraulic systems is a wire-mesh device. Such a device is normally referred to as a strainer rather than a filter because of its rather coarse filtration capability (typically around 150 µm).

Strainers are normally designed to be fitted to the end of the pump suction line to prevent large contaminant particles from entering the pump and potentially causing catastrophic damage. Hence they are often referred to as 'suction strainers' (Fig. 4.7). Strainers are a very simple, low-cost filter, but must be removed and cleaned when necessary. The pros and cons of this type of filter are discussed in the next section.

To reduce the likelihood of a strainer restricting the inlet flow to the pump it may be fitted with a bypass valve, which will open when the strainer starts to become clogged. Strainers can also be fitted inside a body external to the tank, in order to improve their accessibility and to enable the incorporation of an indicator for the condition of the strainer element (Fig. 4.8).

DEFINITION

Strainers are sometimes defined by their mesh size (particularly in the USA). The mesh size is simply the number of openings per inch. For example, a '100 mesh' strainer has hole sizes of 1/100 in, which corresponds to 149 µm.

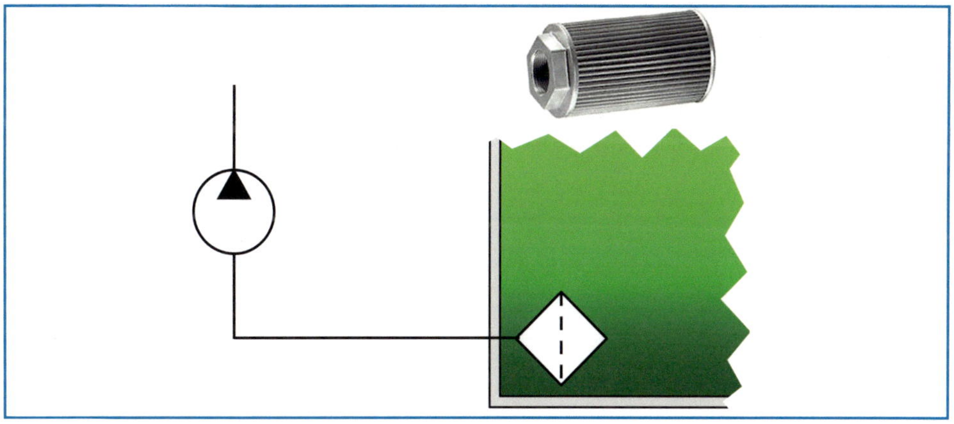

Fig. 4.7 Suction strainer

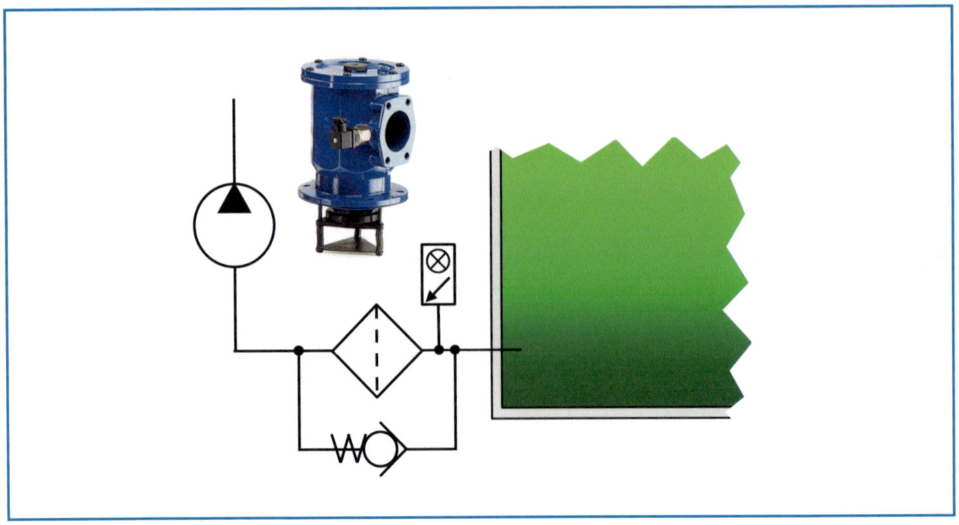

Fig. 4.8 Inlet filter (Photo courtesy of MP Filtri Ltd)

However sophisticated, a suction strainer is basically a pump-protection component only and cannot be expected to achieve the fine levels of filtration needed for the hydraulic system as a whole. This will require a filter element capable of removing much smaller particles of contamination, in some cases down to 3 µm or less in size. Low-cost filters employ paper or cellulose fibre elements, whereas high-performance elements may consist of several layers of specially formulated filter media.

Two common types of filter are the **spin-on can filter**, which is commonly found on mobile applications and is similar to the filters used on automotive engines, and the **bowl filter**, in which the replaceable element is enclosed within a removable bowl (Fig. 4.9).

TOP TIP

If pump suction filtration is necessary, it is best to ensure that the pump has a positive head of fluid (pump mounted below the fluid level) and that filters are located outside the tank, where they can be fitted with a bypass valve and a condition indicator.

Fig. 4.9 Spin-on can and bowl filters (Image courtesy of MP Filtri Ltd)

Both types of filter can be fitted with bypass valves to ensure that the element does not collapse when it becomes clogged with dirt, and often a visual and/or electrical indicator to indicate when the element needs to be changed (Fig. 4.10).

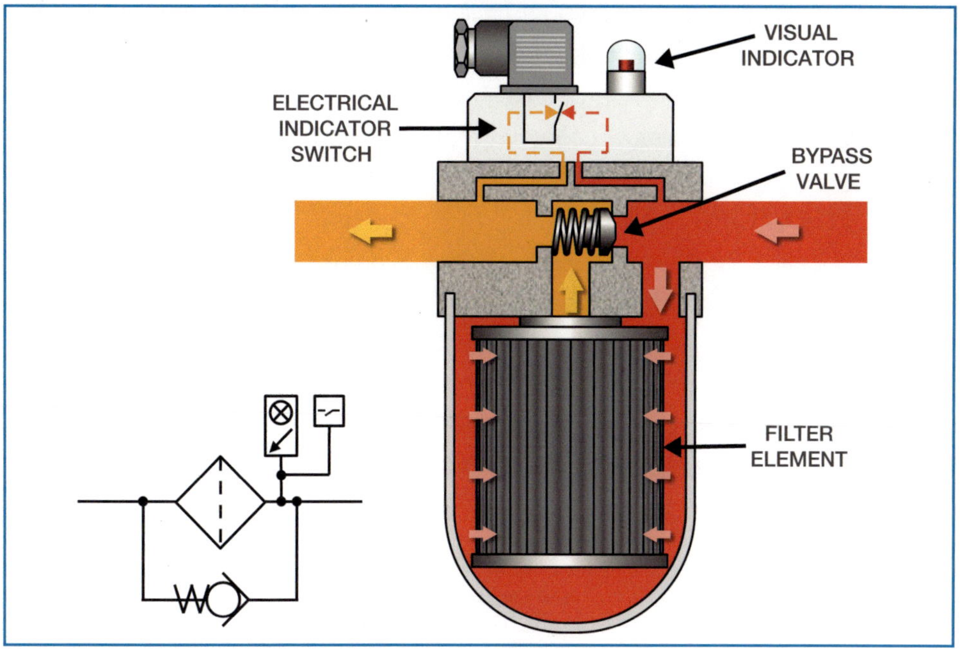

Fig. 4.10 Typical filter construction

In practice, there are many options available for hydraulic filters for different applications, including:

- **Tank-top mounted filter** – the filter element sits inside the reservoir, reducing cost and space requirements.
- **Stack mounted filter** – for mounting in a CETOP size 3 or 5 valve stack.
- **Duplex filter** – this consists of two filters and a changeover valve. If the operating filter becomes blocked, operation can be switched to the second filter, thus enabling the system to continue working while the clogged element is being replaced.
- **High pressure drop element** – an element used with a non-bypass filter to withstand a high pressure difference before the element collapses. These elements may be used with very sensitive components (such as servo valves) where it is important to always have filtration present to protect the components.
- **Water-removal filter** – these are designed to trap and retain water droplets in the hydraulic oil.

FILTER RATINGS

In the past, filter ratings were expressed in terms of their **nominal rating** and **absolute rating**. Whereas the nominal rating is somewhat arbitrary, the absolute rating is defined as the largest spherical glass particle that will pass through the filter under specified test conditions. However, in real life, contamination particles are rarely spherical in nature, and the test conditions may vary considerably from the actual system operating conditions. A better method of rating filter performance is, therefore, to use its **beta ratio** (β), which is determined using a standardised multi-pass test:

$$\beta_x = \frac{\text{Number of particles bigger than } x\ \mu\text{m upstream of the filter}}{\text{Number of particles bigger than } x\ \mu\text{m downstream of the filter}}$$

POINT OF INTEREST

The β rating of a filter is established under laboratory test conditions and may not necessarily indicate how the filter will perform in reality. It does, however, provide a useful means of comparing the performance of one filter with that of another.

POINT OF INTEREST

The correct design of a filtration system involves determining which components are most vulnerable and where the most likely points of contamination entry are.

For example, for a filter described as having a β_3 ratio of 1000, only one particle larger than 3 µm will pass through for every 1000 which do not. Again, this may not be an absolutely true reflection of what happens under working conditions, but it does provide a good means of comparing one filter with another.

FILTER LOCATIONS

The best location for a filter or filters within a hydraulic system is something that requires careful consideration. There is no single answer to where a filter should be located. There are several options, and each has advantages and disadvantages (Fig. 4.11). The choice of location will depend on a number of factors, such as which are the most vulnerable components and where the contamination is most likely to enter. The requirements of the system and its operating environment will therefore determine which filter location or locations are the most suitable.

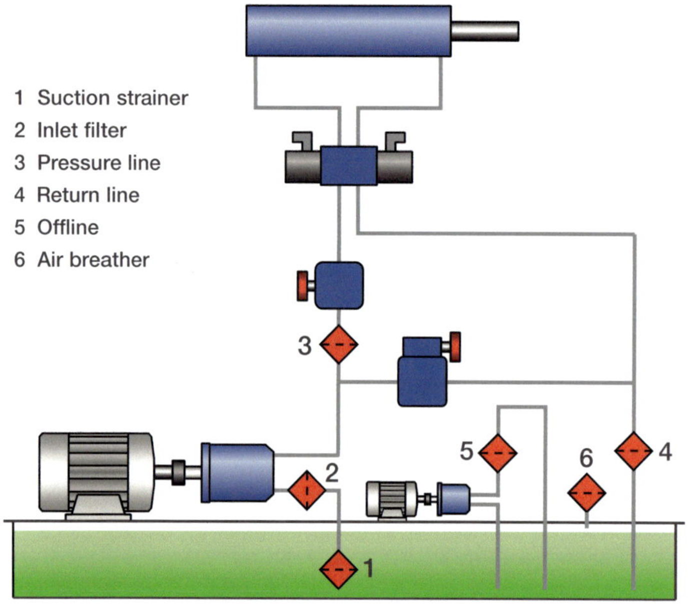

Fig. 4.11 Potential filter locations (courtesy Eaton Corp)

Basically, the location choices are the suction line, pressure line, return line or off-line.

Suction line

Any component in the suction line of a pump has the potential to restrict the inlet flow to the pump and thus cause cavitation damage. In particular, a component such as a suction filter where the restriction increases as it gradually becomes more blocked can be very damaging to a pump. Often, suction strainers are fitted to the ends of suction pipes inside the reservoir, where they are difficult to access for cleaning or replacement, and all too often get forgotten until cavitation damage to the pump occurs.

On the positive side, however, they are low-cost components that will protect the pump from large particles, which could cause catastrophic damage. On a multi-pump reservoir, for example, unless suction filters are used a sudden failure of one pump could cause a large volume of debris to enter the tank (via the pump case drain line) and cause a consequent failure of the other pumps.

If suction filters have to be used, the pump(s) should preferably have a boosted inlet, achieved either by means of a low-pressure charge pump or a positive-head reservoir, in order to reduce the risk of pump cavitation damage. In addition, the filters should be fitted external to the reservoir, both for convenient access and so that they can be fitted with a bypass valve and indicator.

In summary, therefore, the best solution is to avoid the use of suction strainers or filters where possible (by preventing contamination entering the reservoir). If it is not possible to guarantee a contamination-free reservoir, filtration at the pump inlet may be necessary, but it must be designed and maintained in such a way that it does not cause pump cavitation.

Pressure line

Filters for use in the pressure line tend to be expensive because they have to be constructed to withstand the full system pressure. They also have the potential to introduce air into the system whenever an element is changed, thus requiring the system to be re-bled. They may also experience variations in flow and pressure drop, depending on the system operation, which tend to detract from the element's capability to trap and retain dirt particles.

However, the pressure line is a much less sensitive part of the system than the suction line, and so pressure-line filters can be constructed to achieve very fine levels of filtration, which may be required for sensitive components. They will also trap contamination created by the pump as it starts to wear.

Return line

Although filters located in the return line do not have to withstand the full system pressure, they may have to handle higher flow rates than just the pump flow (due to flow intensification across cylinder pistons or the use of accumulators). Therefore, they may not be significantly less expensive than pressure line filters. As with pressure-line filters, they may be subject to flow and pressure variations, which detract from the filter performance. In addition, dirt generated by the pump has to travel all the way round the system before it can be caught in the return-line filter.

On the positive side, filters in the return line do not introduce air into the system when an element is changed, and they are in a good location to catch contamination drawn into the system on cylinder rods, for example. They are thus a popular choice on many mobile machines operating in dirty or dusty environments.

Off-line

Off-line filtration systems (sometimes referred to as 'kidney loop' systems) use a separate pump to draw fluid from the reservoir, pass it through a filter and then return it to the reservoir. Although this arrangement does involve additional components, they are low-pressure items and therefore relatively inexpensive. However, there is no guarantee that an off-line filtration loop will catch all the dirt particles that could cause damage to particular components.

The significant benefits of off-line filtration include:

- The filter is working under ideal conditions, with a steady flow and constant pressure drop across the element.
- The separate drive to the off-line pump can be left running continuously (especially in industrial applications), independently of the main system.
- The separate drive allows the filter element to be changed while the main system is still in operation.
- The filter can be located in a convenient position for access.
- Air is not introduced into the system when changing a filter element.
- The off-line loop can be used to fill and empty the reservoir, thus ensuring that only filtered fluid is added to the reservoir.
- The off-line loop can include a cooler if necessary, and this is an ideal location for a cooler.

MONITORING FLUID CLEANLINESS

FURTHER READING

www.mpfiltri.co.uk/the-knowledge-centre

Whatever target cleanliness level has been established for a particular machine, it is obviously necessary to determine if that level has been achieved and whether it is being maintained over the lifetime of the machine. In critical applications, online fluid-condition monitors can be permanently installed in the system to constantly monitor fluid condition (dirt and water content). More often, however, fluid-condition monitoring involves taking a sample of fluid from the system and analysing it, either by using a portable particle counter (Fig. 4.12) or via a laboratory-based service.

Whichever method is used, the important part of the process is to obtain a representative sample of fluid from the system. Simply unscrewing the drain plug from

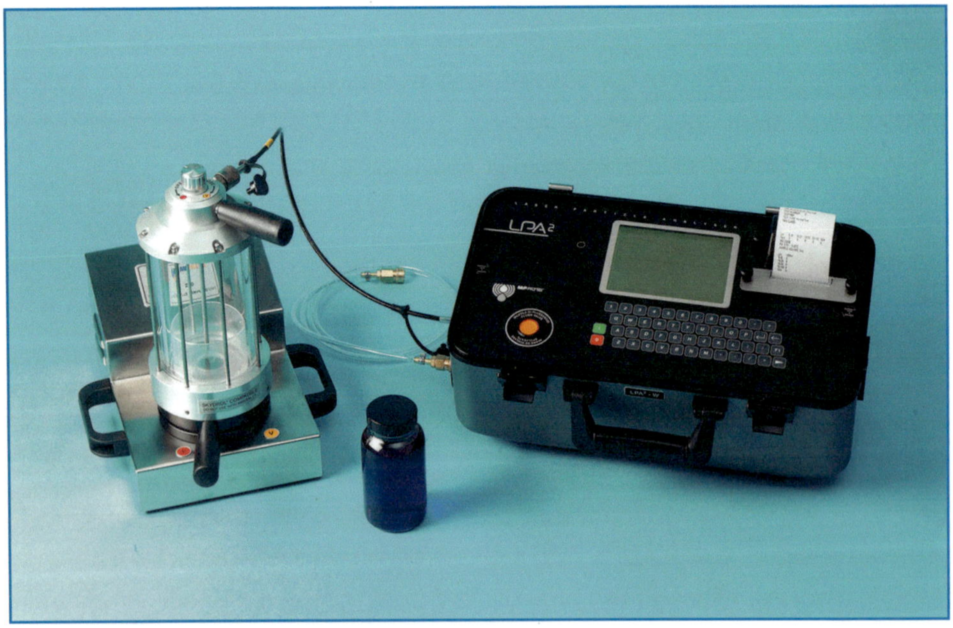

Fig. 4.12 Portable contamination monitoring equipment (Image courtesy of MP Filtri Ltd)

the bottom of the reservoir (where much of the dirt and water tend to accumulate) is unlikely to provide a true indication of the fluid's condition.

Fluid samples should be taken with the system at normal operating temperature and, ideally, from the pressure line via an inbuilt sampling valve. To take a sample, a volume of fluid is first run through the sample valve (to flush out any accumulated dirt), and then a specially cleaned sample bottle is held under the fluid stream, without touching the valve. The bottle is then sealed and sent for analysis as required. If a sample from the fluid reservoir is required, a similar process can be used. In this case, a sampling valve and internal pipe are used to ensure the sample is taken from the middle of the reservoir (Fig. 4.13).

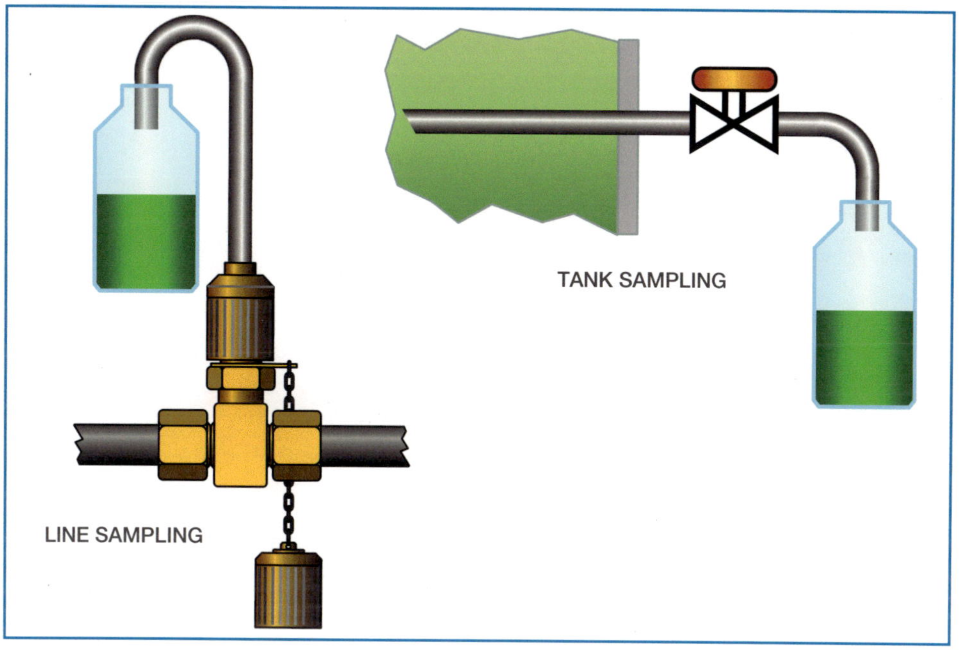

Fig. 4.13 Fluid sampling

Whichever method is used, it is important to ensure that the sample itself is not contaminated either by the action of taking the sample or via the sample bottle.

> **WARNING**
>
> Take special care not to contaminate a fluid sample during the sampling process.

Monitoring the cleanliness level of a system over a period of time will not only confirm that the target level has been achieved (or not) but can also provide information on the wear rates of components and indicate when preventive maintenance may be required.

FLUID-TEMPERATURE CONTROL

Apart from dirt and water contamination, the other major factor that determines the system life and reliability is the control of the operating temperature. In some applications the fluid may have to be heated to a minimum temperature for satisfactory operation. Mineral oil fluids tend to become more viscous as their temperature drops, so pump suction becomes critical at very low temperatures.

More often, however, the requirement is to cool the fluid to maintain it within its optimum temperature range, which will depend on the type of fluid being used. As explained in Chapter 3, water-based fire-resistant fluids and biodegradable fluids have a lower maximum temperature capability than mineral oils.

POINT OF INTEREST

Fluid temperature will increase by approximately 1°C for every 17.5 bar, which is approximately equivalent to 1°F per 140 psi.

The need to cool a hydraulic fluid normally arises from the fact that virtually all of the inefficiencies within the system are turned into heat. If the flow from a fixed-displacement pump is passing across a relief valve, for example, all the power being used to drive the pump is being turned into heat. As a rule of thumb, every 17.5 bar pressure drop across a component heats up the fluid by 1°C if no mechanical work is done.

In order to determine the amount of heat dissipation required, it is necessary to determine the power input to the system (i.e. pump drive power) and subtract from this the mechanical power output from the cylinders and motors (Fig. 4.14). Normally this will be averaged over a typical machine cycle or operating period. The difference between input power and output power will be the heat that the system must dissipate in order to maintain a stable fluid temperature.

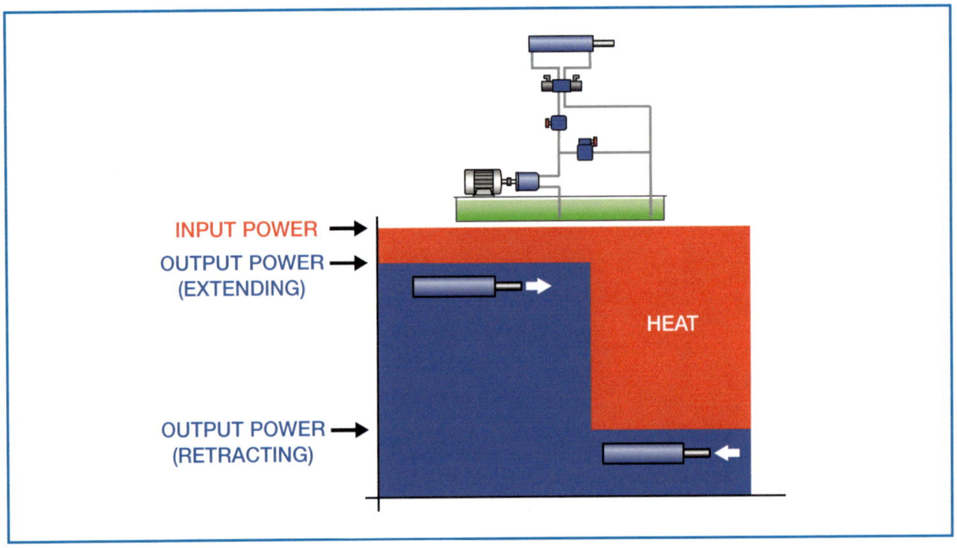

Fig. 4.14 Heat generation

While some heat will radiate naturally from the system components, pipework and, especially, the tank, any requirement over and above this amount will require some form of cooler. Coolers will normally use water as the cooling medium, if available, or air if not (e.g. on vehicle applications).

WATER COOLERS

Water coolers normally consist of a series of pipes inside a casing, with hydraulic fluid flowing through the inside of the pipes and water along the outside (Fig. 4.15). The water supply to the cooler will normally be controlled by a thermostatically controlled valve, which will only switch the water supply on when the oil temperature reaches a certain value. Best results (i.e. the greatest cooling effect) are obtained when the water and oil flow through the cooler in opposite directions (known as 'counterflow').

The size of cooler is determined by the amount of heat it is required to dissipate and the temperature of the cooling water available. However, coolers are often fitted into the return lines of systems (to cool the oil after it has been heated up in the system), where the flow rate may be greater than just the pump flow alone (as with return-line filters). The peak return-line flow rate may, therefore, also determine the size of cooler required.

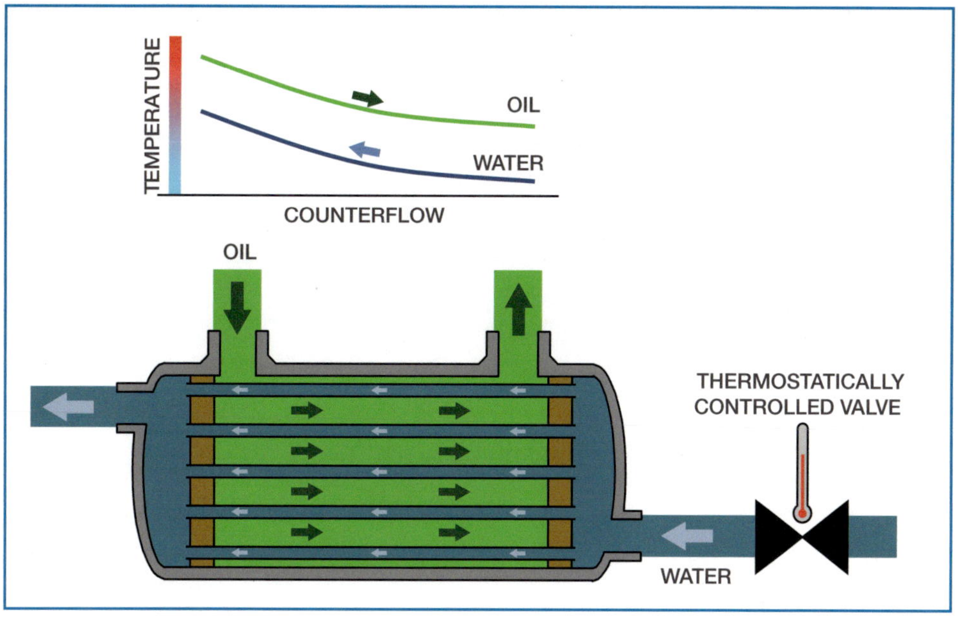

Fig. 4.15 Water cooler

As mentioned previously, an off-line filtration system, if one is used, is an ideal location for the cooler, because it has a steady and predictable through-flow under all conditions. In such arrangements the cooler is usually placed after the filter so that the oil is filtered while hot and then cooled down.

Any internal damage to a water cooler will create the possibility of water leaking into the hydraulic fluid, or vice versa. A protective low-pressure relief valve (often a spring-loaded bypass check valve) is normally fitted around the cooler to guard against sudden flow surges or pressure peaks.

When no cooling water supply is available, an air blast cooler is the usual alternative (Fig. 4.16). Here, hot oil is passed through a cooling element while air is blown through by means of an electric or hydraulic motor driven fan. On some vehicles the hydraulic cooler is mounted alongside other heat exchangers (e.g. main engine, turbocharger or air-conditioning unit) that all share a common cooling fan.

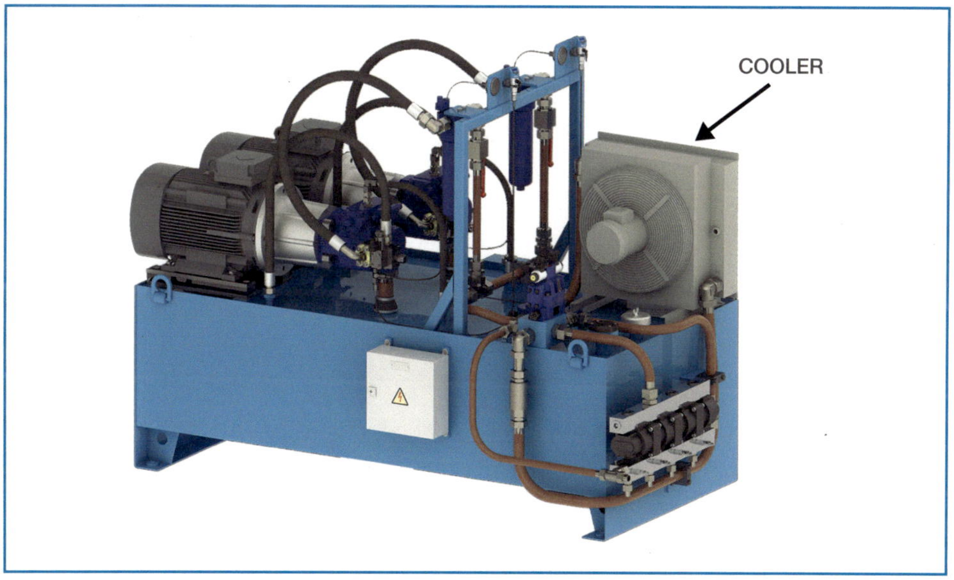

Fig. 4.16 Air-blast cooler fitted to a power unit

As mentioned previously, the reason why a cooler may be required in a hydraulic system is to dissipate the heat created by system and component inefficiencies. Careful design of the system, however, can often reduce the amount of inefficiency (and therefore heat), but this may sometimes involve a higher initial cost. For example, variable-displacement pumps or variable-speed drives can often provide a more efficient system than one using simple fixed-displacement pumps, but usually involve an initial cost penalty.

The design of the system therefore always involves finding the best compromise between capital cost and running cost. But, again, operating conditions will influence this decision; a system that is used for 10 minutes every day will have different priorities to one that is operational 24 hours a day.

Energy-recovery arrangements involving accumulators are now also a practical means of increasing system efficiency, but again may involve a higher initial cost.

POINT OF INTEREST

Traditionally, power-transmission efficiency has not been a strong point of hydraulic technology. However, times change, and now the emphasis in many applications is on making the hydraulic system as efficient as possible in order to reduce running costs, emissions and waste-heat generation.

CHAPTER

5 ELECTRIC POWER AND ELECTRONIC CONTROLS

INTRODUCTION

In Chapter 1 a hydraulic system was defined as a means of transmitting power from a prime mover to the various functions of a machine where the power is required. In most industrial applications, and some mobile applications also, the prime mover is an electric motor of some description. The electric motor provides input power to drive the pump, which converts the mechanical shaft rotation of the motor into hydraulic power in the form of flow and pressure.

For many years, electrically operated directional valves have been used to control and distribute the hydraulic power to where it is needed. In many cases, these are simple on/off devices, but increasingly there is a demand for proportional controls, where valves are capable of modulating their output, as opposed to simply switching it on or off. An analogy can be made with a simple on/off light switch and a dimmer switch. The first is only capable of switching the light on and off, while the second can vary the amount of light produced, from fully off to fully on and anywhere in between.

Once proportional valves had been developed to respond to analogue electronic signals, it was then a relatively small step to introduce digital electronic controls, thereby adding some 'intelligence' to the control function. As a result, hydraulic components are now available with built-in electronics that not only sense what the component is doing but also can make control decisions about how to react and then communicate their actions back to a centralised control station.

ELECTRIC MOTORS

Most electric motors used in industrial systems are **three-phase induction motors**. These are sometimes referred to as **squirrel-cage motors** due to the configuration of their rotating component. They are typically powered by a three-phase AC supply at a relatively high voltage of around 400 V (480 V in the USA), which is the normal industrial supply voltage. By contrast, the domestic electricity supply is around 220 V (110 V in the USA) and single phase.

A single-phase supply involves just two wires, where the voltage in one is constantly changing relative to the other in a sinusoidal manner. The rate at which the voltage changes is known as the supply **frequency** (i.e. the number of complete sinusoidal cycles per second). In Europe, the supply frequency is a nominal 50 cycles per second (50 hertz (Hz)), whereas in North America and some other countries it is 60 Hz. A three-phase supply involves three wires, where the voltage in each varies sinusoidally relative to the other two (Fig. 5.1).

In an induction motor, the electrical supply is connected to windings in the stationary part of the motor known as the **stator**. The cyclical nature of the voltage and the

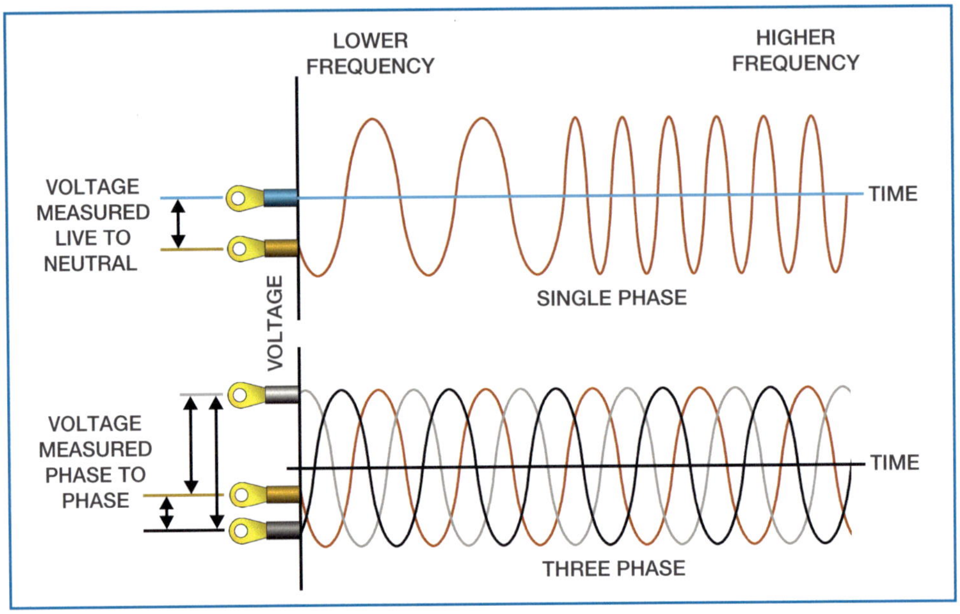

Fig. 5.1 Single-phase and three-phase AC supplies

resulting current in the windings generates a magnetic field that rotates within the stator. The speed of rotation of the magnetic field is determined by the frequency of the supply and the number of windings that can be arranged as, typically, two pole, four pole or six pole. The speed of rotation of the magnetic field (known as the **synchronous speed** of the motor) can be determined using the following formula:

$$N = \frac{120 \times f}{P}$$

where N is the rotational speed (rpm), f is the supply frequency (Hz) and P is the number of poles. For example, a four-pole motor operating with a 50 Hz supply voltage would have a synchronous speed of 1500 rpm.

As the magnetic field rotates within the stator, it induces a current within the bars of the 'squirrel cage' located within the rotor (Fig. 5.2). This induced current also

DEFINITION

Pole – a set of windings in the stator of a motor. The greater the number of poles the slower the motor will rotate for a given frequency.

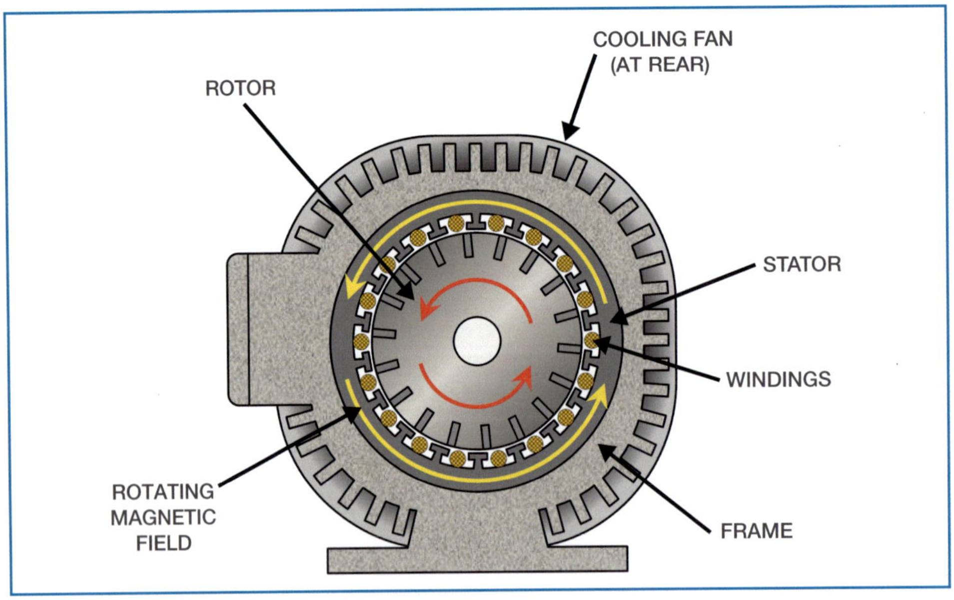

Fig. 5.2 Squirrel-cage electric motor

creates a magnetic field, which interacts with the one created in the stator to cause the rotor and motor shaft to rotate.

In order to create a rotational force (torque), however, there must be a slight difference between the speed of rotation of the stator magnetic field and that of the rotor (known as slip). The difference is typically around 3–5% of the synchronous speed. So, at its rated torque output, a 1500 rpm synchronous speed motor would actually rotate at around 1440 rpm.

Inevitably there will be losses internally in the motor due to friction, wire resistance, etc., so heat will be generated within the motor. Normally, a fan is attached to the motor shaft and incorporated in a cowling on the rear of the motor. The purpose of the fan is to force air across cooling fins on the motor frame in order to dissipate this heat. Most induction motors are able to rotate in either direction, and usually incorporate a bi-directional cooling fan to allow for this.

The direction of rotation is determined by how the three phases are connected (i.e. reversing any two will reverse the direction of rotation). Unfortunately, it is not always easy to tell which way the motor will rotate when first connected, and this often has to be determined by trial and error. Ideally, in a hydraulic system the motor shaft should be disconnected from the pump shaft during this process, but in practice this is unlikely to be the case. When checking motor rotation, therefore, it must be ensured that there is sufficient hydraulic fluid available in the pump in order to lubricate it for the short period of time it takes to determine the rotation of the motor. In particular, the cases of piston pumps *must* be filled with fluid before the pump is operated. A pump that is started up dry can be damaged in just the few revolutions it takes to determine the motor rotation.

The three-phase induction motor has been the standard method of driving hydraulic pumps in industrial applications for many years. In the majority of applications, fixed-speed motors have been used (1000 or 1500 rpm in 50 Hz supply regions, and 1200 and 1800 rpm in 60 Hz regions). Where a variable flow was required in a hydraulic system the designer had a choice of using either multiple fixed-displacement pumps or a variable-displacement pump.

Today, however, a third alternative exists, which is to use a variable-speed electric motor in conjunction with either a fixed- or a variable-displacement pump. Variable-speed DC motors have been available for many years but these were generally too expensive for use in hydraulic systems. However, the advent of **variable-frequency drives** for AC induction motors has made this alternative much more viable from a cost point of view.

As explained previously, the synchronous speed of an induction motor is dependent on the frequency of the AC supply. Therefore, if the frequency of the supply can be varied, it follows that the speed of the electric motor can be varied accordingly. In fact, in many situations few, if any, modifications may be necessary to the standard fixed-speed motor itself. All that is required is a **variable-frequency controller** to drive it (Fig. 5.3).

DEFINITION

The **synchronous speed** of a motor is the theoretical speed of rotation if there is no load on the motor shaft. In practice, there will always be some load (caused by bearing friction, fan resistance, etc.), so even with no external load the speed of the motor will be slightly lower than the synchronous speed.

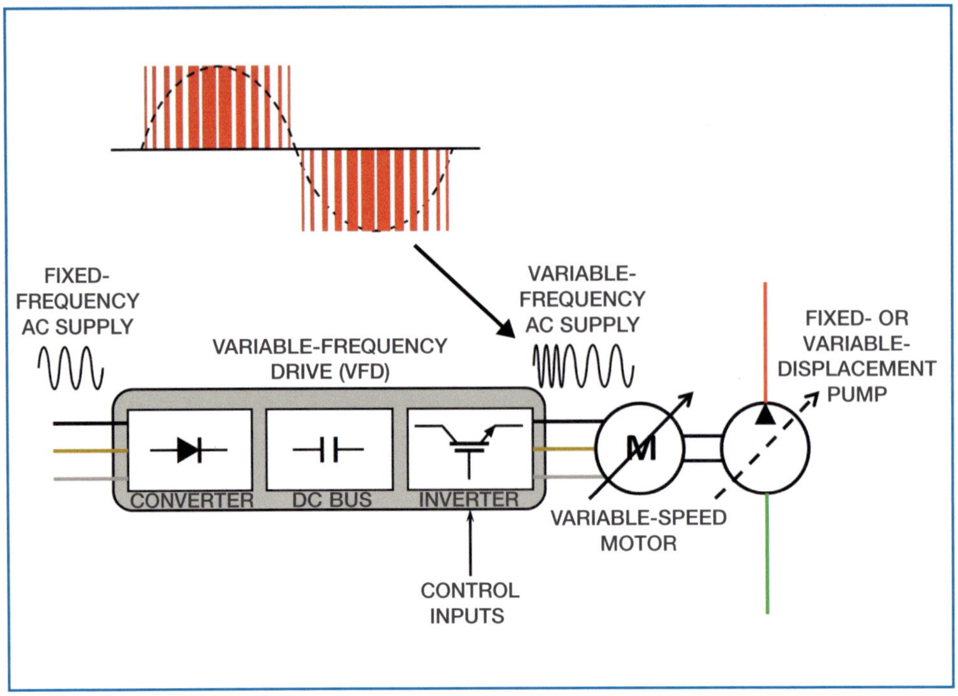

Fig. 5.3 Variable-frequency motor drive

DEFINITION

Rectification is the process that converts the bi-directional current flow of an AC supply to a single-direction current flow.

The fixed-frequency AC supply is first rectified to convert it to DC, and then converted back to AC at a variable frequency, as determined by the control input signal. The output AC is in the form of a series of on/off pulses of varying duration, which approximate to a sine wave. This is achieved by simply switching the output on or off at the required frequency using high power capacity solid-state switching devices. This technique, known as **pulse-width modulation (PWM)**, can also be used to control the current flow through proportional valve solenoids, as will be explained later.

The variable-speed drive described so far is a relatively low-cost method of achieving a variable-flow pump. However, a drive of this type has limitations if it is required to operate at very slow speeds or under high dynamic conditions (e.g. rapid speed changes). Although the motor itself will have a maximum drive speed, the maximum speed capability of the hydraulic pump will normally be much lower, and this will therefore be the factor that limits the capability of the arrangement from this point of view.

DEFINITION

An **open-loop control system** varies the output of a device or system by varying an input control signal. However, any disturbances or variations in the system may cause the output to change.

A **closed-loop control system** adds a **feedback device** so that any variations which tend to cause the output to change unintentionally are automatically corrected.

Where a more dynamic performance is required, or where speeds need to be accurately controlled (down to very slow or even zero speed), a closed-loop arrangement can be used (Fig. 5.4). In this case signals indicating the pump drive speed and pressure can be fed back to the motor controller in order to provide a more accurate and responsive control system. The type of electric motor used in such applications may also be uprated to one that incorporates permanent magnets in the rotor, which is more suited to this type of control arrangement.

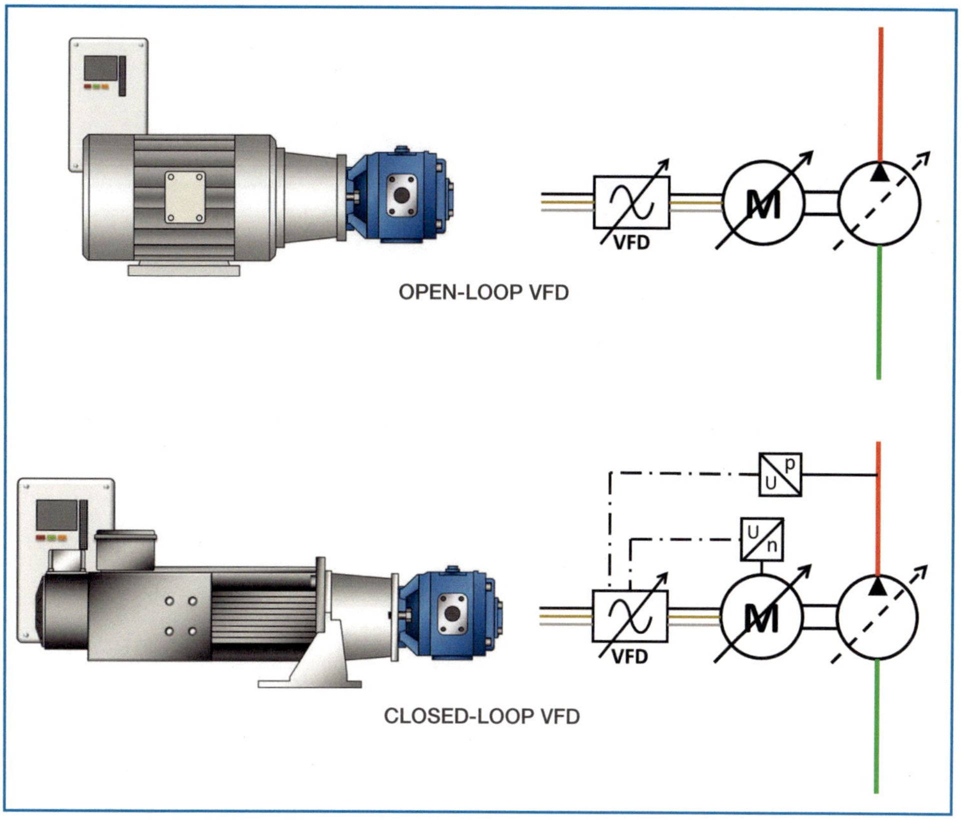

Fig. 5.4 Open- and closed-loop variable-frequency drives (VFDs)

ELECTRIC AND ELECTRONIC CONTROLS

In contrast to the high-voltage AC electrical supply required by electric drive motors, most control systems in mobile and industrial applications now use a low-voltage (typically 12–48 V) DC supply to operate control components. In industrial applications, 24 V DC is most common, whereas mobile systems may operate with a 12, 24 or 48 V control supply.

Where the control voltage is derived from an AC power network, the first requirement is for a component that reduces the voltage down from 110 or 220 V and converts it from AC to DC. This component is generally known as a **power supply** (Fig. 5.5). Power-supply units are available in many forms to provide either a fixed or variable DC output with varying degrees of 'smoothness' (i.e. free from any residual AC ripple). However, all power supply units have a maximum current output, which if exceeded is likely to cause a drop in output voltage and a potential malfunction of the components they are operating.

Mobile systems normally generate electrical power from an engine-driven **alternator** or externally charged battery. The alternator again produces an AC output that has to be rectified (converted to single-direction current flow) and then fed to an on-board battery, which smooths out the AC ripple.

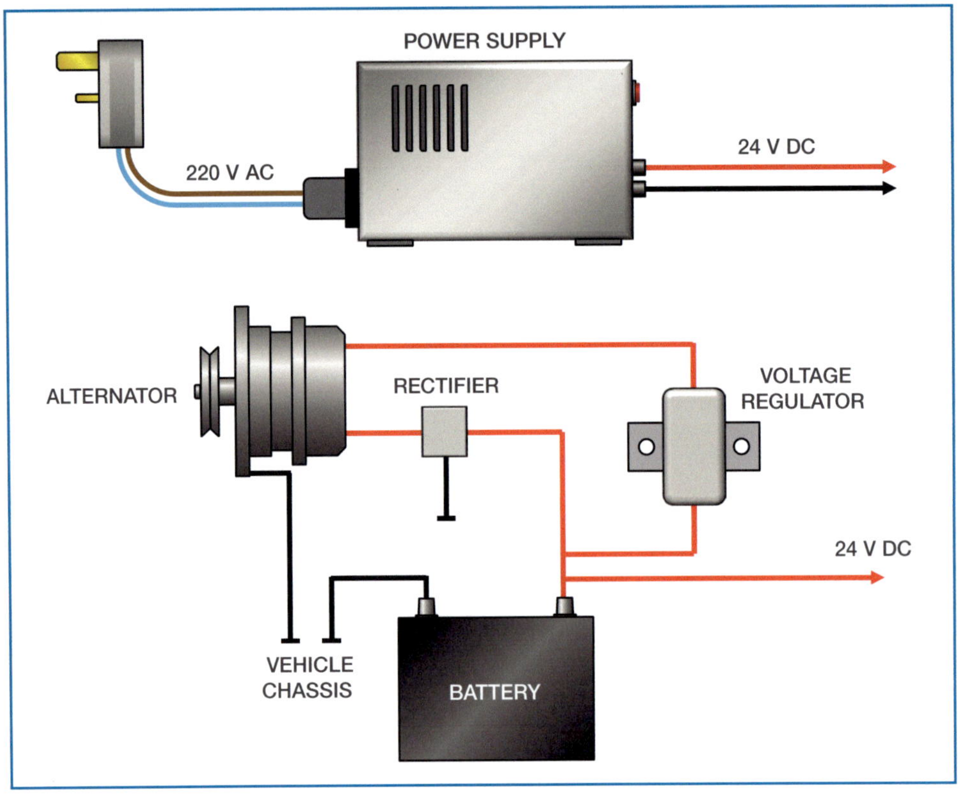

Fig. 5.5 Power supplies

SOLENOID VALVES

Solenoid-operated directional control valves are probably the most numerous of all hydraulic components. Unlike a manually operated valve, moving a valve spool or poppet by means of an electrical solenoid enables the valve to be operated remotely and also to be controlled by the output of an electronic control system of some description. In its basic form, a solenoid simply needs to be switched on or off, but this process is not as straightforward as it may first appear.

The current flowing through a solenoid coil will depend on the voltage difference across it and the resistance of the wire in the coil. The relationship is governed by Ohm's law, which states that

$$\text{Current (amps)} = \frac{\text{Voltage (volts)}}{\text{Resistance (ohms)}}$$

The resistance of the wire is determined by the conductivity of the wire material (how good an electrical conductor it is) and its physical dimensions (cross-sectional area and length). In addition, the wire temperature will also determine its conductivity. For the copper wire used in solenoid coils, the higher its temperature the greater its resistance.

When current flows in a wire, a magnetic field is created. Winding the wire into a coil concentrates this magnetic field until it is strong enough to operate components within valves (Fig. 5.6).

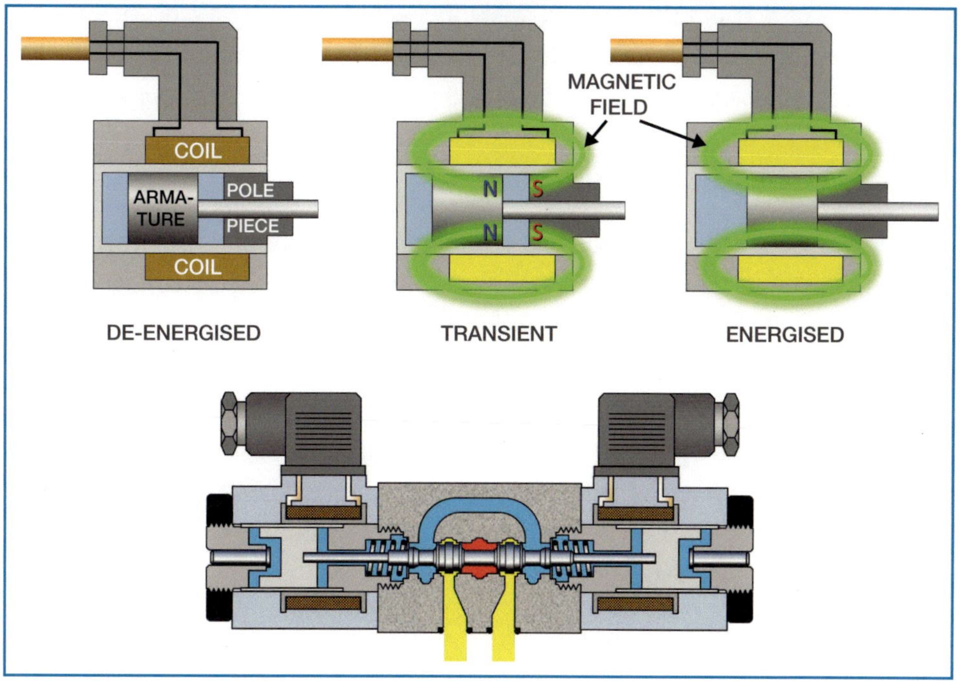

Fig. 5.6 Solenoid operation

When current flows through the solenoid coil, the magnetic field created magnetises the **pole piece** and **armature** with opposite poles, thus creating a magnetic force that acts to pull them together. As the armature moves towards the pole piece it operates a spool or poppet within the valve and thus changes the fluid flow path. The magnetic field created by the electrical current within the coil acts in a similar way to inertia in a mechanical system. This property is referred to as the inductance of a coil, and its first effect is to delay the build-up of current in a coil when the current is first switched on. It is therefore analogous to starting up a flywheel from rest, where the inertia of the flywheel determines how fast it can accelerate up to its maximum speed (Fig. 5.7).

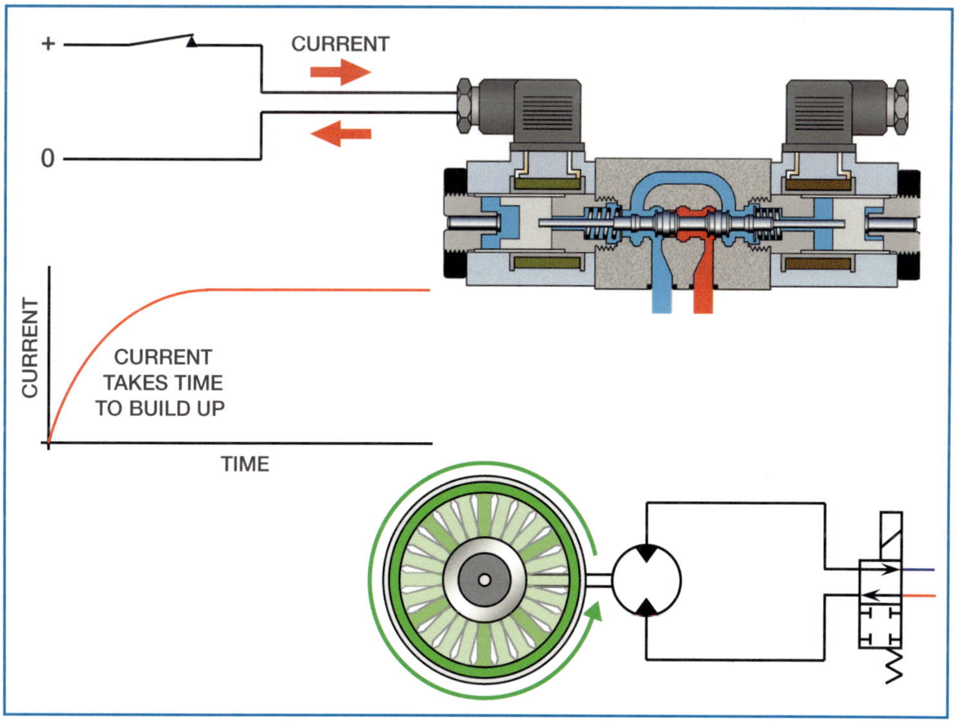

Fig. 5.7 Analogy between a DC solenoid and a flywheel

In the case of a hydraulic solenoid valve this delay may be only of the order of 20–30 ms, which may or may not be significant in the operation of the valve. Just as flywheels are used on internal combustion engines to smooth out the torque fluctuations caused by the cylinder firing, so inductive coils can be used in electrical circuits to smooth out current fluctuations.

The 'inertia' effect of the magnetic field in a solenoid coil has a more significant effect, however, when the coil voltage is switched off. Having established the magnetic field within the coil, if the current is then suddenly switched off, by opening a switch or relay contact for example, the magnetic field attempts to keep the current flowing in a similar way to how the inertia of a flywheel tries to keep it rotating when the drive is removed. It does this by building up a very high voltage across the switch contacts, which often results in arcing (sparking) and thus physical damage to the switch. If solid-state switches are used (e.g. transistors), a complete breakdown of the switch can result from this high voltage.

POINT OF INTEREST

Just as the energy stored in a rotating flywheel cannot be dissipated instantaneously, so the magnetic field energy built up in a solenoid coil cannot be switched off suddenly without causing damage to components.

A similar situation will arise in a hydraulic system if a flywheel is being driven by a hydraulic motor controlled by a solenoid-operated directional valve. A sudden movement of the solenoid valve spool to the blocked port condition will attempt to stop the flywheel almost instantaneously, resulting in a very high pressure build-up in the motor exhaust line, thus causing possible damage to components or pipework (Fig. 5.8).

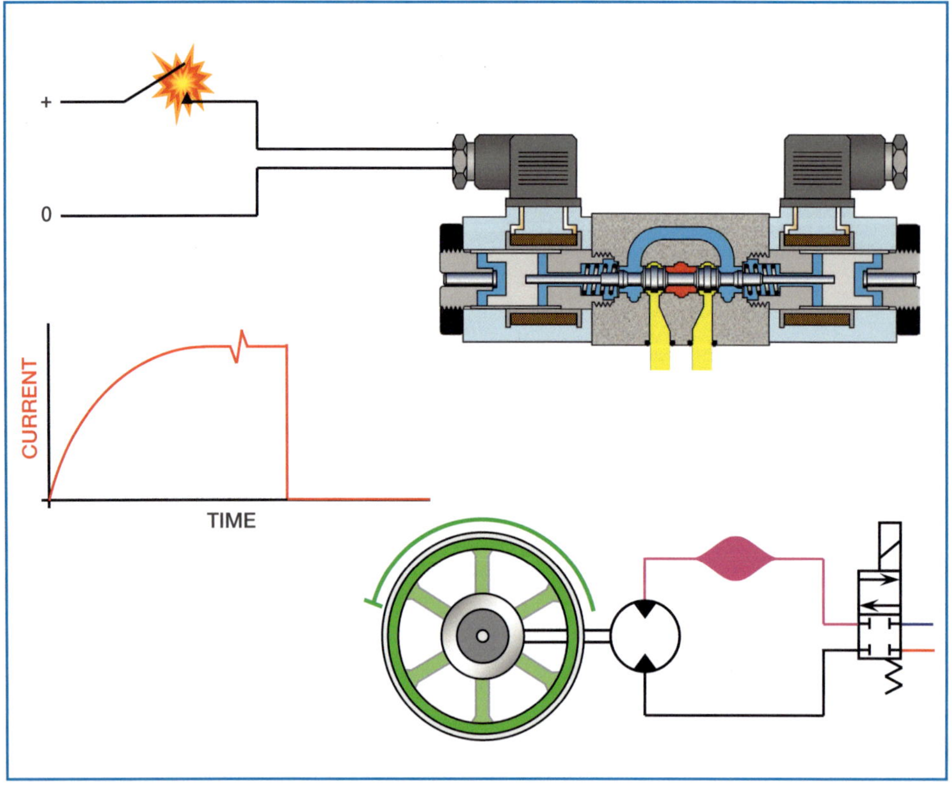

Fig. 5.8 Voltage spikes

One solution to this in a hydraulic system is to add a check valve across the motor ports. When the solenoid valve is de-energised, fluid can circulate freely around the motor so that the flywheel can come to a stop gradually without building up a damaging pressure peak in the system (Fig. 5.9).

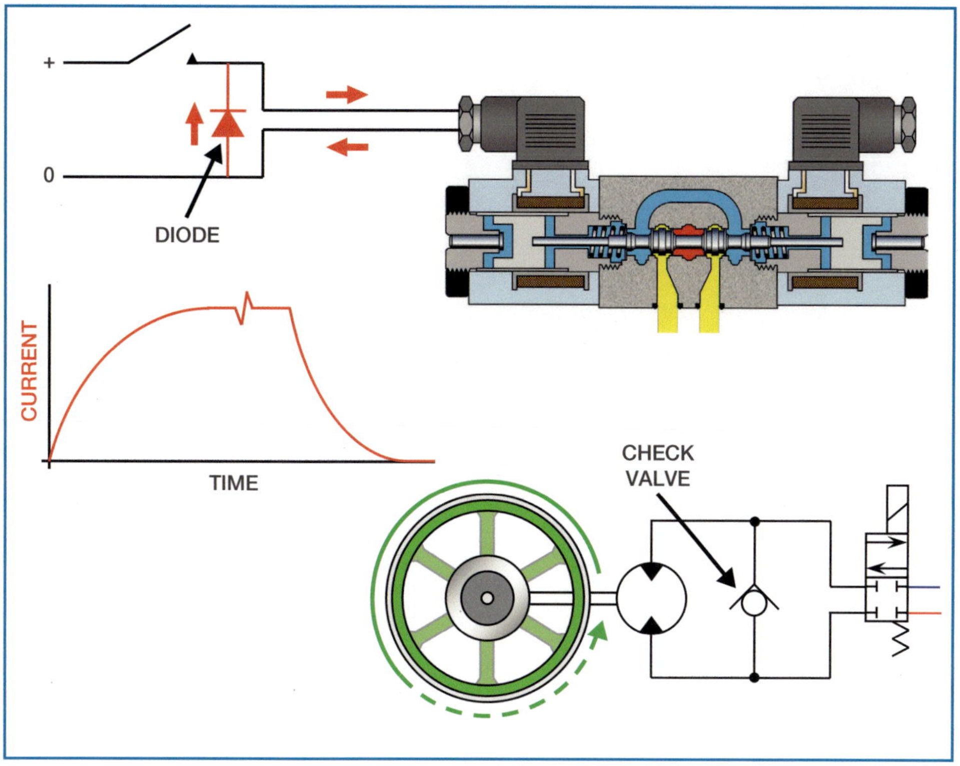

Fig. 5.9 Flywheel diode

The electrical equivalent of a hydraulic check valve is known as a **diode**, which allows a current flow in one direction but blocks it in the reverse direction. Connecting a diode across the coil connections therefore avoids the potential damaging effect of voltage spikes in the circuit. In fact a diode used for this purpose is sometimes referred to as a **flywheel diode**. In practice, the diode is often incorporated in the solenoid valve coil.

DIGITAL ELECTRONIC CONTROLLERS

The control of most modern machinery is now carried out (at least in part) by some form of digital electronic device, often referred to as a **programmable logic controller (PLC)**. Terminology with this technology is still evolving, however, and as PLCs become more advanced in their capabilities the description **programmable automation controller (PAC)** is also used.

In their simplest form these controllers take the signals from a range of input devices (stop/start buttons, limit and pressure switches, temperature sensors, etc.) and produce appropriate outputs to operate hydraulic and other components. Outputs are generated according to the inputs received following a particular logic (e.g. pressing the 'start' button may energise a directional valve to operate a cylinder until a limit switch is triggered). This logic is entered into the device by means of software (rather than being hard-wired) through a **human–machine interface (HMI)**. The HMI typically consists of a screen and series of buttons or keyboard (Fig. 5.10).

The solenoids of hydraulic valves often require relatively large electrical currents (1–2 amps), which is normally too much for the PLC to switch directly. Therefore, some form of power switch is required between the PLC and the valve. It can either be a separate component or be built into the PLC (or even built into the valve).

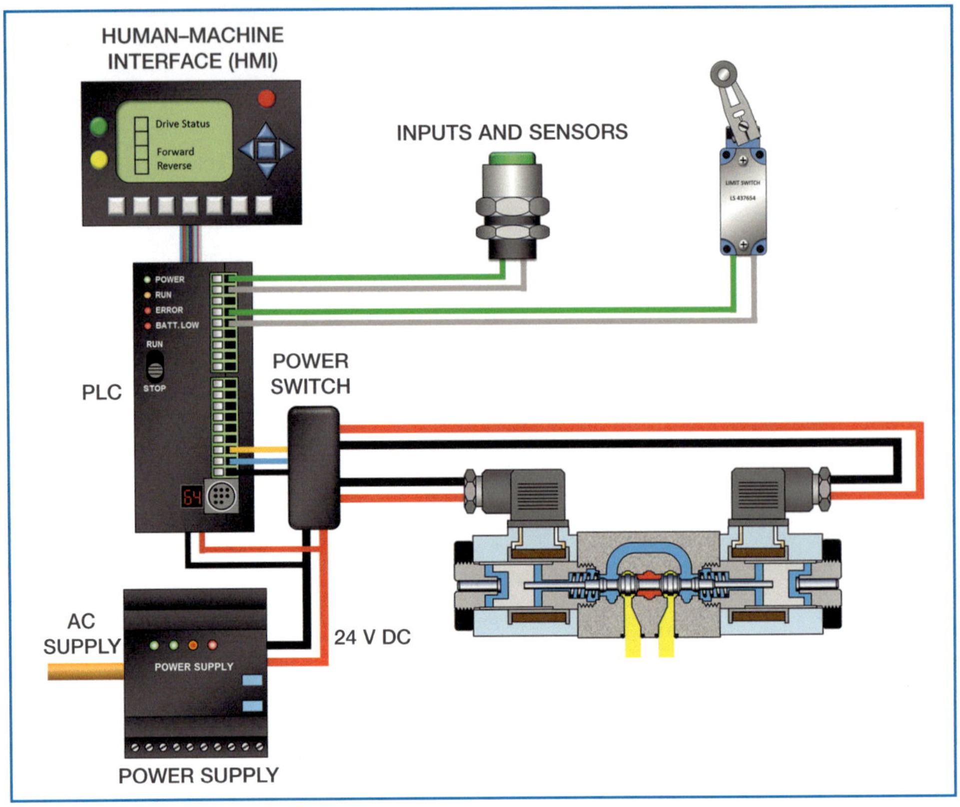

Fig. 5.10 PLC control arrangement

PROPORTIONAL VALVES

While simple on/off-type solenoid valves are very numerous in hydraulic systems, their function is purely to act as a hydraulic 'switch', and their rapid operation can sometimes create shock in the system when heavy loads are stopped and started quickly. For this reason they are sometimes referred to as **bang-bang valves**.

However, by suitably modifying the valve it is also possible to provide it with the capability to control the flow rate as well as the direction of flow. In other words, the amount of spool opening can be controlled electronically, both in terms of the amount of opening and the rate at which it opens and closes. This type of valve is then referred to as a **proportional directional valve**, because the amount of valve opening (and therefore the valve flow rate) is proportional to the input signal it receives.

As can be seen from Fig. 5.11, modifications can be made to the valve spool by adding notches to the edges of the spool lands to provide a gradual opening and closing of the flow path. Modifications are also made to the solenoid to provide low-friction bushes or bearings (to reduce 'stick–slip' effects) and to linearise the magnetic characteristics. As before, the flow direction can be controlled by energising one solenoid or the other, but now the amount of flow can also be controlled by varying the level of solenoid current. By controlling the rate of change in the solenoid current (i.e. a **ramped** change), the speed at which the spool opens and closes the flow path can also be controlled. This has the effect of determining the acceleration and deceleration of an actuator, enabling system shocks to be reduced to acceptable levels. As mentioned, therefore, the comparison between an on/off solenoid valve and a simple proportional valve is similar to that of an on/off light switch and a dimmer switch.

POINT OF INTEREST

A **proportional directional control** valve controls both the amount and the direction of flow through it in response to a variable electronic signal.

A **proportional flow control valve** controls only the amount of flow passing through it (i.e. not the direction of flow).

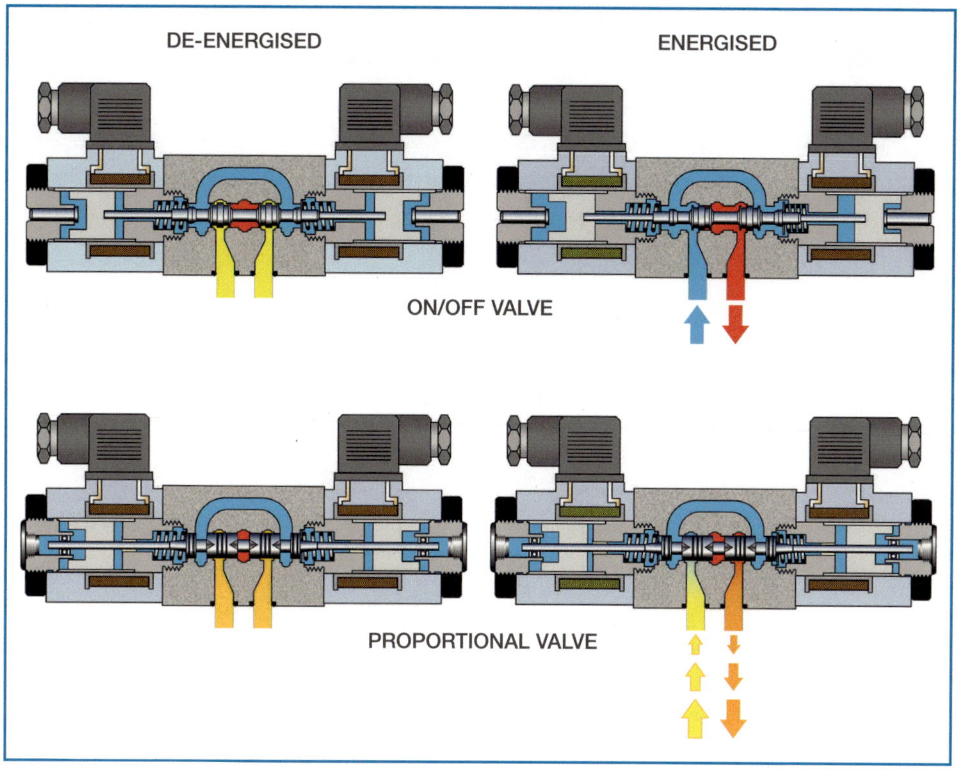

Fig. 5.11 On/off and proportional solenoid valves

In theory, a simple variable resistor could be used to vary the current flow through the proportional valve solenoid. However, in practice the problems of heat generation that this would create, together with the variation in the solenoid coil resistance with temperature, means that the degree of control would be very imprecise. In practice, therefore, the coil current, and hence the valve opening, is normally controlled by a simple electronic amplifier arrangement, which not only provides controllability but also additional control features (Fig. 5.12).

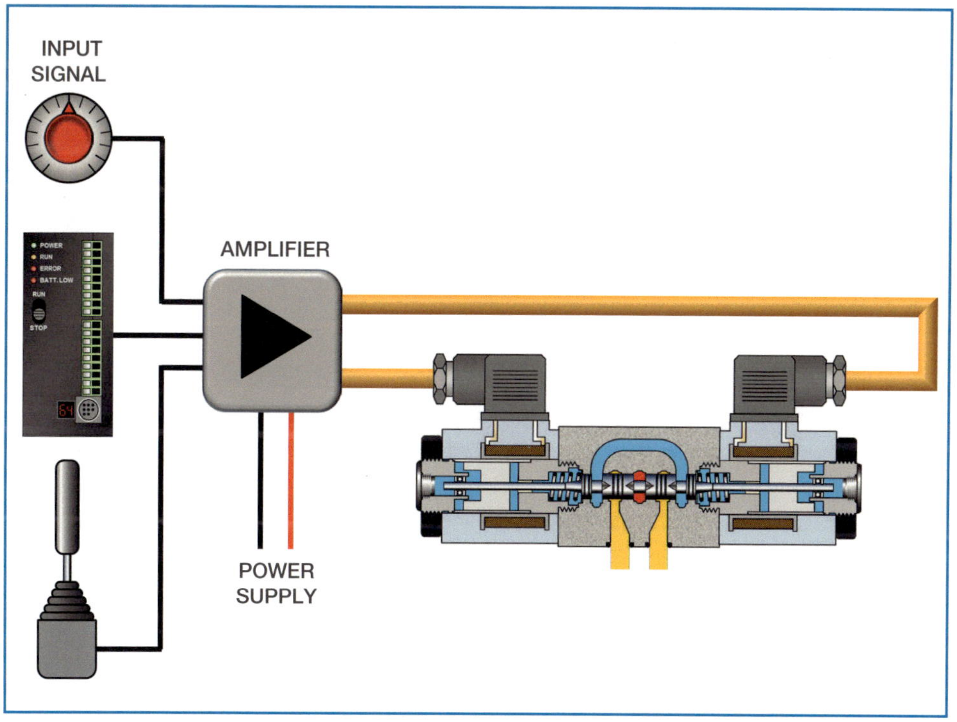

Fig. 5.12 Arrangement for proportional valve control

POINT OF INTEREST

The proportional valve amplifier may be a separate component (mounted in an electrical control cabinet, for example) or be built onto the hydraulic valve itself. Simple amplifiers may also be mounted in the solenoid plug connectors.

DEFINITION

Deadband refers to the overlap region between a spool and its port. As the spool moves within this region, the port is not opened to flow, hence the term 'deadband'. The reason for the overlap is often to reduce leakage in the de-energised position. The unwanted effect of the deadband can be reduced significantly by increasing the amplifier gain in this region in order to move the spool rapidly to a position where flow starts to pass through the valve.

The input signal to the amplifier can be created by, for example, a simple rotary potentiometer, the output from a PLC or a joystick-type potentiometer. Input signals can be either a variable voltage (often 0–10 V or ±10 V) or a variable current (typically 4–20 mA). Current signals tend to be less susceptible to interference (electrical 'noise') and to variations in signal level when transmitted over long signal lines, and so are more suited to some environments. Figure 5.13 shows a block diagram of a typical proportional valve amplifier.

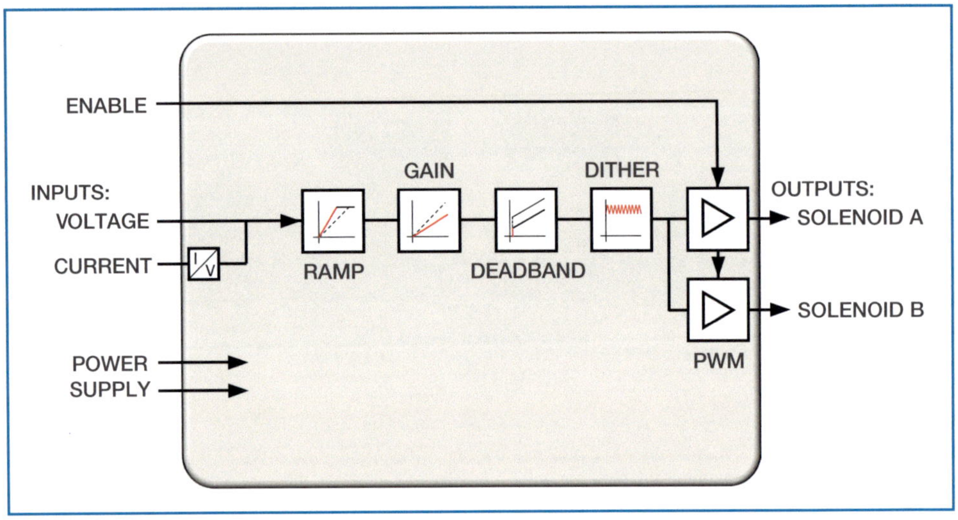

Fig. 5.13 Block diagram of a proportional valve amplifier

An adjustable **ramp function** can be included in the amplifier. This will provide a gradually increasing or decreasing output even though the input may change suddenly from one value to another (step change). This is, therefore, a very useful feature for controlling actuator acceleration and deceleration.

The **gain adjustment** determines the amplifier output for a given input signal (similar to the volume control on a radio, for example), and the **deadband compensation** effectively removes the effect of spool overlap (i.e. the portion of the spool movement before the flow path opens).

A **dither signal** is often included, and this is sometimes adjustable and sometimes factory set. The dither signal is a small high-frequency AC signal superimposed on the DC signal to keep the valve spool vibrating very slightly. This helps to remove the effect of 'stiction' in the valve (i.e. the tendency of the spool to stick due to friction or contamination).

Finally, the output signal is converted to a pulse-width modulated signal which, as mentioned earlier, is a series of on/off pulses at high frequency, where the signal level is determined by the duration (width) of each 'on' pulse relative to the duration of each 'off' pulse. This technique is used to reduce heat generation in the amplifier.

An **enable function** is normally included in the amplifier circuit to disable the amplifier output unless the signal is present. This is often used as part of the system safety control or connected to the machine emergency stop system.

Simple proportional valves control the spool opening, and therefore valve flow rate, by balancing the solenoid force against the spool spring force. This may be perfectly adequate for less demanding applications (such as shock reduction). However, when a more precise or faster valve operation is required, a **spool position sensor** can be added to the valve to provide a feedback signal to the amplifier proportional to the spool position (Fig. 5.14).

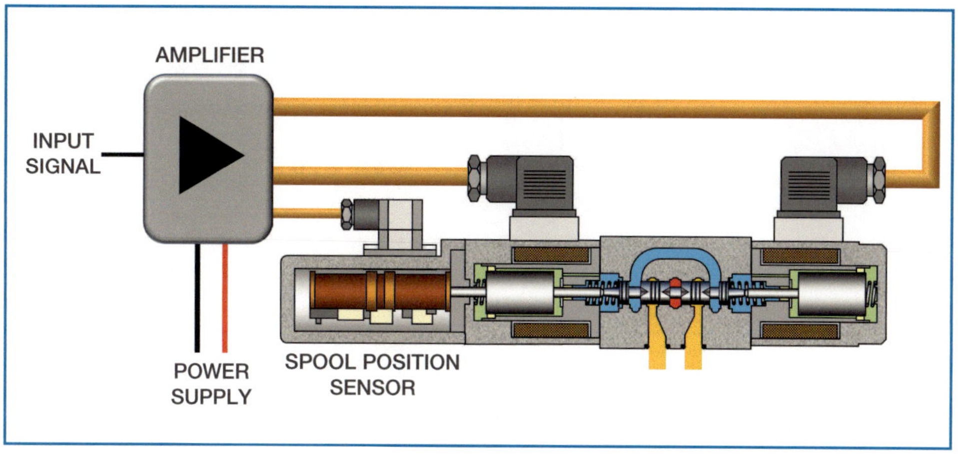

Fig. 5.14 Feedback-type proportional valve (Image courtesy Eaton Corp.)

The spool position sensor is often a non-contact inductive-type sensor (e.g. a **linear variable differential transformer (LVDT)**) and provides a feedback signal to the amplifier. The feedback signal is dependent on the spool position either side of the central position. The amplifier then compares the input signal with the feedback signal and acts to correct any error between the two, thus providing a much more precise level of spool control.

BUS SYSTEMS

In order to complete an electrical circuit, control components such as sensors, switches, solenoids, etc., need at least two electrical connections. In addition, where electromagnetic interference is likely (caused by high-current devices, powerful radio-communication signals, etc.), sensitive electronic cables have to be **screened**. This involves surrounding the cable with a protective wire-mesh sheath, which also needs to be connected to ground.

On vehicle systems, the vehicle chassis can sometimes act as one of the connections, i.e. one of the battery connections is connected to the chassis (referred to as **negative earth** or **positive earth**). In complex systems with long distances between components the amount of electrical wiring required can be significant, and the potential for errors due to incorrect cabling or bad connections can be quite high (Fig. 5.15).

In addition, any modifications required to the electrical control system will require physical changes to the wiring, which may be difficult and/or time consuming to carry out. An alternative to this traditional approach is to use a digital bus system to transmit control signals. This involves running just two signal wires around the machine and connecting control components locally. A further two power supply wires may also be required to provide the electrical current to operate higher power components such as valve solenoids. All components connected to these 'bus' wires are provided with a small piece of electronic circuitry that allows them to send and receive messages along the bus connections (Fig. 5.16).

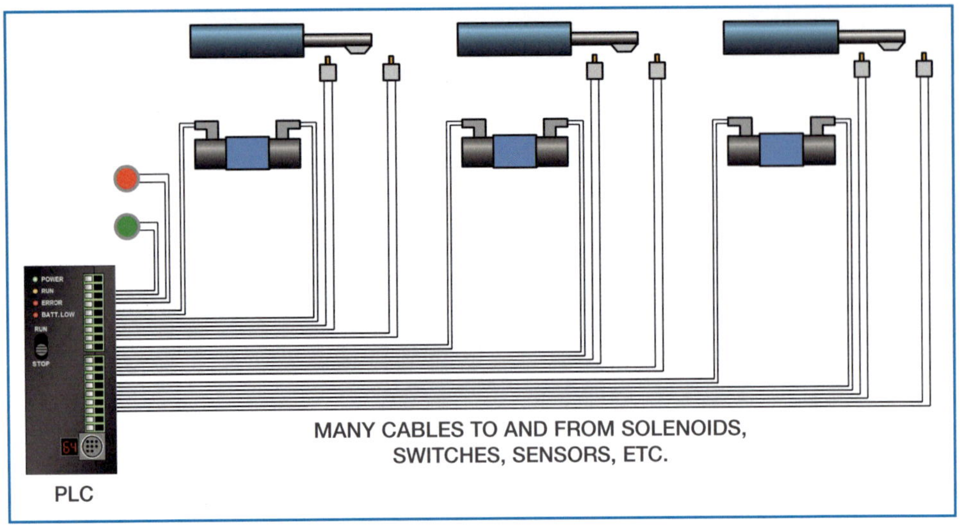

Fig. 5.15 Conventional wiring (Image courtesy of Eaton Corp.)

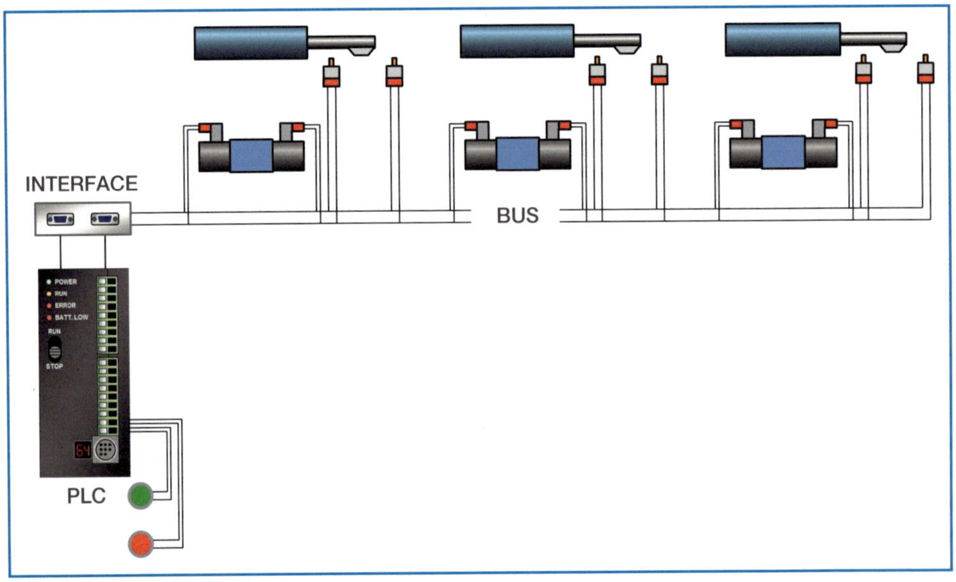

Fig. 5.16 Bus system wiring (Image courtesy of Eaton Corp.)

Each device connected to the bus has an individual 'address' (similar to a specific phone number). Each device will only react to messages addressed to it, and the central controller or PLC can recognise which device a message has been sent from. Using a bus communication system therefore means that the amount of wiring can be reduced significantly, error detection can be incorporated and modifications can often be made via software rather than hard wiring. Over the years, different bus protocols have evolved (such as CAN bus and Profibus), which all use a similar principle but differ in the way they code messages, etc.

FURTHER READING

For a white paper on best practice in cabling and 10 steps to 'not letting the gremlins in' – how to minimise the effects of EMI – go to www.webtec.com/education

CHAPTER

6 MOUNTING AND CONNECTING COMPONENTS

INTRODUCTION

Once the components of a hydraulic system have been selected, it is then necessary to connect them together and build them into a system. Several different methods of mounting components are used in practice, depending on the application. The connections between components can be made either by pipes or flexible hoses attached by means of end fittings.

VALVE MOUNTING OPTIONS

The simplest method of connecting two or more components together is to use screwed ports in the bodies of the components and to make the interconnection via pipes or hoses attached to the component via a fitting of some description (Fig. 6.1).

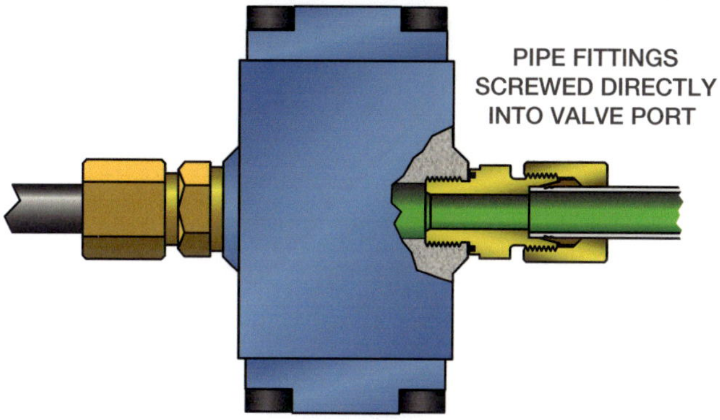

Fig. 6.1 Pipe-mounted valve

This is often referred to as a **pipe-mounted** or **line-mounted** component and is a common method for connecting pumps, actuators, filters, etc., into a system. However, where several components have to be mounted in close proximity (such as groups of control valves) this method can take up a lot of space, may be prone to leakage and can be difficult to access for maintenance purposes. Also, the number of individual pipes or hoses required in complex systems can make such arrangements impractical from a cost or space point of view.

Panel-mounted valves use a **sub-plate** or **manifold** to mount the valve, with the pipe or hose connections being made to the sub-plate rather than to the valve directly (Fig. 6.2). Valves are mounted onto the sub-plate by bolts and sealing is achieved by O-rings recessed into the valve body. This type of valve mounting is sometimes referred to as **gasket mounted**, because the early arrangements used gasket-type seals rather than rubber O-rings. Although the number and complexity of pipework and fittings are no different from those of pipe-mounted valves, this approach has the advantage of making the valves much easier to access, either for

DEFINITION

An **O-ring** is a circular rubber seal (shaped like the letter 'O') that normally has a circular cross-section.

maintenance or replacement. Also, many valves (in particular solenoid directional valves) now have standardised interfaces, making valve replacement even simpler.

The original directional-valve interfaces were defined by the European committee on fluid power known as **CETOP** (Comité Européen des Transmissions Oléohydrauliques et Pneumatiques), and such valves are sometimes known as 'CETOP valves', although the original standards have now been extended and incorporated in the international standard ISO 4401. Typical examples of ISO 4401 interface layouts are illustrated in Fig. 6.3.

POINT OF INTEREST

The American NFPA standard is basically the same as the ISO standard for valve interfaces.

Solenoid valves are also sometimes referred to by their nominal port diameter (e.g. NG6 denotes a 6mm port, which is equivalent to ISO 03, and NG10 denotes a 10mm port, which is equivalent to ISO 05).

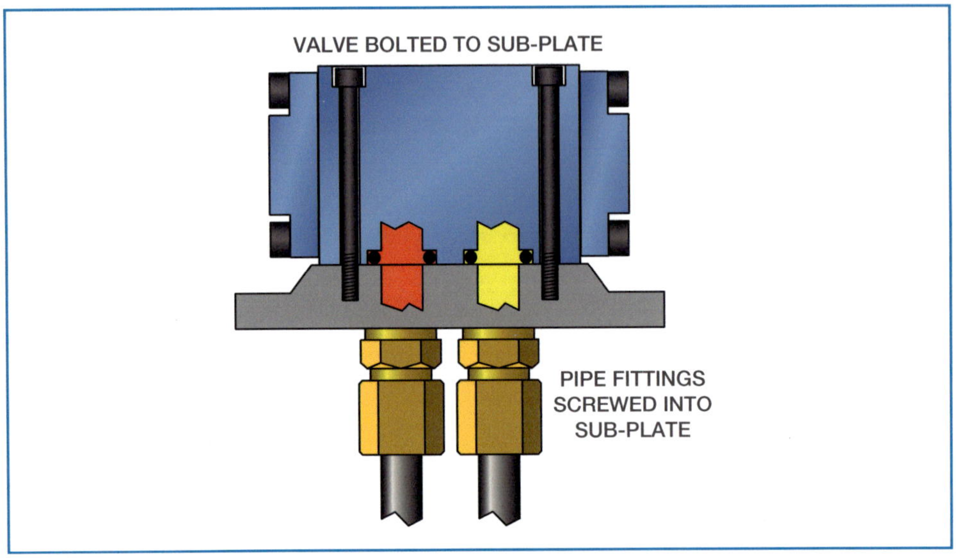

Fig. 6.2 Gasket-mounted valve

For smaller systems (ISO 03 and 05), mounting the valves in a stacking arrangement is now a very popular method of assembling a number of control valves (Fig. 6.4). In this case, valves are mounted onto a sub-plate or manifold but are designed

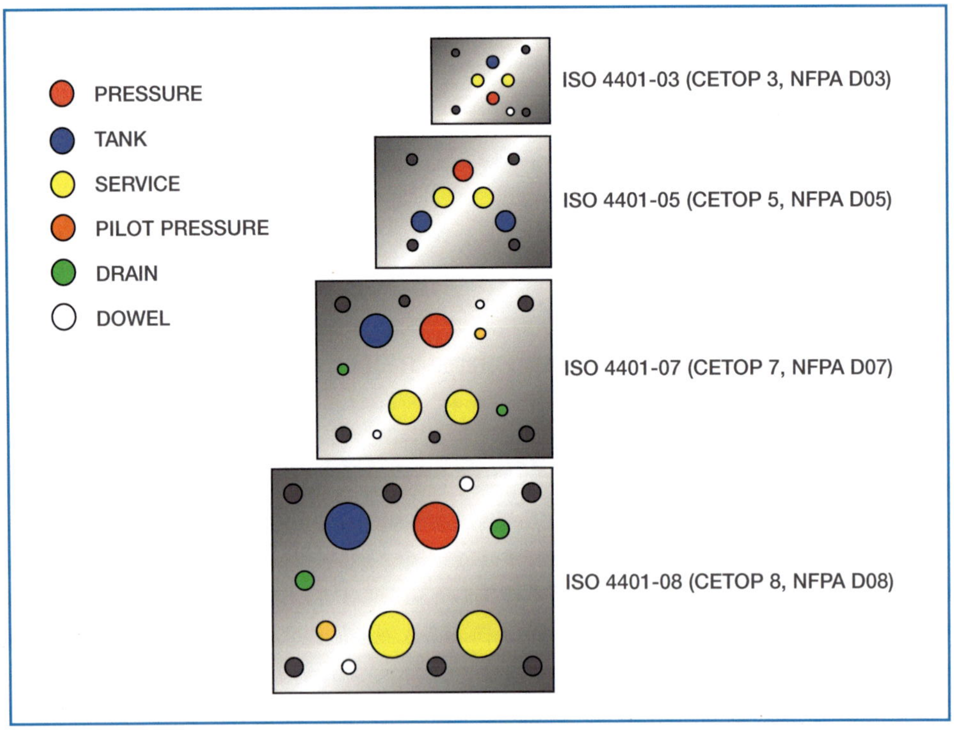

Fig. 6.3 ISO (CETOP) interfaces

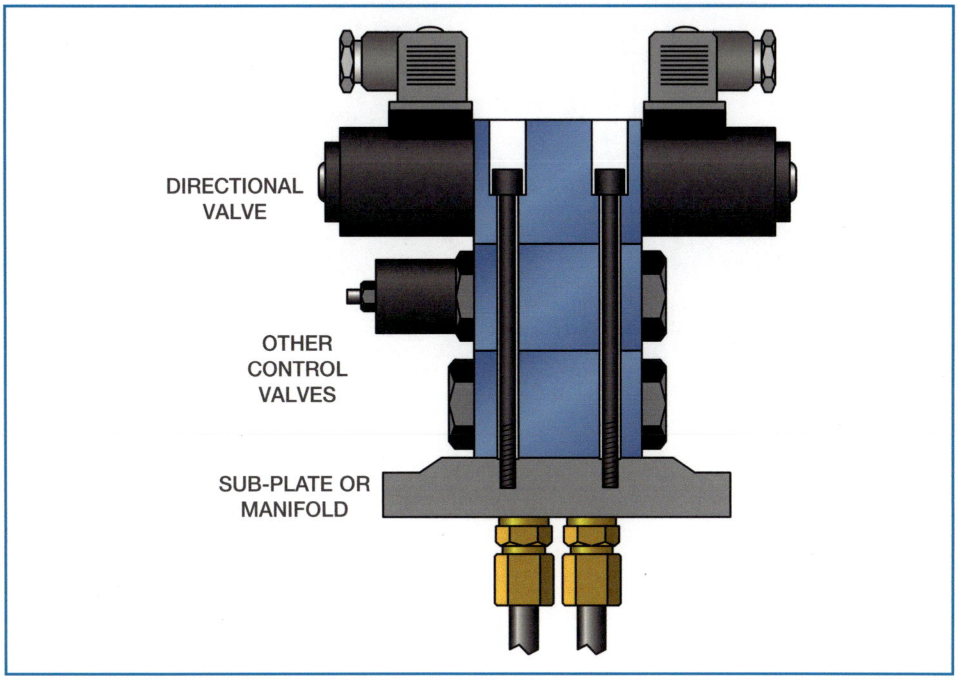

Fig. 6.4 Stacking valves

to stack one on top of the other rather than being mounted individually. Sealing between valves is again achieved by means of recessed O-rings, and the whole stack is held together and attached to the sub-plate by bolts passing through the complete stack. The directional valve normally is the top-most valve, so all the other valves use the same interface layout (port positions, etc.) as the directional valve.

Stack mounting provides a very compact and low-cost method of mounting valves together, especially for smaller valve sizes (typically up to approximately 120 L/min (32 gpm)). However, the relatively tortuous fluid pathway through the valve stack has to be taken into account in the rating of the components. For example, fluid entering the pressure (P) port of the stack has to travel upwards to the directional valve and then down again to the service port (A or B) before flowing on to the actuator. Similarly, return fluid from the actuator has to pass up and down the stack before returning to tank, all of which adds to the internal pressure drops within the system and a consequent reduction in efficiency.

For higher flow rate systems, therefore, and in order to reduce the space requirement of pipe-mounted valves, a manifold block construction can be used, where passageways within the manifold provide the interconnections and valves are gasket mounted onto the surfaces of the manifold (Fig. 6.5).

This approach provides a very compact system with minimal potential for leakage, yet retains the ease of serviceability of the components mounted on the block. System fault-finding can sometimes be more difficult with this approach, but pressure test points can be incorporated in the manifold to minimise this disadvantage.

The other main disadvantage with manifold systems in general is the design cost of the manifold itself. Although modern CAD (computer-aided design) software has simplified the design process, there is still a significant amount of skill and work involved in designing a bespoke manifold block, which obviously increases the cost

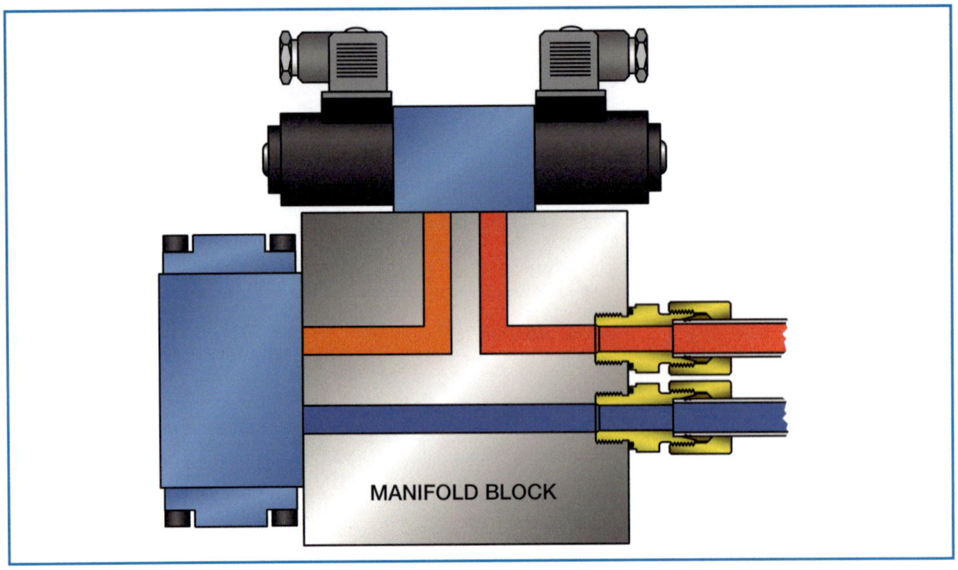

Fig. 6.5 Manifold-mounted valves

of the system. Where several or many systems are being manufactured to the same design, this initial cost can be distributed and is less of an issue, but for one-off systems the manifold design cost can be significant.

The next step on from surface mounting valves onto a manifold block is the **cartridge valve system**, where the manifold acts as the valve bodies as well as the interconnecting passages. This then provides an even more compact system, because much of the volume of the components is mounted within the manifold rather than on its surface.

There are two basic types of cartridge valves: **screw in** and **slip in**. Slip-in cartridge valves are also known as **DIN cartridges** (after the original German standard that defined their dimensions) or sometimes **logic elements** (as their operation is in some ways similar to electronic logic circuitry).

As their name suggests, screw-in cartridge valves are held in place within the manifold by means of a screw thread on the neck of the valve, and interconnections from one valve to another are via drilled passageways within the manifold (Fig. 6.6). Although

POINT OF INTEREST

The abbreviation SICV is usually taken to mean 'screw-in cartridge valve', although there is potential for confusion with the term 'slip-in cartridge valve'.

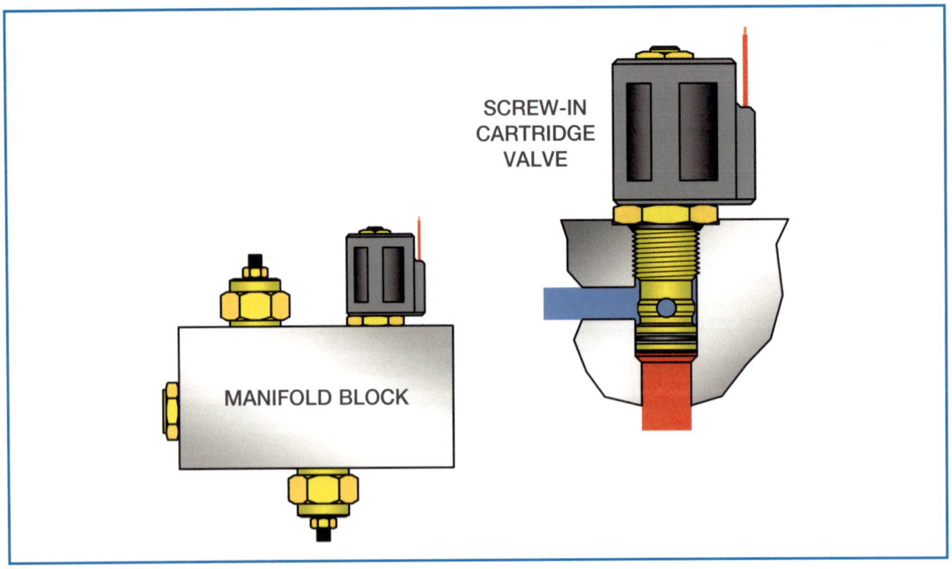

Fig. 6.6 Screw-in cartridge valves (Image courtesy of Eaton Corp.)

there is no single industry standard for the cavity dimensions, port positions, etc., some manufacturers' valves are interchangeable with others. Screw-in cartridge valves are commonly used for low to medium flow rate systems (up to a few hundred litres per minute), and are very popular in mobile systems in particular due to their compact nature.

Slip-in cartridge valves also fit into a recess in a manifold block, but in this case are retained in place by means of a cover plate on the top, which is bolted to the manifold (Fig. 6.7). In this case the international standard ISO 7368 (German standard DIN 24342) defines the cavity sizes and layouts, so that most manufacturers' components will be interchangeable from a mounting point of view (but not always functionally). Slip-in cartridge valves tend to use seated poppets rather than sliding spools to control the flow, and thus leakage within valves can be almost zero. In addition, very large flow rates can be controlled (several thousand litres per minute) by relatively small components (compared with spool valve equivalents).

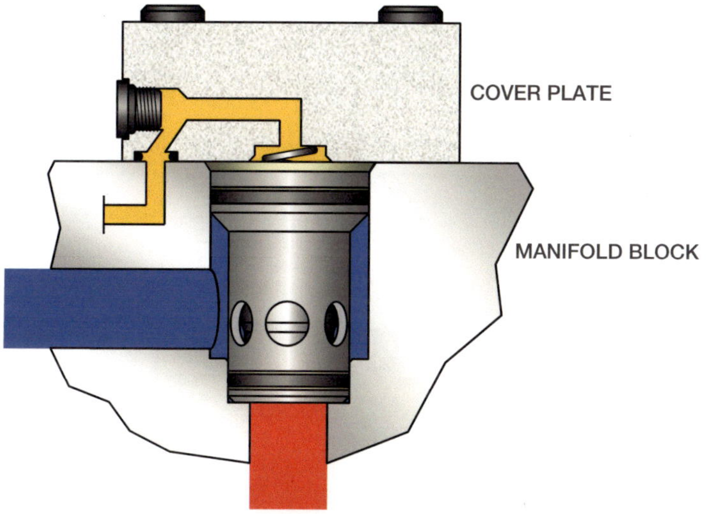

Fig. 6.7 Slip-in cartridge valves (Image courtesy of Eaton Corp.)

With manifold systems it is possible to combine several different types of valve-mounting arrangements. Combinations of slip-in and screw-in cartridge valves, together with gasket- or stack-mounted valves on the same manifold block are not uncommon.

PIPES AND PIPE FITTINGS

Cold-drawn seamless steel tubing is the general-purpose material used in hydraulic systems for conveying fluid flow over distances. Where corrosion is a significant factor, however, or a higher strength material is required, stainless steel tubing is an alternative, albeit at a higher cost.

Hydraulic pipe is specified by quoting its outside diameter (OD) and wall thickness. For example, a 25 mm OD pipe with a wall thickness of 3 mm has an internal diameter of 19 mm. The wall thickness selected depends on the pressure the pipe is required to withstand. For example, the pipe wall thickness on the inlet side of a pump will be considerably less than that on the outlet (pressure) side.

In the early days of hydraulic systems, tapered threads were sometimes machined onto a pipe so that it could be screwed directly into the female threaded port of a valve body. Sealing was then achieved by means of the metal-to-metal contact of the male and female thread forms, often aided by a sealant material such as PTFE (polytetrafluoroethylene) tape. However, this approach, together with tapered threads in general, has largely been replaced by more convenient and reliable methods of connecting pipes to valves and other components.

Connectors used for joining pipe lengths or for fitting between pipes and the ports of components basically have two functions to fulfil. First, they have to mechanically secure the pipe to the component body, counteracting the pressure inside; and, second, they have to provide a leakproof seal between the pipe and the component. This can be achieved in several different ways, as described below.

Compression fittings

Compression fittings (sometimes called **bite fittings**) function by 'squeezing' a metallic compression ring (often referred to as a **ferrule** or **olive**) onto the outside of the pipe. As shown in Fig. 6.8, as the backing nut is tightened onto the body of the fitting a taper drives the compression ring into the pipe to provide the mechanical joint and the fluid seal.

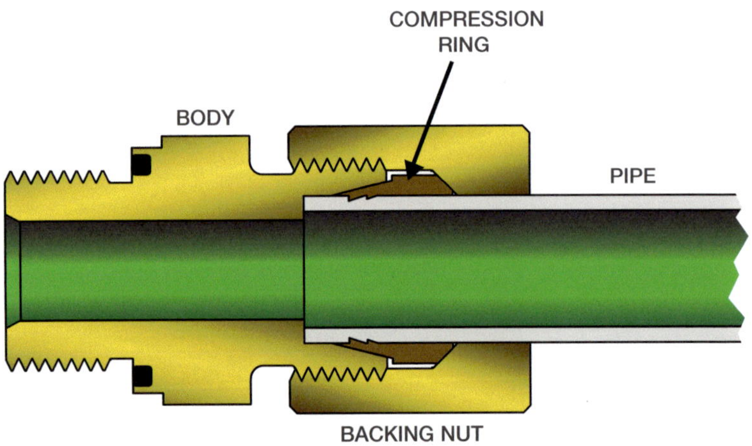

Fig. 6.8 Compression fitting

Provided the fitting is assembled correctly, especially with respect to the tightening torque applied to the backing nut, the design is adequate for general-purpose systems where very high pressures, mechanical vibration or frequent removal are not significant factors. Compression fittings have the advantage that no special tools are required to assemble the fitting. However, as the size of the fitting increases, the effort necessary to tighten the backing nut to the required torque also increases. For pipe sizes above approximately 25 mm (1 in), therefore, other types of fittings begin to become more practical.

TOP TIP

Always consult the manufacturer's specification for the correct tightening torque for compression fittings.

Flared fittings

As their name suggests, the principle of the **flared fitting** is to flare the end of the pipe into a funnel shape to provide the mechanical joint, and sometimes the fluid seal also. This is done by means of a separate flaring tool (i.e. not by using the fitting itself). Different tools are available with different angles of flare (Fig. 6.9).

Flared fittings incorporating O-ring seals have advantages over compression fittings when conditions are difficult (wide temperature ranges, high pressures, mechanical vibration or shock, etc.), but have the disadvantage that a separate process and tooling is required to produce the flare in the end of the pipe.

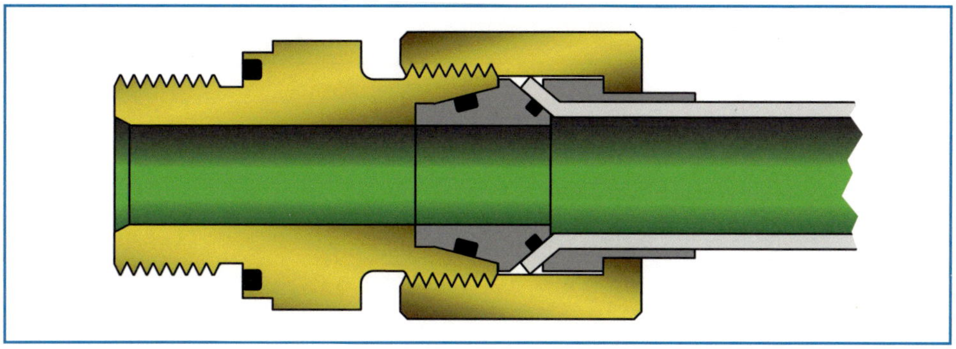

Fig. 6.9 Flared fitting

Form fittings

Form fittings also require the tube end to be formed in a separate dedicated machine, but this time the profile is more complex than a simple flare (Fig. 6.10).

This type of fitting is suitable for heavy-duty, high-pressure applications. Although more costly than compression-type fittings, the torque required to assemble the fitting is considerably less, especially for larger sizes.

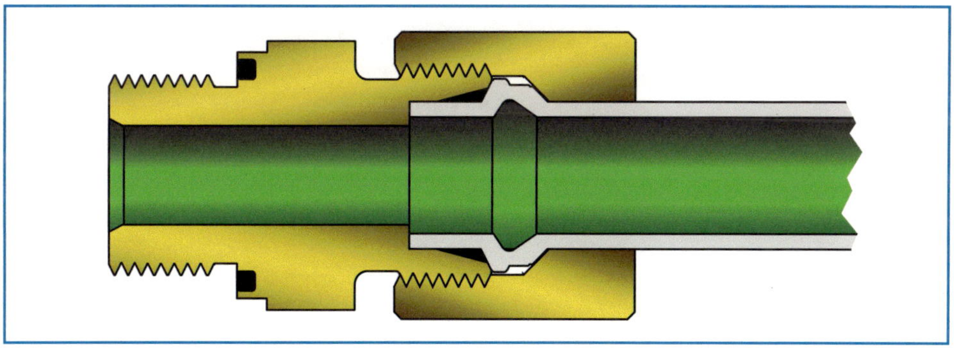

Fig. 6.10 Form fitting

Welded fittings

Socket weld fittings provide a means of permanently joining pipes together, either end to end or as T-joints. Where connections will need to be disassembled in the future, the pipe can be welded to a 'nipple', which is then retained in a fitting body and sealed with an O-ring type seal (Fig. 6.11).

Care has to be taken in the welding process with regard to integrity and scale formation, although specialist machines are available for carrying out the process.

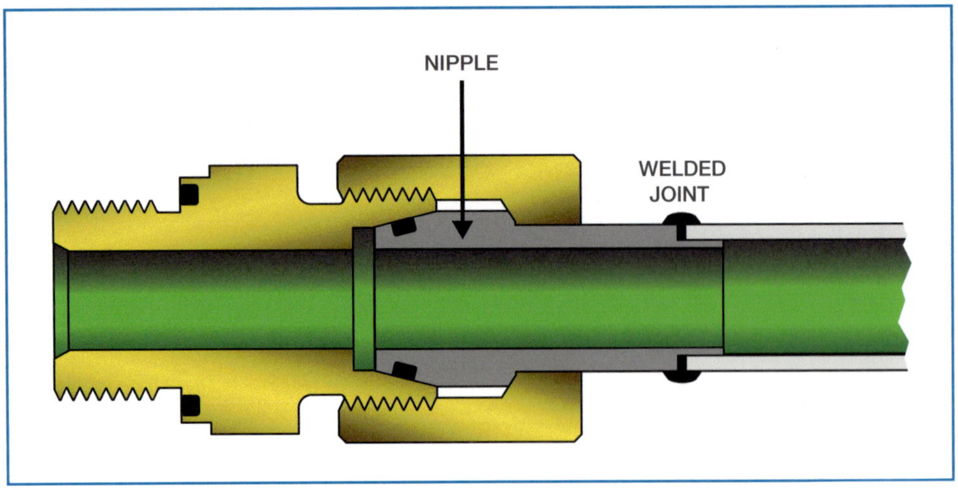

Fig. 6.11 Welded fitting

Screw threads

When connecting pipes or hoses to components such as pumps, valves or manifolds, one side of the fitting has to connect to the component port. This can be either a screwed or a flanged connection, with flanges tending to be used for larger port sizes. As mentioned above, tapered screw threads seal by means of the metal-to-metal contact between the male and female threads (Fig. 6.12).

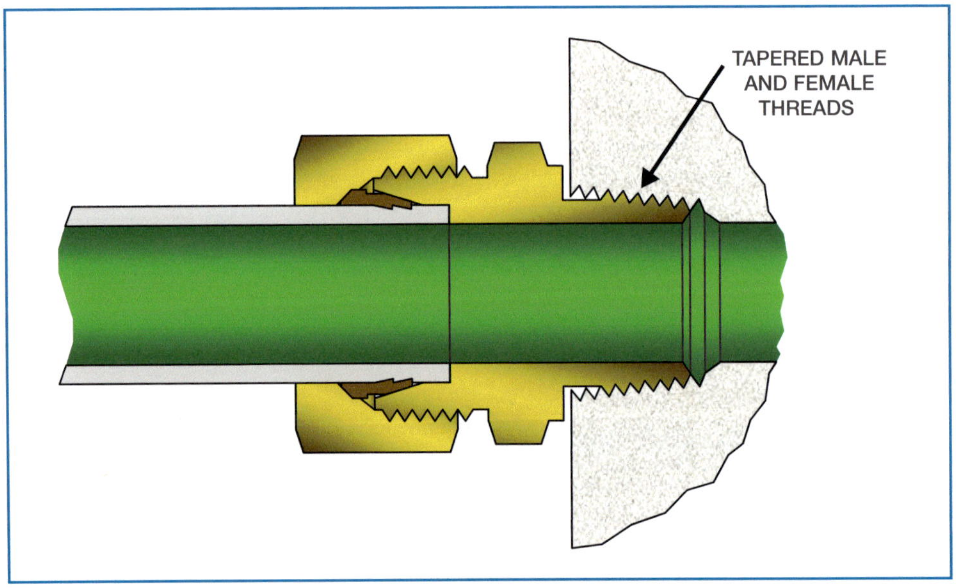

Fig. 6.12 Tapered thread

TOP TIP

Always ensure male and female screw threads are compatible with each other. In smaller sizes especially it may be possible to assemble mismatched thread forms, but the required connection strength will not then be obtained.

However, apart from very low-pressure areas of a hydraulic system (such as the pump inlet), tapered threads are no longer used due to their limited sealing capability and the potential for under- or over-tightening of the connection. Virtually all screwed fittings now utilise parallel screw threads fitted with some type of sealing component.

In the USA, Society of Automotive Engineers (SAE) standard ports are in common use. These ports have a unified fine (UNF) thread and an O-ring seal fitted to the inside shoulder of the fitting. As the fitting is screwed into the port, the O-ring becomes trapped in the chamfer on the outside edge of the port to provide the seal (Fig. 6.13). This type of connection is sometimes referred to as an **O-ring boss** (ORB), and is generally available with thread sizes up to 2½ in. The metric equivalent of the SAE

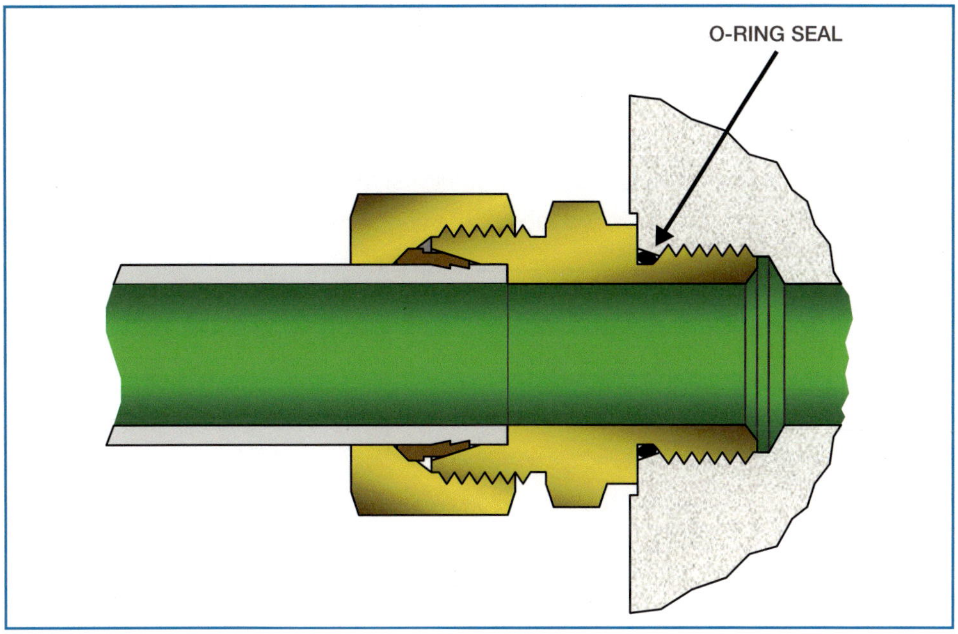

Fig. 6.13 SAE thread O-ring boss

fitting is the ISO 6149 standard fitting, which uses the same basic principle (with a trapped O-ring) but with a metric dimension instead of UNF screw thread.

The **British Standard Pipe (BSP)** thread is based on an imperial Whitworth thread form (dimensioned in inches) and, due to its popularity, it is also an ISO standard thread. Sealing of this type can be achieved by using a soft metal (e.g. copper) washer or bonded washer (metal and rubber) in between the fitting and the face of the port. Alternatively, an O-ring seal can be recessed into the fitting. This seal contacts the face of the port as the fitting is tightened (Fig. 6.14).

The German standard DIN 3852 fitting uses a similar principle, with a variety of sealing components, but with a metric (rather than Whitworth) thread form.

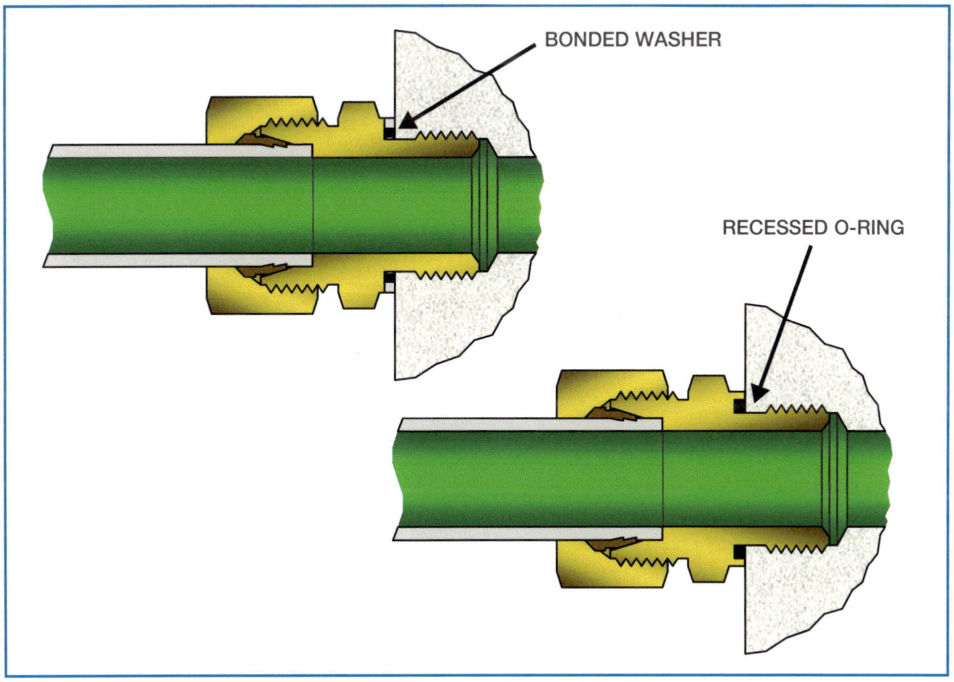

Fig. 6.14 Parallel thread fittings

POINT OF INTEREST

British Standard Pipe (BSP) threads are often denoted by the letter G. For example, G3/4 denotes a 3/4 in BSP (parallel) thread form.

Bonded washers are commonly known as **Dowty washers**, after the company that first produced them.

Flanges

Flanged connections offer an alternative to screwed fittings, and in some cases may be more robust or easier to install. In a flange fitting the pipe or hose is attached to a metal flange plate (normally screwed or welded), which is then clamped to the flat face of the component by four (sometimes eight) bolts. Sealing is achieved by means of a recessed O-ring in the face of the flange plate (Fig. 6.15).

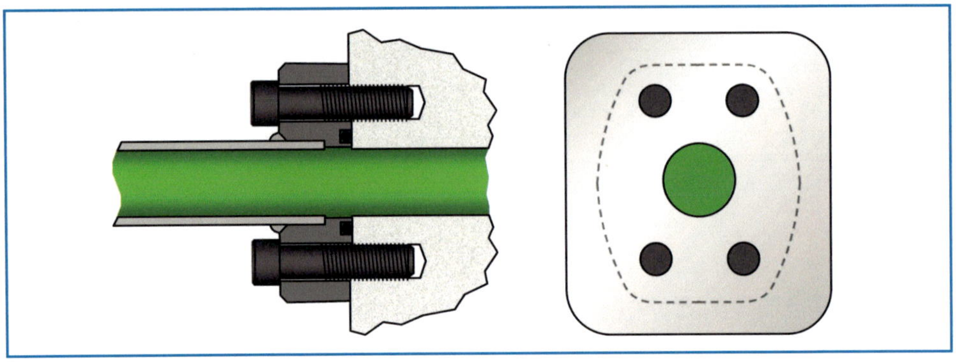

Fig. 6.15 Flange fitting

As with other connectors, different standards for flange dimensions and ratings have evolved over the years, with the American SAE and German DIN standards being the two most common for hydraulic use. In addition, two SAE standards, Code 61 and Code 62, are also in common use. Although similar in dimensions, these two standards have different pressure ratings, so care must be taken not to confuse the two types.

Hoses

Flexible hoses are used in hydraulic systems for a number of purposes, such as:

- *To enable components to move relative to the rest of the system*. Obviously, cylinders mounted by means of flexible mountings (such as trunnion or clevis mounts) may have to pivot as the piston rod extends and retracts. Other actuators, such as wheel motors, may be flexibly mounted to the frame of a vehicle via the vehicle suspension, and therefore have to move relative to other components. Similarly, pumps mounted on anti-vibration mountings will move slightly relative to other components, and therefore should have flexible inlet, outlet and drain connections.
- *To reduce the transmission of noise from the components that generate it*. Typical components that generate noise are pumps and motors, and typical components that transmit or radiate noise are flat-sided reservoirs.
- *For convenience*. In some situations flexible hoses are used simply for convenience, for example to avoid having to bend solid pipes to follow a particular route.

Apart perhaps from those used on the inlet side of pumps, all flexible hoses have to be constructed to be able to withstand internal hydraulic pressure. In addition, the material they are constructed from must be compatible with the fluid they are to carry and their operating environment (temperature range, required flexibility, resistance to abrasion and mechanical damage, etc.).

In general, flexible hoses consist of an inner tube, a reinforcement layer of some description, and an outer protective layer (Fig. 6.16). The inner tube has to be compatible with the system fluid over the expected temperature range of operation. The outer layer has to have the strength characteristics demanded by the working environment in order to resist abrasion, mechanical or chemical damage, etc. The reinforcement then has to provide the strength required to resist the internal hydraulic pressure, while remaining sufficiently flexible for the hose to perform its task.

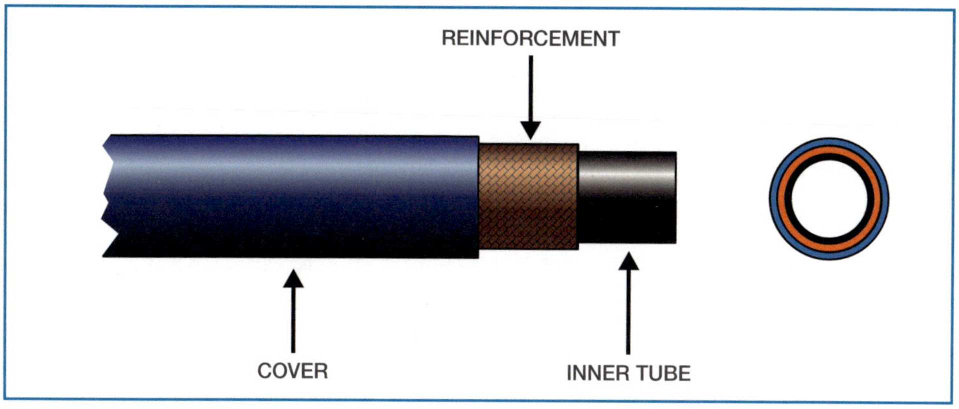

Fig. 6.16 Hose construction

As a result, many different hose constructions and materials are in use, depending on the particular function they have to provide and their working environment. For example, the two most common types of reinforcement media used for hydraulic hoses are known as **wire braid** and **wire spiral**.

As its name suggests, **braided hose** is reinforced by one or more layers of woven thin wire mesh similar to that used on screened electrical cables (Fig. 6.17). For lower-pressure applications, textile braid can be used instead of wire mesh to provide even greater flexibility. In general, however, wire-braided hoses are used for medium-pressure applications, although smaller sizes can be used in higher-pressure parts of the system.

Where high pressures and large hose diameters are called for, wire-wound **spiral hoses** are normally required. In these hoses, one or more (typically up to six) layers of thicker wire are wound spirally around the inner core of the hose to provide the hydraulic strength (Fig. 6.18).

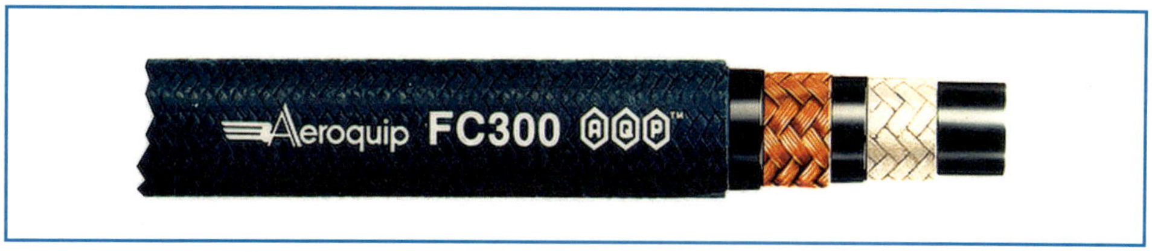

Fig. 6.17 Braided hose (Image courtesy of Eaton Corp.)

Fig. 6.18 Multi-spiral hose (Image courtesy of Eaton Corp.)

Inevitably, however, as the hose construction becomes stronger the flexibility of the hose reduces. Flexibility is usually specified in terms of the **minimum bend radius** of the hose (i.e. the minimum radius to which the hose can be flexed without damaging its construction). The flexibility of a hose must be taken into account not only for the condition when the machine is at rest but also when the machine is operational and the hose is flexed to its greatest extent.

When installing hoses in a hydraulic system, therefore, precautions have to be taken to ensure that the hose:

- is fit for purpose in terms of its pressure rating, material compatibility, temperature range capability etc.
- has been assembled with the correct end fittings
- cannot flex tighter than its minimum bend radius under all operating conditions
- cannot rub or chafe against other components likely to damage it during operation
- cannot become kinked, crushed or stretched during operation
- is protected from mechanical damage caused by machine components, falling rocks, etc.
- is not twisted longitudinally.

TOP TIP

It is vital that personnel who manufacture, repair or fit hydraulic hoses are fully trained in choosing, assembling and fitting the correct components.

Although hoses are designed to flex in one plane, any twisting of the hose along its length is likely to damage the hose and considerably reduce its life. To avoid this, most hoses have an embossed or printed line along their length known as the **lay line**, which makes any twisting of the hose visually obvious when it is installed.

Also printed onto the hose (often along the lay line) is important information such as:

- the hose manufacturer and its designation for the hose type
- the hose part number and/or size. (Hose diameters are specified by their inside diameter (ID). The most common method is to specify the hose ID in terms of a **dash number**, which is the hose ID in terms of sixteenths of an inch. For example, a -6 (dash six) size hose has an internal diameter of six-sixteenths of an inch (3/8 in).)
- the maximum working pressure rating of the hose
- the industry standard(s) to which the hose conforms
- a date code – rubber hose will have a definite life (typically 6 years) and should not be used beyond its recommended life.

In some situations it may be necessary to restrain the hose in order to prevent it causing a potential 'whiplash' injury to equipment or personnel in the event of its failure. This is normally achieved by wire retainers attached to clips on the hose.

Hose-end fittings

Although there are many ways of attaching hoses to end fittings, two basic fitting types cover the majority of hydraulic pressure line applications. The simplest type is the **reusable fitting** (or field-attachable fitting), which as its name suggests requires no special tools to attach it to the hose (Fig. 6.19).

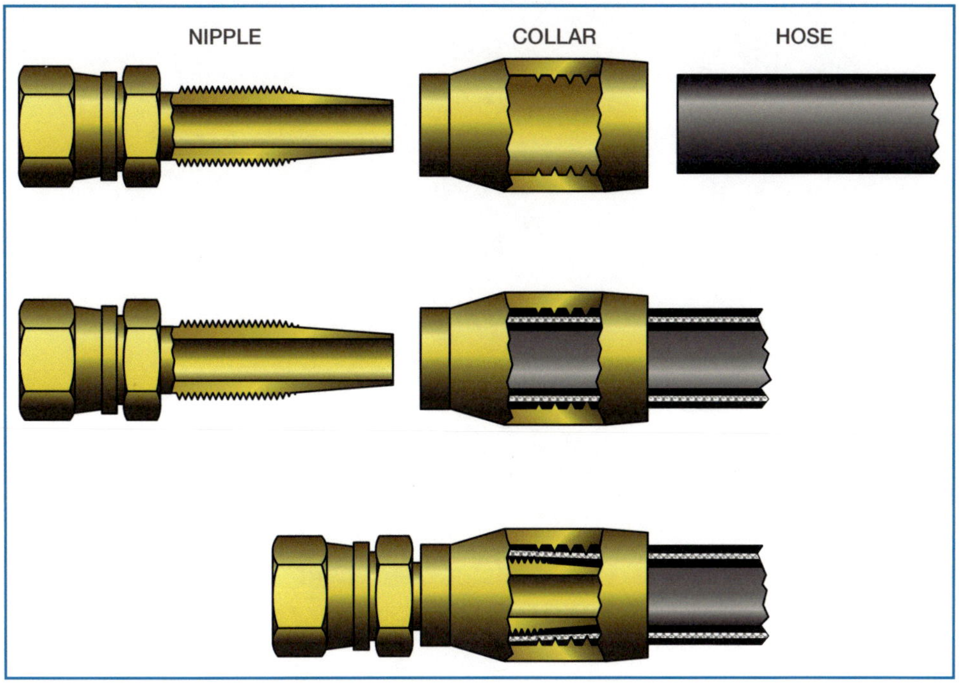

Fig. 6.19 Reusable hose-end fitting

With this type of fitting a left-hand threaded collar is first screwed onto the outside diameter of the hose. A tapered nipple is then screwed into the collar, which then squeezes the hose material and reinforcement tight into the threads of the collar to provide the mechanical joint and the fluid seal.

The connection between the hose and the fitting is a potential failure point for a hose assembly, but the term 'reusable fitting' does not mean that in a failure situation the end of the hose can be simply cut off and the fitting reattached. In fact the practice of 're-ending' hoses, as it is called, is now strongly discouraged for the following reasons:

- a re-ended hose will be shorter than the originally specified length and may therefore be vulnerable to further damage as it flexes
- the hose may have deteriorated in other ways and may be past its useful life
- fittings and hoses should be matched together (i.e. from the same manufacturer's specification), and a new fitting may not be compatible with an existing hose.

The second type of fitting involves crimping or swaging the fitting onto the end of the hose (Fig. 6.20). In this case the fitting is permanently squeezed onto the end of the hose and cannot be reused.

Swaged fittings are pushed through a die to reduce the full circumference of the fitting, whereas in **crimped fittings** a set of radial dies is squeezed against segments of the fitting. In both cases a relatively large amount of force is required to carry out the process, so it is normally done using a hand- or power-operated hydraulic swaging or crimping machine.

In some cases it is necessary to remove part of the hose covering (known as **skiving**) in order to correctly assemble the fitting to the hose. Again it is strongly recommended

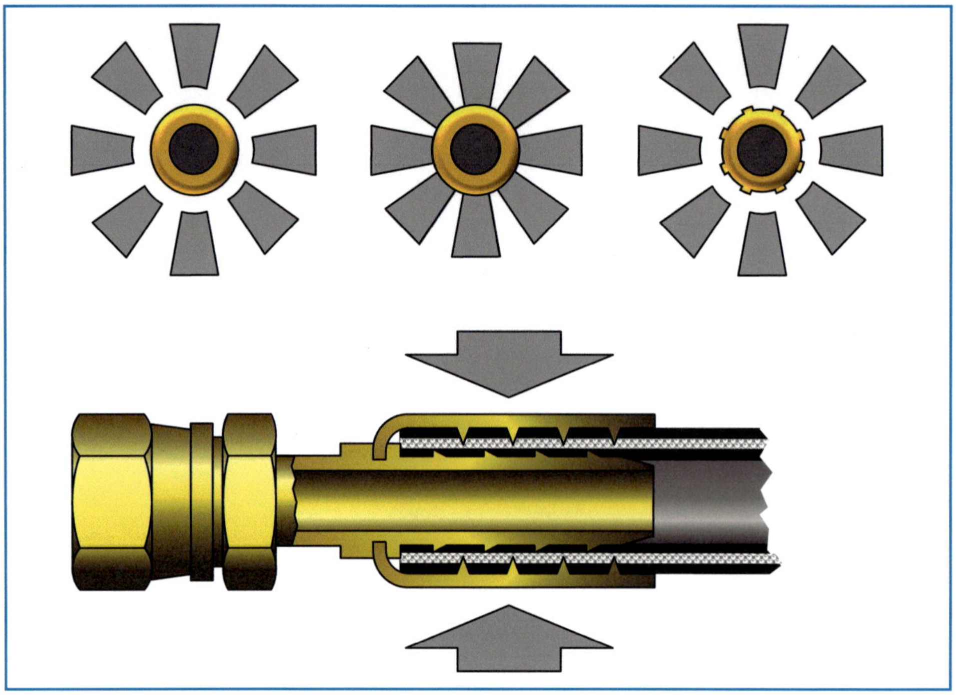

Fig. 6.20 Crimped fitting

not to use a hose and a fitting from different manufacturers, as fittings are often specifically designed to suit a particular hose construction. Also, the manufacturer's specifications for the crimping or swaging process must be accurately followed in order to achieve the correct assembly of the hose and the fitting.

FURTHER READING

For further information on hose training courses and tools for selecting hoses, visit www.webtec.com/education

CHAPTER

7 HYDRAULIC SYSTEM SAFETY

INTRODUCTION

Legislation exists in most industrialised countries to ensure that machinery is designed, manufactured and operated according to recognised safety standards in order to protect machine operators. Even stricter legislation may apply to hydraulically powered machinery designed to operate in the vicinity of the general public, such as road sweepers, refuse vehicles and movable bridges.

However, when troubleshooting or maintenance work has to be carried out on machinery, some of the operational safety features may need to be overridden or bypassed in order to carry out such work. For example, a machine guard may have to be temporarily removed in order to gain access to an actuator for test purposes. It therefore goes without saying that, for anyone working within hydraulic system maintenance, the number one consideration must be safety. No matter how diligently safety standards have been applied, ultimately the safety of personnel maintaining or troubleshooting hydraulically powered machinery depends strongly on both the competence of the individuals involved and the adoption of safe working practices and procedures.

By their nature, hydraulic systems are used to move heavy loads and generate large forces, and they sometimes use potentially dangerous fluids, so the potential for hazardous situations to arise is significant.

THE RISKS

Mechanical loads

As mentioned above, hydraulic systems are used to lift or move heavily loaded components. Before any work is carried out on a system, loads must be lowered to a position where they cannot move under their own weight (Fig. 7.1). If this is not

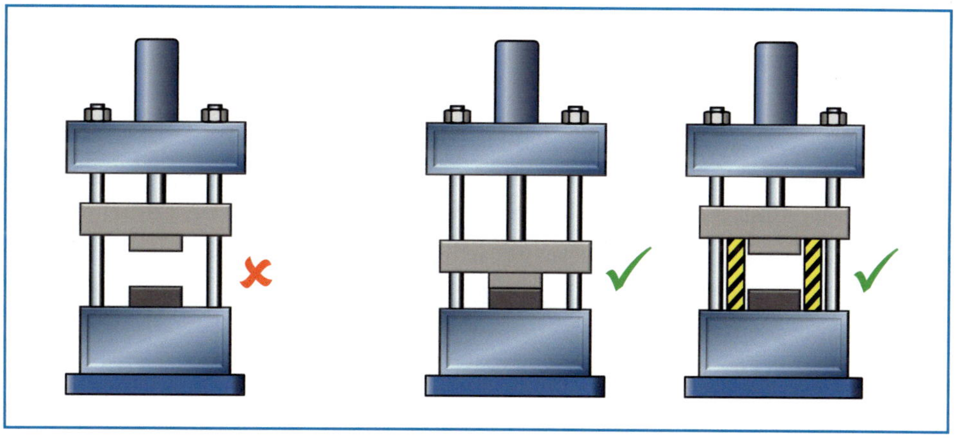

Fig. 7.1 Ensure loads cannot move

possible, the loads must be mechanically secured in such a way that they are unable to move under their own weight.

Similarly for vehicles, before carrying out any work it must be ensured that the vehicle is parked on level ground or chocked to ensure that it cannot move (Fig. 7.2).

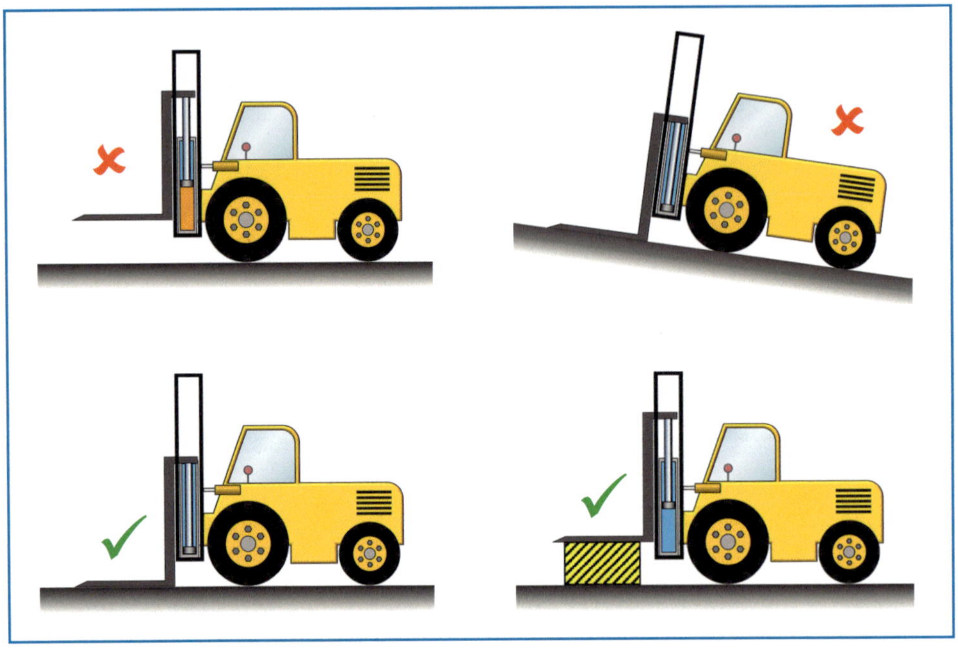

Fig. 7.2 Ensure vehicles and their components are secured

In some applications, hydraulic cylinders and motors can be fitted with mechanical braking devices, which are normally spring applied and hydraulically released. Typically these are used as 'parking' brakes. That is, they prevent movement (due to fluid leakage) when the actuator is stationary. Hydraulic motors operating a winch, for example, are often fitted with a spring-applied brake, and cylinders holding vertical loads can use a **rod lock** device to prevent creep when the cylinder is parked.

TOP TIP

When carrying out maintenance work on a hydraulic system do not rely on the system itself to prevent dangerous movement of machine components. *Always* mechanically secure moving parts of the machine that could create hazards to maintenance or other personnel.

While these devices may provide adequate safety during normal machine operation, a higher level of safety will be required during maintenance work on a machine, particularly where work on the actuator itself may be required. Any failure of the mechanical connection between the actuator and load, for example, could still cause a dangerous situation, as could inadvertent release of the brake due to an incomplete release of pressure in the system. The same also applies to hydraulic components fitted to control loaded actuators. For example, pilot-operated check valves or counterbalance valves must not be relied on to provide adequate safety during maintenance operations on the machine or hydraulic system.

Knowledge of the machine function is therefore vital to determine which parts of the machinery could move in potentially dangerous ways. Practices must then be adopted in order to prevent such movements occurring, and ideally these should be documented in machine maintenance manuals.

Fluid pressure

Pressures up to 350 bar (5000 psi) are not uncommon in modern hydraulic systems, and many operate at higher pressures still. A pressure of 350 bar is approximately

100 times the pressure of a domestic water supply (which as plumbers know can cause a lot of problems if not adequately contained). High pressures acting on even small areas can create very large forces. For example, a pressure of 350 bar acting on a piston as small as 100 mm (4 in) diameter will create a force of almost 28 tonnes (31 US tons) (Fig. 7.3).

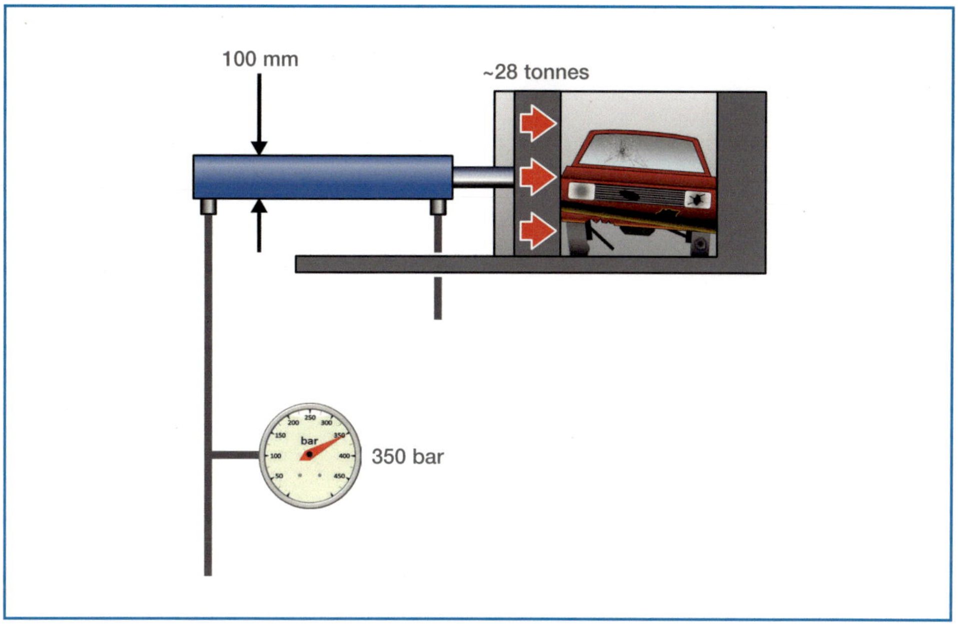

Fig. 7.3 Hydraulic power creates large forces

This means that maintenance work on a hydraulically powered machine should ideally be carried out with no pressure in the system. When this is not possible (e.g. when setting valve adjustments), it must be ensured that all personnel are clear of any moving parts of the machine.

Shutting down the drive to system pumps will not necessarily reduce the pressure throughout the system to zero, as pressure can sometimes become 'locked in' to parts of the system. Any unsupported load will create pressure in the system, even when the drive is shut down. Also, if accumulators are used in the system, the system pressure may be maintained unless the fluid is drained from the accumulators on shutdown.

A further danger of high fluid pressure is caused by small 'pin-hole' leaks (typically in flexible hoses), where the fluid escapes at very high velocity, as described in the next section.

Hydraulic fluids

As mentioned in Chapter 3, certain fluids used in hydraulic systems are potentially dangerous to humans because of their chemical composition. Fire-resistant phosphate esters, for example, can cause skin irritation, and much more serious problems if they are ingested or come into contact with the eyes. Mineral oil fluids obviously carry the risk of fire if they come into contact with hot surfaces, especially when sprayed as a mist from a high-pressure leak. Water-based fluids have the potential to form bacterial growth, which again can be harmful to health, and just about any hydraulic fluid when spilled on the floor can create a slip hazard.

It should also be remembered that hydraulic systems, and the fluids within them, often operate at relatively high temperatures. Therefore, carrying out maintenance work on a machine immediately after it has been shut down can result in burns from hot components or fluids. Even a temperature as low as 44°C, which is less than the operating temperature of many hydraulic systems, can cause burns to the skin. Mineral oil fluids are flammable, so care must be taken to ensure that any fluid spillages cannot come into contact with anything hot, such as welding or cutting equipment.

A much more serious danger with hydraulic fluids is the effect they have when they enter the bloodstream. This can happen if a high-velocity jet of fluid, caused by a pin-hole leak, comes into contact with the skin. Such injuries can result in blood poisoning and bacterial infection, potentially requiring amputation of fingers or limbs. Therefore, hands must not be used to check for leaks, even if gloved. If such an injury does occur, even though it may appear to be quite minor at the time, medical attention must be obtained immediately.

WARNING

Medical personnel treating **high-pressure fluid injection injuries (HPFII or HPII)** will need to know the type of fluid involved, so the **material data safety sheet (MSDS)** is a vital piece of information in such circumstances.

HPFIIs can be very serious, and sometimes fatal. For further safety information on this subject, consult the available information on the internet (such as the UK Health and Safety Executive (HSE) website www.hse.gov.uk).

Component failure

Fortunately, it is very rare for hydraulic components to fail in a dangerous manner. As fluids are only very slightly compressible, an external failure of a component usually results in a rapid loss of pressure. Although this may result in a significant fluid spillage (and the hazards that this may result in), the sudden loss of pressure normally means that further mechanical damage is limited.

By contrast, compressed gases will have to expand significantly before their pressure reduces, and so are much more likely to create a damaging, explosive-type failure. Normally, compressed gases are only present in hydraulic systems in accumulators, and these therefore require special safety precautions, as will be described. However, trapped air in a system, caused by ineffective bleeding, can also create a potential hazard.

One type of component failure that may require special attention is the separation of a flexible hose from its end fitting. In this situation the hose tends to flail around, which in itself can cause injury, and hot, flammable hydraulic fluid may be sprayed out in the process. Where such a potential danger exists, **hose restraints** or covers must be fitted to reduce the potential risk.

SAFE WORKING PRACTICES

Before carrying out any work on machinery the most important thing to ensure is that the people who will carry out the work are competent to do so. This means that those concerned will have knowledge about the operation of the machine and have been trained in the operation of hydraulic systems and components. They must also be competent to carry out general workshop practices, and must have been made aware of the particular hazards and risks involved on each job. Ideally, their competence should have been assessed and certified.

In addition, a **risk analysis** should be carried out by suitably qualified people, and appropriate safety procedures determined. Where machinery is located on customer's premises, additional safety requirements may have to be determined and complied with.

All maintenance personnel should have a good understanding of both the particular machine and its hydraulic system operation, and be provided with the necessary circuit diagrams, manuals, procedures, etc. All recommendations and instructions provided by the machine or system manufacturer must be followed. Also, the safety data sheet for the fluid being used should be consulted if necessary (e.g. if the fluid is an unfamiliar one).

Routine maintenance tasks (such as filter element changes, accumulator pre-charge checking and hose replacement) should have detailed written work procedures. However, no matter how exhaustive the documentation is on a piece of machinery, there will always be times when a maintenance technician has to work on their own initiative. Even in undocumented or unforeseen situations the rigorous adoption of safe working practices means that accidents can be avoided. Such working practices include the following:

- *Procedures and equipment*. Have a clear understanding of what work or diagnostic procedures need to be carried out and what instrumentation, tools, etc., will be required. Having the right tools for the job is something common to all maintenance activities (Fig. 7.4). If more than one maintenance technician is involved, one individual should manage the process and be designated to coordinate or lead all activities.
- *Personal protective equipment*. Relevant personal protection equipment must be worn at all times. This may include safety shoes, hard hats, safety glasses, gloves, protective skin creams, ear defenders and high-visibility clothing (Fig. 7.5). If working in unfamiliar surroundings, the location of first-aid facilities should also be established.
- *Put the machine in a safe condition*. The machine must be put into a safe condition for maintenance work to be carried out. Wherever possible, loads that are supported by the hydraulic system should be placed in a position

Fig. 7.4 Hydraulic test equipment

Fig. 7.5 Ensure personal protective equipment is worn at all times

where they cannot move under the action of outside influences such as gravity, wind or waves. If this is not possible, loads should be mechanically secured or supported to ensure that they cannot move when maintenance work is being carried out.

- *Switch off all pumps*. The drive to the pump or pumps should be switched off and appropriate methods used to ensure that pumps cannot be restarted inadvertently. This could involve removing ignition keys from vehicles, locking motor starter switches in the off position, removing fuses and using appropriate signs (Fig. 7.6). The lead maintenance technician should be the only person designated to restart pumps after a shutdown.

TOP TIP

Remember that some systems incorporate more than one pump (such as a standby pump), so all pumps need to be isolated before maintenance work is carried out.

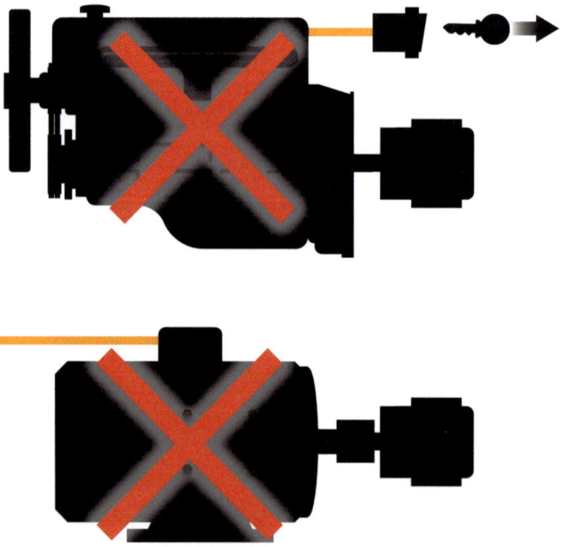

Fig. 7.6 Lock out prime movers

- *Drain all accumulators*. If accumulators are incorporated in the system, they should be drained of fluid by opening the drain-down valve and checking with a pressure gauge that the fluid pressure has been released (Fig. 7.7). In some systems, automatic valves are fitted to accumulators to drain the fluid each time the system is shut down. Even if this is the case, the manual drain valve should

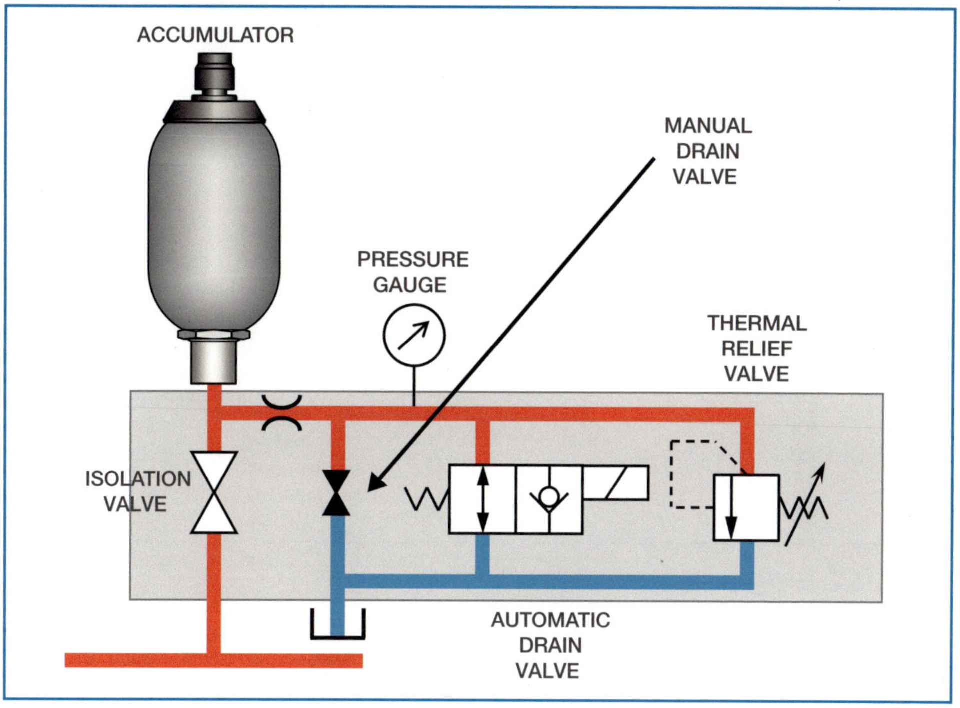

Fig. 7.7 Manually drain all accumulators

also be opened and a check made (by means of a pressure gauge) that the drain-down has occurred. (Note that some pressure gauges are fitted with a **push-to-read valve**, which requires a button to be pushed in before the gauge is connected to the system.) As an extra precaution the isolation valve can be closed to isolate the accumulator from the system. However, this isolation valve must not be used on its own as a safety precaution (i.e. the accumulator should *always* be drained of pressurised fluid before carrying out maintenance work).

Unless it is required to remove or work on the accumulator itself, it is not normally necessary to release the gas pressure from the accumulator, as this will be retained within the accumulator vessel itself by means of a rubber bag or piston. It may, however, be necessary to check the gas pressure in the accumulator (pre-charge pressure), and this must be done with the accumulator drained of fluid.

- *Release any pressure in the system*. Any pressure that may be locked into certain parts of the system can often be released by operating directional valves either electrically or via manual overrides. Checking machine functions using solenoid valve manual overrides must be carried out with extreme caution, however, as this may bypass programmed machine safety sequences.

  Again, if pressure gauges or test points are included in the system, check that the pressure has decayed. Pressure bleed points should be incorporated in parts of the system where pressure is likely to be locked in (e.g. between a cylinder and a counterbalance valve). As fluids are only slightly compressible, only a relatively small release of fluid is normally necessary to relieve the pressure. If the pressure does not drop rapidly, this implies that something is creating pressure other than just the compressibility of the fluid. This could be an unsupported load, spring-loaded actuators or compressed gas in the system (from an accumulator or trapped air) (Fig. 7.8).

TOP TIP

Accumulators are potentially dangerous components and should be treated with extreme care.

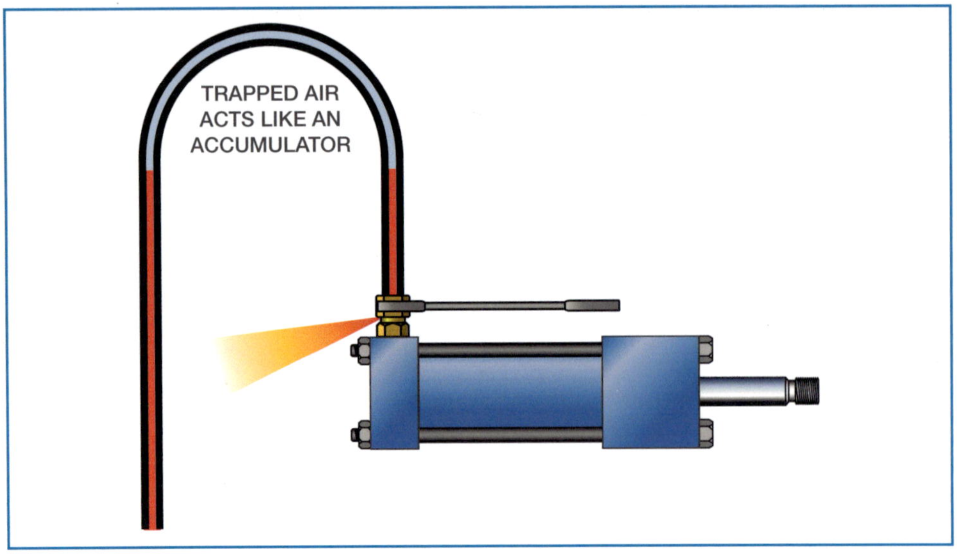

Fig. 7.8 Trapped air may be pressurised

- *Shut off the electrical control supply*. Isolate the electrical control voltage from the machine (i.e. the supply to electrically or electronically controlled valves, etc.) (Fig. 7.9). This may require the services of a qualified electrician.

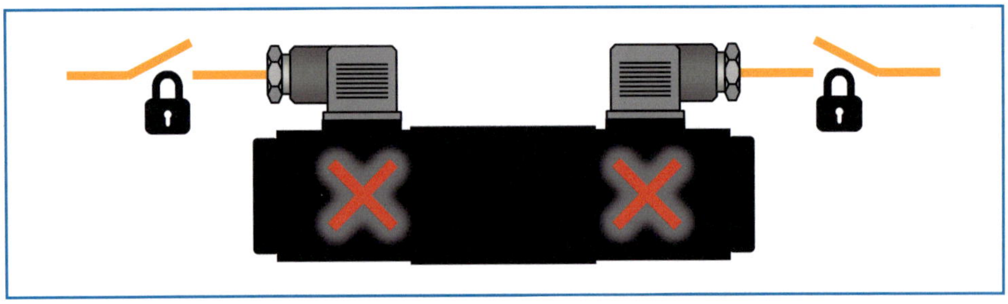

Fig. 7.9 Lock-out electrical control supply

- *Use a temperature-sensing gun*. When checking for heat generation or hotspots in a hydraulic system, people are tempted to use their hands as a temperature sensor. Needless to say this is not a good idea, for obvious reasons, and neither does it provide particularly quantifiable data. A far better and safer solution is to use an infrared temperature-sensing gun (Fig. 7.10), which can be pointed at various components or locations to discover where the heat is being generated.

TOP TIP

Remember – hydraulic systems often operate at temperatures that could cause injury to unprotected skin.

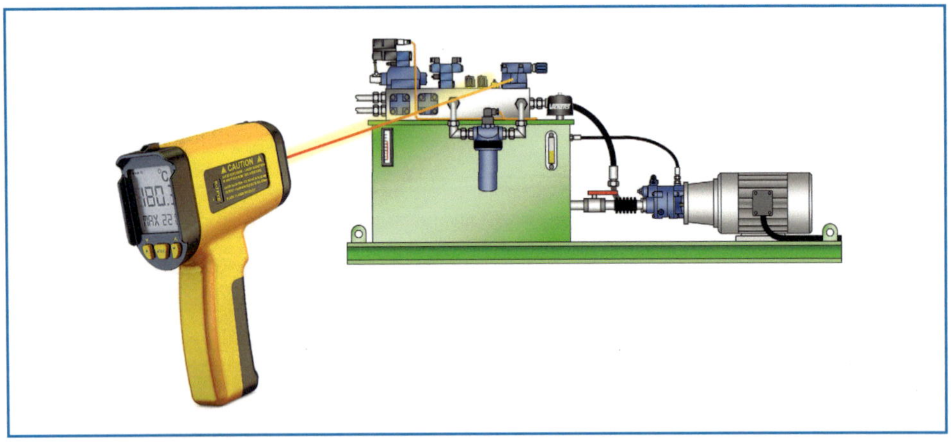

Fig. 7.10 Temperature sensing gun

- *Avoid fluid contamination and collect spillages*. Carry out the diagnostic and/or rectification work as required, ensuring that there is minimal opportunity for contamination to enter the system. Any fluid spillage that is likely to occur should be captured in a drip tray or other suitable container (Fig. 7.11).

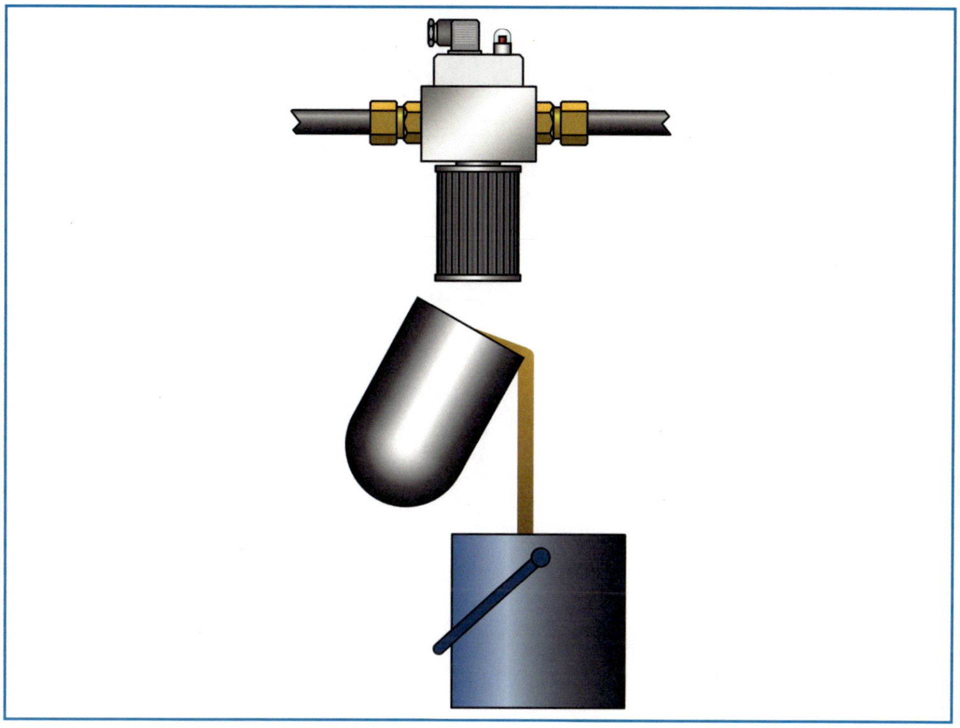

Fig. 7.11 Empty filter bowls into a container

- *Use the correct components and adjust the settings appropriately*. If a component is being replaced, ensure that the new component is the correct one in all respects (pressure rating, solenoid voltage, mounting arrangement, thread form, etc.). Wherever possible use genuine replacement parts recommended by the machine manufacturer. Manufacturer's recommendations for installation must be followed with regard to alignment, torque levels, air bleeding, etc. If an adjustable component is being replaced, if possible set the adjustment to approximately the same setting as the original component (Fig. 7.12). If this is

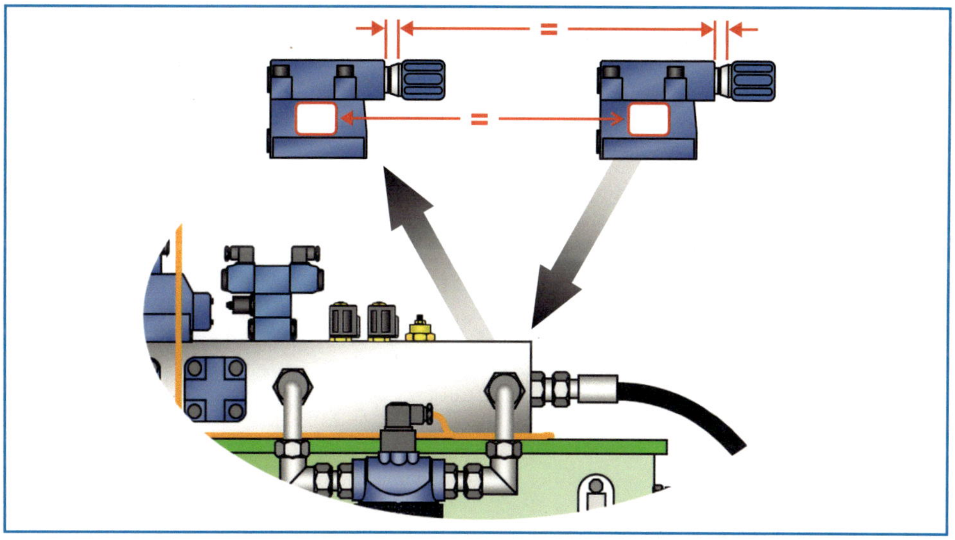

Fig. 7.12 Adjust new components to the approximate previous setting

TOP TIP

Changing a filter element is always likely to create some fluid spillage, so provision should be made for this in work instructions.

TOP TIP

Be aware that not all adjustments operate in the same manner (i.e. screwing an adjuster in one direction may have opposite effects on different components). If in doubt the component data sheet should be checked.

not possible, set the adjustment to a safe condition for a restart. For example, when replacing a relief valve adjust the setting to a low pressure, whereas when replacing a counterbalance valve it is safer to adjust the valve setting higher.

- *Replacing hose assemblies*. Pay particular attention to the replacement of flexible hose assemblies. Hoses are one of the commonest components of a hydraulic system that require routine replacement, while at the same time offering one of the most likely opportunities for mistakes to be made. Hose assemblies must be chosen, manufactured and installed correctly, without twisting and undue bending, and be clear of anything likely to cause mechanical damage. Figure 7.13 illustrates some of the common errors made in the installation of flexible hoses.

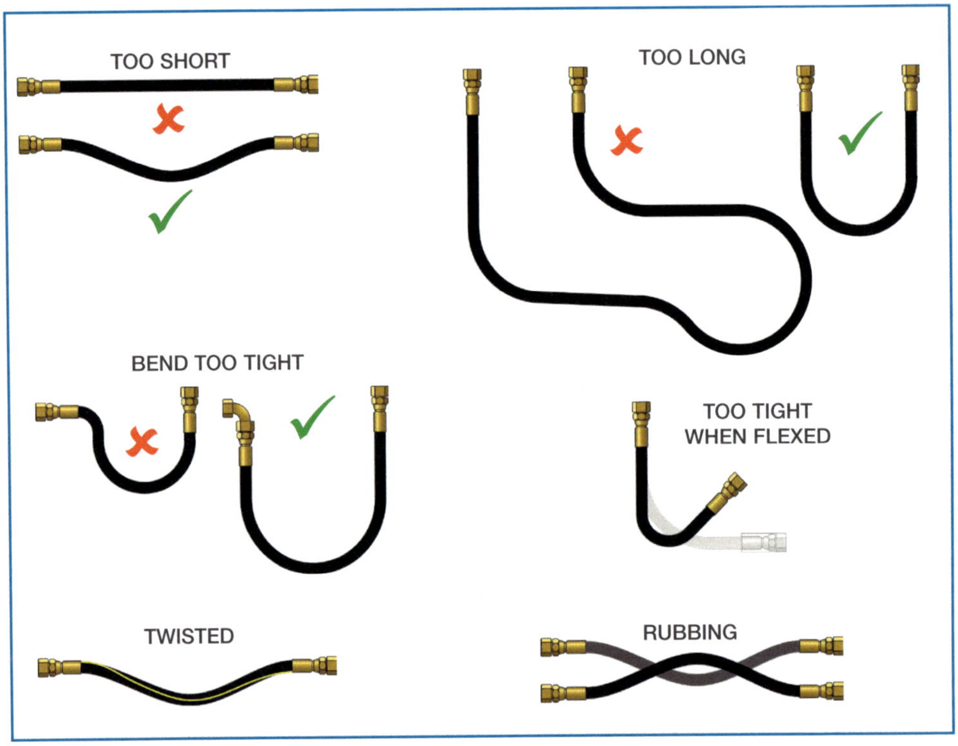

Fig. 7.13 Correct and incorrect hose installations

  Avoid mixing hoses and fittings from different manufacturers (as they are often matched together) and ensure that the correct thread forms are used. Some BSP fittings, for example, will screw into a metric ISO 6149 threaded port (and vice versa), but the required connection strength will not be obtained.

  Non-conductive hoses must be used for equipment (such as man lifts) that may operate close to high-voltage power lines. Care must be taken that any replacement hoses fitted are to the original specification in this respect.

- *Do not assume new fluid is clean*. When replacing or topping up the system fluid, always filter the fluid to an appropriate level before it is put into the system reservoir. If the system is not equipped with a means of filling the reservoir via a filter, a portable filter cart should be used (Fig. 7.14). If changing the type of fluid (e.g. from mineral oil to fire resistant), seek advice from fluid and component suppliers.

Fig. 7.14 Fluid-transfer cart (Image courtesy of MP Filtri Ltd)

- *Maintenance of electrical or electronic components*. If maintenance is required on electrical or electronic components within the hydraulic system, only people certified to carry out such work must be allowed to do so.
- *Do not work on a hydraulic system while it is running, unless absolutely necessary*. In general, no work should be carried out on the hydraulic system while it is running, unless this has been allowed for in the design of the system. For example, it could be acceptable to change a filter element in a duplex-type filter (Fig. 7.15) fitted with a changeover valve designed for this purpose, but

Fig. 7.15 Duplex filter (Image courtesy of MP Filtri Ltd)

a leaking fitting should not be tightened with the system running and under pressure.

However, there will inevitably be occasions when maintenance work has to be carried out with the machine in operation, or at least with the hydraulic system running. This could involve setting adjustable components, taking fluid samples, checking and recording pressure levels, etc. In such situations the risk levels are greater, and additional safeguards have to be taken to protect personnel from moving parts of the machine, etc.

As mentioned, checking machine functions using solenoid valve manual overrides must only be carried out with extreme caution when the hydraulic system is operational.

- *Restarting the machine*. Once the maintenance work has been completed, valves that were opened should be closed and those that were closed should be opened, interlocks should be removed, and everyone in the vicinity should be warned by appropriate means that the machine is about to restart.
- *Leave the work site in a clean and tidy state*. Once final adjustments have been made, if necessary, and the machine performance confirmed, all tools and instruments should be cleared away, any fluid spillages should be cleaned up, and the machine left in a clean and tidy state. Contaminated materials such as used filter elements, cleaning cloths, etc., should be disposed of appropriately and contaminated clothing removed.
- *Record the work done in the machine log and reassess risk and safety*. Any work carried out on the machine should be recorded in a log for the benefit of other personnel at a later date. Wherever possible, the root cause of unexpected failures should be established and the risk analysis for the machine modified if necessary. Note that any significant modifications that are made to a machine or its hydraulic system may require it to be reassessed from the point of view of conformity to relevant safety legislation.

TOP TIP

Wherever possible think about the causes of the failure and any consequences of it before restarting the machine.

To a large extent, sudden catastrophic failures of hydraulic systems that may create potentially dangerous situations can be significantly reduced by establishing a routine preventive maintenance procedure, as described in Chapter 8. Anticipating failures before they occur avoids the stress of breakdown situations, when safety precautions are easily overlooked.

Finally, it is worth repeating that the number one requirement to ensure the safety of hydraulic systems is to ensure that all personnel involved are well trained, well informed and competent to carry out the tasks required of them.

FURTHER READING

ISO 4413. *Hydraulic fluid power. General rules and safety requirements for systems and their components.*

For further information on fluid power training courses available, visit www.webtec.com/education

CHAPTER

8 INSTALLATION, MAINTENANCE AND TROUBLESHOOTING

INSTALLATION

The installation of hydraulic systems in industrial applications normally involves the connection of a centralised power unit to actuators mounted on the machine itself (Fig. 8.1). However, on larger machines, control valve manifold block stations may be mounted in between the power unit and the actuators. In the main, interconnections are made using solid pipework, with flexible hoses only being used where actuator movement is involved or noise or vibration isolation is required.

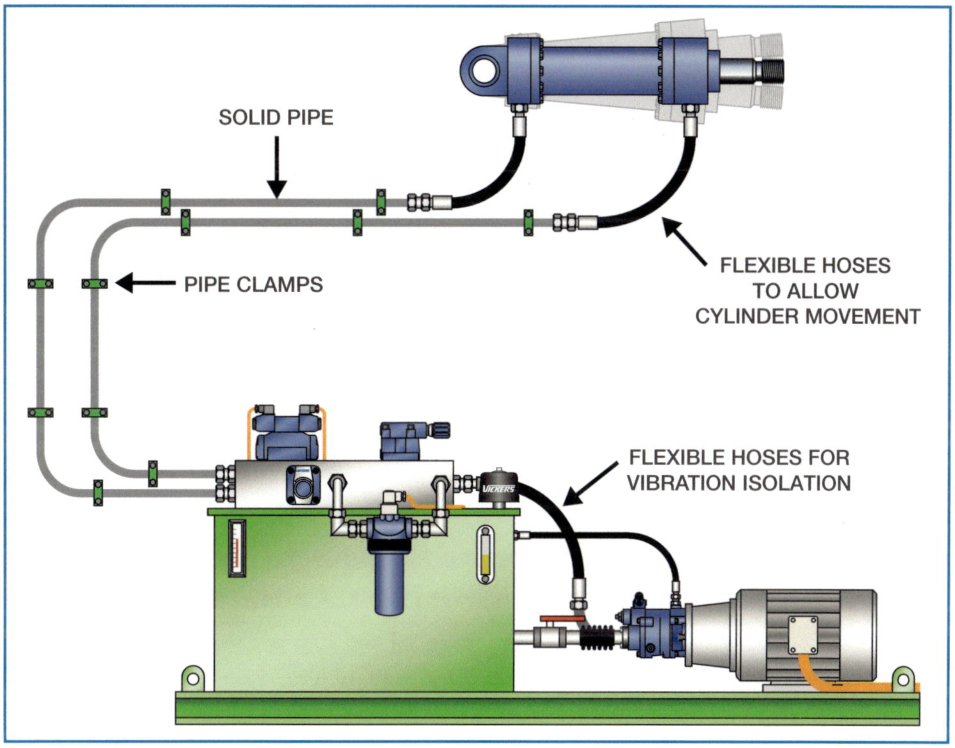

Fig. 8.1 Typical industrial system layout

In mobile applications, hydraulic components tend to be more widely distributed around the vehicle (Fig. 8.2). Pumps are mounted close to the prime mover (diesel engine or electric motor), directional control valves (or their remote controls) are mounted close to the operator's position, and other control valves and ancillary components tend to be mounted wherever space allows on the vehicle. For ease of routing and to account for the greater degree of movement generally experienced on mobile machinery, flexible hoses are much more frequently used on mobile machinery than on industrial systems.

The installation of solid pipework should follow the design specification, with pipe clamps used at the recommended intervals to provide mechanical support, minimise vibration, etc. Swept bends (achieved by bending the pipe) tend to produce less

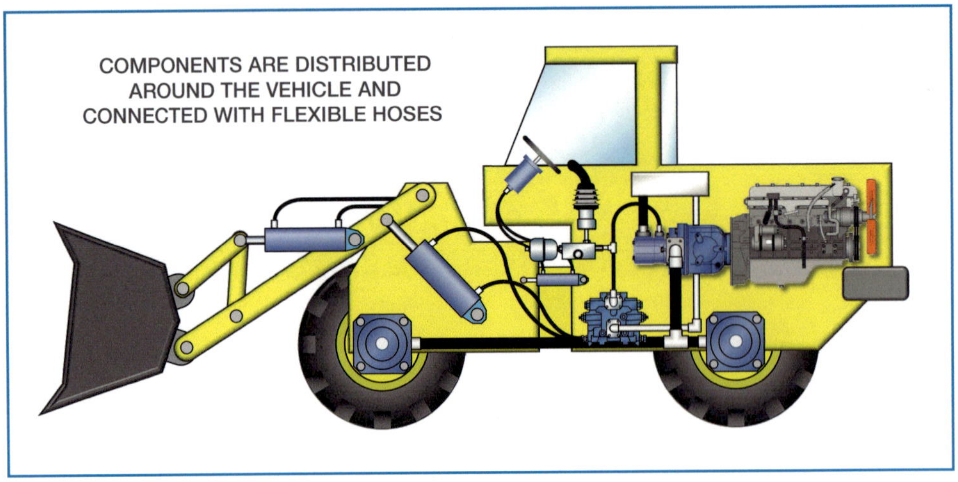

Fig. 8.2 Typical mobile system layout (Image courtesy Eaton Corp.)

turbulence, and hence lower pressure drops, than right-angle fittings, but the space available will normally dictate which approach is most feasible.

TOP TIP

Although they often appear similar, flexible hoses come in many different types and ratings, so always refer to the manufacturer's data sheet for information on the correct hose assembly and fitting.

As mentioned in Chapter 7, the installation of flexible hoses has to be carried out with great care to ensure that:

- their bend radius is above the minimum specified by the hose manufacturer under all operating conditions
- they cannot rub or chafe against other hoses or components during operation
- they are not installed twisted along their length and cannot twist axially during normal operation
- the correct combination of hose and fittings has been used.

It is normally necessary to fill the cases of pumps and motors having external drain connections with clean fluid before they are started, in order to provide initial lubrication for their moving parts (Fig. 8.3). This should be done even before the electric drive motors are jog-started, to check whether the direction of rotation is correct. A pump started dry can be damaged in just a few revolutions, thereby shortening its useful life. Once operating, internal leakage will ensure the cases are maintained full of fluid, provided the drain connections have been piped correctly.

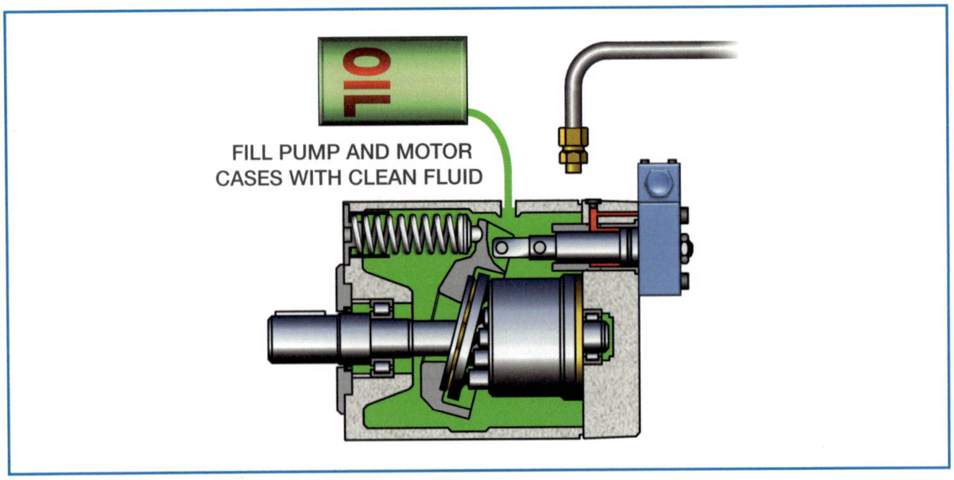

Fig. 8.3 Prefill pump and motor cases with clean fluid (Image courtesy of Eaton Corp.)

If fitted, accumulators can then be **pre-charged** with nitrogen to the required pressure (Fig. 8.4). This will normally involve connecting the accumulator to a nitrogen bottle via a charging kit, and opening the bottle tap, accumulator charging valve and the accumulator fluid drain valve. The accumulator bag can then be charged with gas until the required pressure is achieved. In some cases, two or more nitrogen bottles may be required.

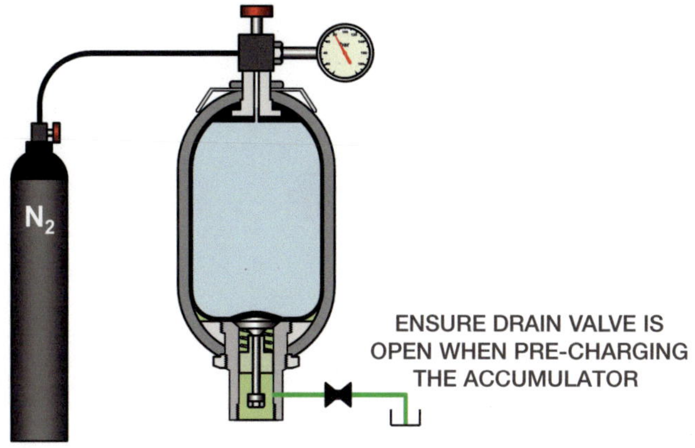

Fig. 8.4 Pre-charging an accumulator

Whatever type of system is being installed, keeping the installation as free from contamination as possible is vital. Whereas it is possible to control the environment of factory-assembled mobile machines, this is often much more challenging when installing industrial systems. Where other construction work is going on at the same time, dirt and moisture in the atmosphere tends to find its way into any exposed openings in components, pipes or hoses and, as explained in Chapter 4, such contamination can cause serious problems.

There are, therefore, two aspects to achieving a clean installation of a hydraulic system on a machine: preventing dirt from entering the system (prevention) and removing any dirt that has entered before the system is put into operation (cure). While prevention may be the best approach, it may never be 100% possible to keep all contamination out of the system during installation, and in practice both methods are usually required. However, placing greater emphasis on prevention will make the subsequent removal of contamination much simpler.

Preventing entry of contamination

It can generally be assumed that hydraulic components purchased from reputable suppliers are clean to an acceptable level and supplied with all ports plugged or capped by some means to prevent the entry of dirt. Screwed ports, hoses and pipework are normally fitted with plastic caps, while gasket-mounted components have shipping plates attached to them (Fig. 8.5).

These protection devices should remain securely attached until the last possible moment (i.e. just before the component is installed in its final location). The same also applies to fittings and other small components, which should remain sealed in their packaging until just before they are used. Similarly, any openings in reservoirs should

POINT OF INTEREST

Note that most manufacturers recommend that bag-type accumulators are mounted vertically to avoid damage to the bag caused by the closing of the poppet valve when draining down.

TOP TIP

It is sometimes stated that removing contamination from a hydraulic system is twice as difficult as preventing it from entering in the first place.

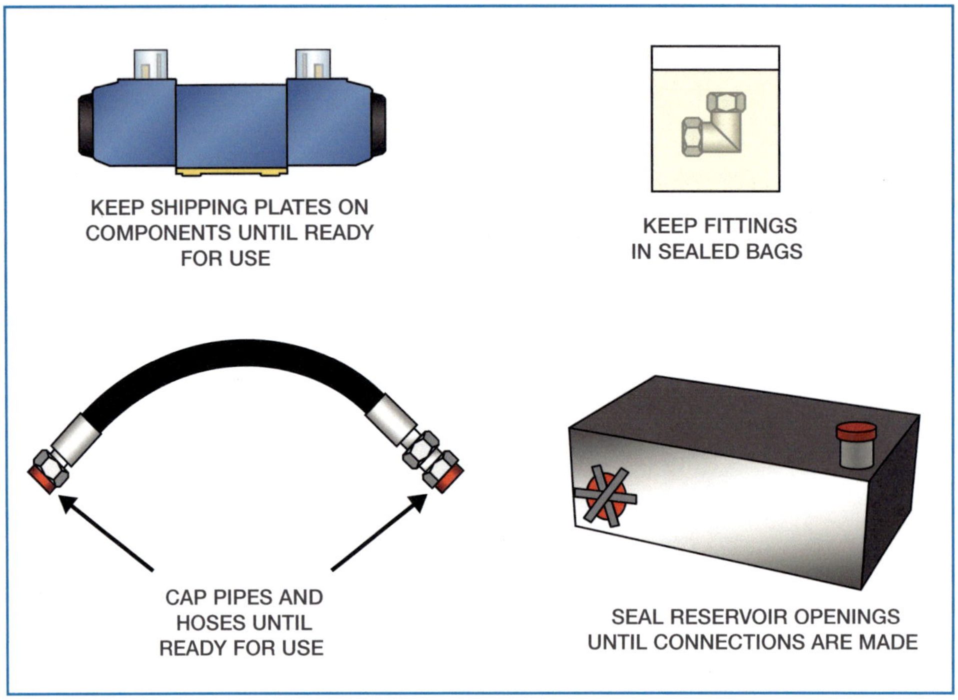

Fig. 8.5 Keep components sealed (Image courtesy of Eaton Corp.)

remain closed until the final connections have been made. Wherever possible, sub-assemblies, such as manifold blocks and reservoirs, should be assembled in a clean environment where airborne contamination can be controlled.

It is often assumed that new hydraulic fluid is clean enough to be used in a hydraulic system. Experience has shown, however, that this is rarely the case, as illustrated in Fig. 8.6.

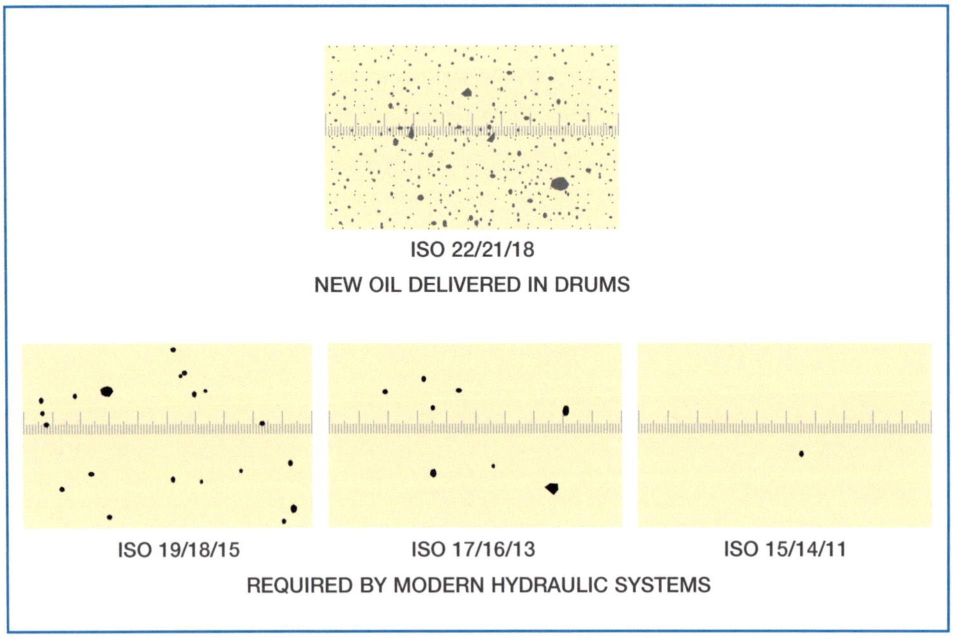

Fig. 8.6 Comparison of fluid cleanliness levels

The initial fill of fluid and any subsequent top-ups should therefore be made via a suitable filter or via a **transfer cart** (see Fig. 7.14).

Removing contamination

Once the system has been installed it will inevitably require some degree of **flushing** to remove inbuilt contamination before it is put into service. However, it may be possible to remove some of the contamination generated during the installation process even before flushing is undertaken.

Where flexible hose assemblies are being manufactured on site, contamination particles will inevitably be generated by cutting the hose. A possible way of removing such contamination is to use a plastic foam pellet 'fired' through the hose by means of compressed air (Fig. 8.7). The tightly fitting pellet then pushes out contamination as it is forced along the hose. The process is normally carried out at least twice (once in each direction), or as many times as is necessary to remove the contamination. Care must be taken, however, to ensure that sharp edges on the end fittings do not 'shave off' particles from the foam pellet and thus create further contamination. If in doubt, the process should be carried out on the hose alone (i.e. without end fittings), and then the hose assembly flushed through with fluid once the fittings have been attached. As before, if the hose assembly is not used straight away the ends should be capped and, ideally, the hose labelled accordingly.

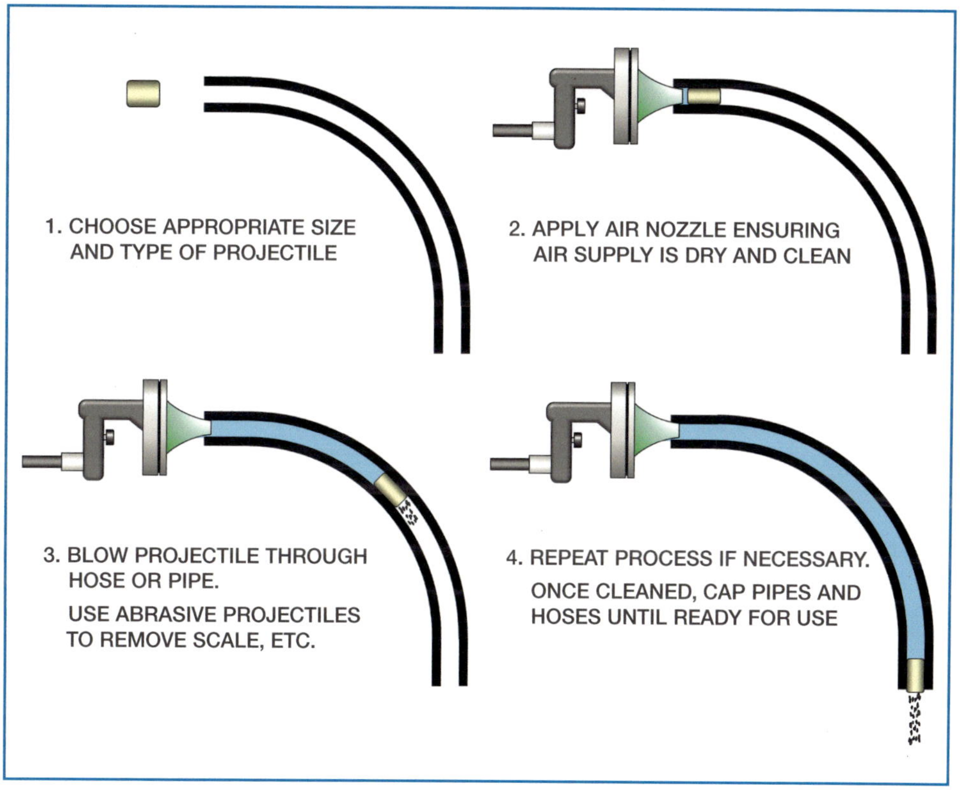

Fig. 8.7 Cleaning flexible hoses

The same process can be used for straight solid pipes. If there is any corrosion evident in the pipes, an abrasive pellet can be pushed through to remove the corrosion, followed by a cleaning pellet to remove the contamination. However, the best solution is to use clean pipes that are free from corrosion.

After the installation is complete, but before the machine is put into operation, most hydraulic systems will need to be flushed to remove any last traces of installation contamination. How rigorous this process needs to be will depend largely on the

> **DEFINITION**
>
> The **target cleanliness level** is the level of cleanliness required to ensure satisfactory life of the hydraulic system. It can be determined at the design stage of a system, and is dependent on operational factors and the type of components used.

target cleanliness level of the system, which in turn will be determined by such factors as:

- the sensitivity of the system components to contamination
- the type of fluid being used
- the operating pressure and temperature levels of the system
- the criticality of the machine operation.

Although the flushing process only takes place at low pressure, it is often recommended to remove very sensitive components (such as servo or high-performance proportional valves) from the system while flushing takes place. Normally, such components are manifold or sub-plate mounted, so manufacturers often supply special connection blocks that can be fitted in their place while system flushing takes place.

In order to create a continuous flushing loop, cylinders should be bypassed by a temporary connection across their ports (Fig. 8.8). The same may also apply to motors if they are likely to be damaged by contamination being flushed through them, albeit at low pressure.

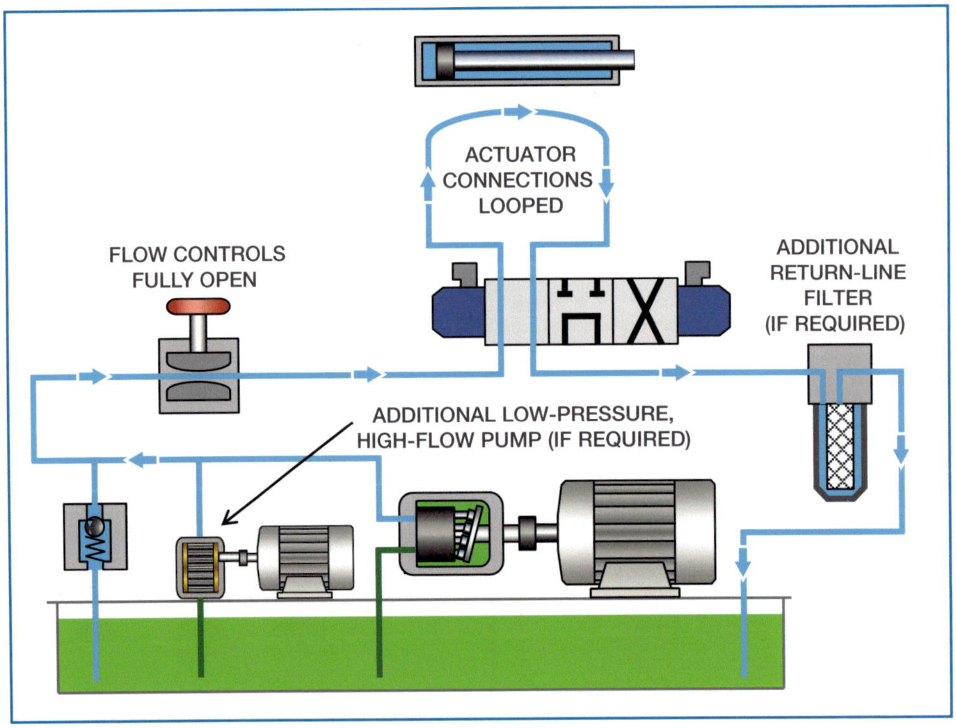

Fig. 8.8 A typical flushing arrangement

The process of flushing requires a high flow velocity through the system, to create as much turbulence as possible in order to remove contamination particles from every corner of the installation. Whether or not a flow is likely to be turbulent can be estimated by calculating a factor known as the **Reynolds number**. Generally speaking, the higher the flow velocity and the lower the fluid viscosity the more likely it is that turbulent flow will result.

In practice, this will mean ensuring that the flushing fluid is fully heated to the maximum practicable temperature and that as much flow as possible is passing through the system. If the normal system pump does not create sufficient volume on its own to generate turbulent flow, additional pumps may be required on a temporary basis to increase the flow velocity. Alternatively, if standby pumps are incorporated in the system, these could be used to augment the normal system pump flow. Similarly with filters, additional large-capacity flushing filters can be temporarily incorporated in the return line of the system if the normal operational filters are not suitable.

Fluid samples should be taken at regular intervals during flushing, and the flushing process continued until the target cleanliness level has been achieved. Wherever possible, the fluid used to flush the system should be the same fluid that will be used when the machine is put into operation. In this way, at the same time as the system is being cleaned by the flushing process the fluid is being cleaned as well.

MAINTENANCE

Maintenance of a hydraulic system, or any mechanical device, involves ensuring that it remains in an operational state whenever it is required. There may, of course, be times when a machine is taken out of service in order for maintenance work to be carried out, but this should be done on a routine or planned basis. However, over a period of time, almost all mechanical machines will wear out, but either through monitoring, forethought or experience the wear process can be predetermined and allowance made for it, either through machine refurbishment or complete replacement. What planned maintenance should be able to do is to eliminate, or at least minimise, sudden unexpected breakdowns, which tend to be costly and sometimes dangerous.

As mentioned before, the component parts of most mechanical machines will wear with use. In most cases this will be a gradual process, and to start with may not noticeably affect the performance level of the machine. As wear increases, however, the machine performance will start to degrade more rapidly. If no action is taken, the performance level will drop off to a level where the machine can no longer perform its function or a catastrophic failure will occur.

There are four main approaches to machine maintenance:

- breakdown maintenance
- preventive maintenance
- predictive maintenance
- proactive maintenance.

These are discussed in turn in the following sections.

Breakdown maintenance

The philosophy of breakdown maintenance (Fig. 8.9) is to repair something when, and only when, it goes wrong. The argument behind this thinking is that sometimes more harm can be done by carrying out the maintenance work than by leaving something alone that is working satisfactorily, as illustrated by the well-known saying 'If it ain't broke don't fix it'.

DEFINITION

The **performance level** of a machine is a measure of how well a machine carries out the task it is designed to do compared with its specification or 'as new' level. It may be measured in terms of speed of operation, efficiency, accuracy, etc., or a combination of several such factors.

DEFINITION

Maintenance is dealing with the expected, whereas **troubleshooting** is dealing with the unexpected.

Fig. 8.9 Breakdown maintenance (Image courtesy of Eaton Corp.)

Whether or not this is a valid approach to maintenance really depends on the consequences of any breakdown of the machine or device. It may be a perfectly acceptable approach for a TV remote control but totally inappropriate in other situations. It is unlikely that many people remove the batteries from their TV remote control once a week to check their voltage level and replace them when they have dropped to a predetermined level, as the consequences of the remote control not working are relatively minor. And perhaps more damage is likely to occur by constantly removing the battery cover or putting the batteries back in the wrong way round. In such a situation, therefore, breakdown maintenance (only replacing the batteries when they no longer work) would be an acceptable philosophy. Adopting the same practice for a nuclear reactor in a power station, however, would obviously be wholly unacceptable.

So, is breakdown maintenance an acceptable practice for hydraulic systems? The answer to that question is almost certainly no. At best, unexpected breakdowns will mean a loss of machine production and quite possibly a high cost of rectification. Depending on the situation, unexpected breakdowns may also be potentially dangerous to personnel. Even something as simple as a burst flexible hose is likely to spray hot hydraulic fluid (which may be flammable) over the surrounding area and potentially cause contact injuries due to unrestrained whiplash movements.

So, almost without exception, it is possible to state that breakdown maintenance is not applicable to hydraulic systems. That is not the same, however, as saying that such maintenance does not happen in practice!

Preventive maintenance

Preventive maintenance (Fig. 8.10) involves predicting when components of a machine are likely to wear out or become unserviceable, and scheduling their replacement before this occurs.

For example, it may be possible to determine, through experience, testing or calculation, that the seals on a cylinder piston have a life of 8 months in a particular

Fig. 8.10 Preventive maintenance (Image courtesy of Eaton Corp.)

application. A maintenance schedule may then be drawn up to strip the cylinder down and replace the seals every 6 months. A 'safety margin' of some degree needs to be built in, as no lifetime predictions are likely to be 100% accurate.

This technique is often used on machinery that has a more or less predictable pattern of usage, such as passenger cars. In the main, passenger cars are used on roads (as opposed to off-road), in winter and summer, and at speeds nominally within national speed limits. Therefore, the interval between necessary oil changes or cam-belt replacements can be predicted reasonably accurately, and recommendations can be made in the service manual to carry out these maintenance activities after a set period of time or mileage. The same may not be the case, however, in relation to race or rally cars, the usage of which is likely to be much more severe and unpredictable.

Preventive maintenance can therefore work well when component lifetime expectancies are well understood and the machine operation is reasonably predictable. A further everyday example is the batteries in a camera. From experience we may know that after a period of use the batteries will have lost a certain amount of charge. So, before taking the camera to an important event, such as a wedding, we would change the batteries for new ones, or give them a precautionary top-up if they are rechargeable.

The disadvantages of preventive maintenance is that it is not always possible to predict component lifetimes or machine usage patterns accurately. If the component life is overestimated, then costly, unexpected breakdowns may still occur before the scheduled replacement time. If the component life is underestimated, perfectly serviceable components, with many hours, days or months of useful life left in them, will be replaced needlessly, thus increasing operational costs.

Preventive maintenance is a technique that is used regularly in relation to hydraulic systems. The most common example is probably the replacement of flexible hoses, where it is recommended that all hoses older than a stated age (typically 5–10 years)

POINT OF INTEREST

Car service manuals are a classic example of where preventive maintenance procedures are used in practice. For example, they specify that an oil change or cam-belt replacement should be done after a certain number of miles or kilometres.

are replaced. In some applications, however, the recommended replacement age may be much less. Another example is the filter or air-breather elements typically used on mobile applications where condition indicators are not fitted. In such situations the recommended period between element changes will be stated in the service literature.

DEFINITION

Predictive maintenance is also known as **condition-based maintenance (CBM)**.

Predictive maintenance

Predictive maintenance (Fig. 8.11) is based on measuring the performance of key components in order to predict when they will approach the end of their useful life. This could involve something as simple as a blockage indicator on a filter, or something much more complex such as an on-line debris analyser.

Fig. 8.11 Predictive maintenance (Image courtesy of Eaton Corp.)

A blockage indicator on a filter or breather senses the pressure drop across the filter element, and triggers a visual indicator or electrical switch when the pressure difference reaches a predetermined value. This pressure difference is somewhat less than the pressure difference required to open the bypass valve in order to protect the element from total collapse. The indicator therefore tells the maintenance people that an element change is required before the filter actually goes onto bypass and becomes inactive as a contamination-removal device.

This is obviously a more precise means of judging when to replace a filter element than simply replacing it after a recommended period of use, but additional cost is added to the system by including the sensor and indicator. In the past, the relatively high cost of incorporating sensors to detect not only filter-element condition but also properties such as fluid cleanliness and condition, component leakage rates and noise and vibration has probably been the main disadvantage of predictive maintenance. In recent years, however, the cost of such sensors and monitoring devices has dropped dramatically. In addition, monitoring devices do not have to be permanently mounted in the system. Provided easy access is built into a system for taking pressure or flow measurements, fluid samples, temperature measurements, etc., then monitoring can easily be carried out on a routine scheduled basis using portable monitoring equipment.

Where machine up-time is critical, the cost of online sensors can normally be justified quite easily. Such devices could include:

- **pressure sensors** for monitoring system pressures
- **flow sensors** for monitoring system flow rates or leakage flows (e.g. drain flows for pump-condition monitoring)
- **temperature sensors** for remote monitoring of the bulk fluid temperature or localised hotspots
- reservoir **fluid-level sensors**
- **fluid-condition monitors**
- **vibration monitors**
- **hose-condition sensors** for monitoring the condition of flexible hoses.

The advantage of basing maintenance decisions on actual measurements rather than experience or best practice are:

- advance warning of impending component failure can be obtained, and rectification work scheduled for convenient times
- replacing a component before complete failure occurs reduces the risk of further failures caused by contamination of the system, fluid spillages or the cost of re-flushing the system
- the cost of carrying out necessary maintenance work can be justified by hard data
- the expense of replacing serviceable components 'just in case' can be avoided.

The key aspect of predictive maintenance is being able to determine trends in the data collected (i.e. to see how a property of a component is changing over a period of time). For example, a temperature reading of 52°C (125°F) on the case of a pump does not really mean much on its own. However, if the temperature was 46°C (115°F) last week and 43°C (109°F) the week before, it is clear that the pump is starting to generate more heat than normal and should be examined or replaced in the near future.

In the days when most machines were controlled by an operator (as opposed to a box of digital electronics), skilled operators could often tell when the performance of a machine started to deteriorate. When a drill or saw blade started to become blunt, for example, the level of force the operator had to exert might change or the noise made would alter slightly, which would tell an experienced operator that the drill or saw blade would soon need replacing. However, since machines have become automated and skilled operators replaced by electronic controls, this capability has largely disappeared. The object of predictive maintenance is to try to replicate the experience and skill of an operator by sensors and software that perform a similar task.

An example from everyday life of predictive maintenance that uses a sensor is the battery alarm in a domestic smoke detector. Obviously the battery condition in such devices is more important than that in a TV remote control, for example. As a result, smoke detectors normally monitor the battery condition and flash a visual alarm or sound a 'beeping' noise when the battery needs to be replaced so that the smoke detector can be kept in an operational condition. Normally this happens around 3 o'clock in the morning!

Proactive maintenance

The final maintenance methodology considered here is generally known as proactive maintenance (Fig. 8.12), or sometimes **design for six sigma (DFSS)**. The aim of DFSS is to reduce the possibility of errors or defects in a process or piece of equipment to a target level of 3.4 or less in every million opportunities (or 1 failure per 300,000 opportunities approximately). 'Six sigma' is a statistical term that defines this number.

Fig. 8.12 Proactive maintenance (Image courtesy of Eaton Corp.)

Mechanical machines tend to follow a predictable degradation curve (Fig. 8.13). The aim of preventive and predictive maintenance is to carry out repairs or replacements during the period between points A and B on the curve in Fig. 8.13, but as close to point B as possible (i.e. before rapid degradation starts to set in). This is especially true in hydraulic systems, as beyond point B wear often accelerates, creating a rapid increase in the number of contamination particles generated.

POINT OF INTEREST

Taking steps to prevent a component or system failure from reoccurring is similar to applying proactive maintenance techniques.

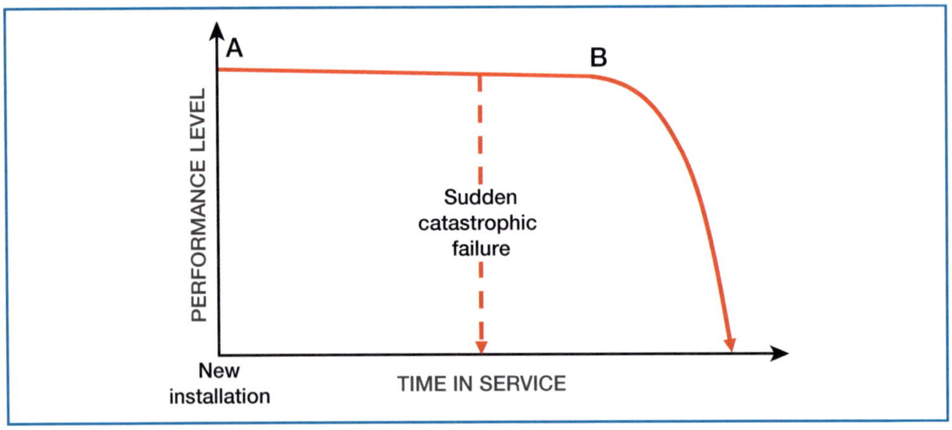

Fig. 8.13 Machine degradation

There are times, however, when a machine or system will fail catastrophically with no prior warning (see Fig. 8.13). For example, this could happen as a result of a component manufacturing defect, operation of the system beyond its design capability, unexpected operational conditions (e.g. a lightning strike) or poor routine maintenance. Proactive maintenance aims to reduce unexpected breakdowns by analysing the design of a component, system or machine from the point of view of what could go wrong with it, the likelihood of it going wrong and the resulting consequences. There are several well-established methods for doing this, and these are similar to the processes carried out for risk analysis.

Having identified all the possible modes of failure of a machine or component, the design is modified to ensure that those modes of failure cannot happen or their likelihood of happening is reduced to a minimum or acceptable level. If this is not possible, back-up devices should be designed into the machine to allow continued operation in the event of a failure of the primary device.

A relatively common example of this approach is to include a standby pump on a hydraulic system, which can be brought into operation in the event of an unexpected failure of the normal running pump. In this case, however, proactive maintenance analysis could also be carried out to reduce the likelihood of the first pump failing unexpectedly in the first place. Such analysis could consider, for example:

- monitoring any shut-off valves on the pump suction or drain lines to prevent the pump being started unless they are fully open
- preventing the pump from starting or shutting it down if the fluid level in the reservoir is below a predetermined minimum
- monitoring the temperature and cleanliness level of the fluid supplied to the pump and shutting the system down if these exceed acceptable levels
- using pre-set (rather than adjustable) components to avoid the possibility of incorrect settings, or using key-lockable adjustments on pump compensator or relief valves
- monitoring the pump inlet and outlet port pressures to ensure they remain within specification
- monitoring the pump drive speed to ensure it remains within specification, and taking appropriate action if not.

In practice, the cost of such sensing devices needs to be taken into account and balanced against the likely cost of an unexpected machine breakdown. The merits of the proactive approach will, therefore, differ from one situation to another. However, there are many examples of where modifications that cost very little can reduce the likelihood of human error or equipment failure, provided the modification is made at the design stage. For example, the addition of a dowel pin to the interface of a gasket-mounted valve costs very little, but prevents the valve from being mounted the wrong way round.

POINT OF INTEREST

Due to the consequences of failure, aircraft hydraulic systems often have at least two back-up systems capable of taking over in the event of a failure of the normal system.

Continuing with the battery analogy for maintenance, remote-control keys for cars are often powered by a conventional silver oxide button-type battery. Such keys will obviously not lock or unlock the car when the battery power is exhausted. Using a rechargeable battery that is topped up whenever the key is placed in the ignition will reduce the likelihood of the key suddenly failing to operate.

Setting up a maintenance schedule

Preventive, predictive and proactive maintenance are not mutually exclusive techniques. That is, two or more of these techniques can be combined to avoid unexpected breakdowns. If all these techniques fail, of course, then breakdown maintenance will still be required, but even so this can be pre-planned to a certain extent (e.g. by having available spare parts or components, prepared procedures and trained personnel).

As mentioned above, predictive maintenance has become increasingly popular in recent times, mainly due to the availability of low-cost sensors and digital communications. It is now possible to monitor machine performance in real time anywhere in the world. Commercial jet engines, for example, are monitored continuously by their manufacturer, no matter where the plane happens to be flying in the world. However, there will always be instances where examination by human beings is still worthwhile, and in such cases it is advisable to establish a maintenance schedule.

TOP TIP

Maintenance schedules need to define not only *what* needs to be measured or checked but also *how* it is to be measured.

When setting up a maintenance schedule for a hydraulic system the first thing to decide is what properties need to be checked. The list could include:

- fluid level
- fluid temperature
- fluid cleanliness level
- fluid chemical analysis
- pump temperature
- pump noise
- pump flow rate
- pump case drain flow rate
- pump outlet pressure
- external leakage
- actuator performance (speed, force, leakage, temperature, etc.)
- flexible hose condition
- any abnormal symptoms.

Based on knowledge and experience, a timescale (daily, weekly, monthly, etc.) for checking each property can then be established. It will also be necessary to establish procedures for checking and recording each property to ensure that data is obtained and recorded consistently.

Consider, for example, the simple task of checking the fluid level inside a reservoir. To check this manually involves reading the level of the fluid visible in the sight glass and recording it. If the level drops from one reading to another, this might indicate a leakage of fluid in an area that might not be readily accessible and therefore has gone unnoticed. However, it could also mean that the second reading was taken with all of the cylinders fully extended (and therefore containing the maximum amount of fluid) and perhaps with accumulators fully charged with fluid, whereas the first reading was taken with the cylinders retracted and the accumulators drained.

The correct procedure for checking the fluid level should therefore include the requirement that the machine state (cylinder positions, etc.) is the same each time a reading is made. Similarly, for fluid or component temperature readings, the readings should only be taken after the machine has been operational for a certain period of time and preferably when performing the same operations.

Good maintenance practice

As stated several times previously, it is vital that the people involved in recording data, taking samples, using test equipment, etc., are properly trained in the jobs they have to perform and the equipment they have to use. No matter how comprehensive the maintenance procedures are, there will inevitably be occasions when maintenance personnel have to use their initiative, so it is important that they are properly equipped to do so.

Examples of good maintenance practice include the following:

- *Pressure test points*. Fit pressure test points into the system where pressure readings are likely to be required (such as when setting pressure-relief or pressure-reducing valves). Gauges permanently connected into the system may be subject to pressure peaks and ripple, and so tend to have a short life. They can be protected by 'push to read' or 'twist to read' valves, which normally isolate them from the system, or they can be removed completely and only connected when required via simple quick-release test points (see Fig. 9.3).
- *Access points*. Include access points for taking fluid samples from the system, normally from the pressure line or from the reservoir. Sampling from the reservoir should ideally be via a pipe, taking fluid from the centre of the reservoir, away from where water or contamination is likely to settle.
- *Flow meters*. Flow meters are sometimes permanently connected into systems but more often need to be connected temporarily for test purposes. This process can be made simpler by including quick-release couplings and/or three-way valves to divert flow through the meter when required (Fig. 8.14). Care should be taken, however, to ensure that the test equipment and

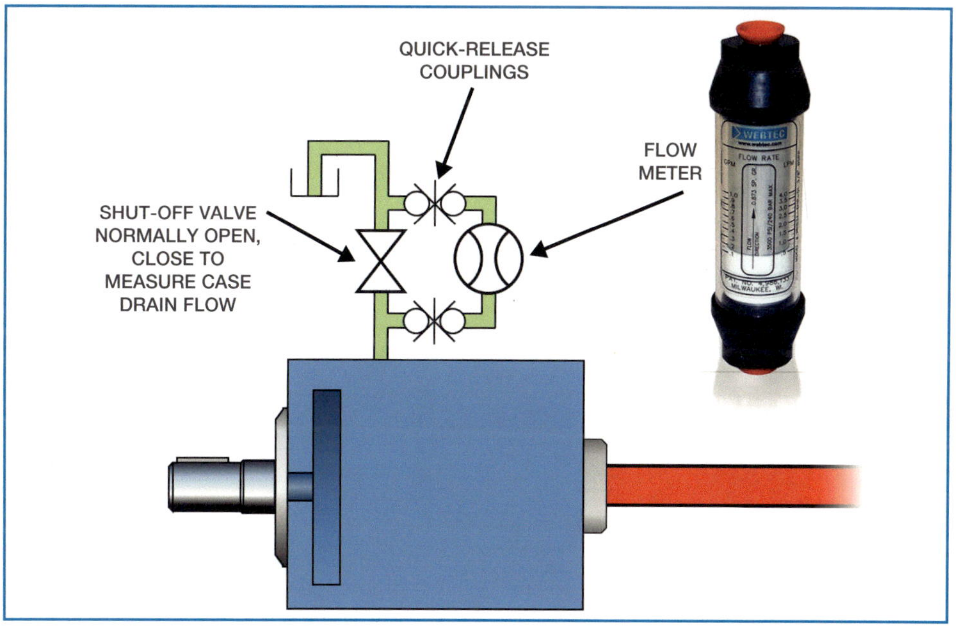

Fig. 8.14 Monitoring pump drain line flow

associated fittings do not create undue restriction (e.g. in the case drain line of pumps and motors).

- *Adding new fluid*. As stated previously, new fluid to be added to a reservoir will normally require cleaning, so provision should be made to incorporate a filter on the reservoir filling connection. If this is not practical, a filling point should be included so that it is not possible to simply pour fluid into the tank. If filling of the tank has to be carried out by pumping the fluid in (e.g. via a quick-release coupling or side connection), it is more likely to be done correctly via a fluid-transfer cart and filter.
- *Label reservoirs*. Colour code or clearly label each system reservoir with the correct fluid to be used for filling and topping up.
- *Mark normal fluid levels*. Mark the normal maximum and minimum fluid levels on the reservoir sight glass.
- *Shut-off valves*. Where shut-off valves are likely to cause damage if set incorrectly, for example on pump inlet or drain lines, they should be monitored via limit or proximity switches that are interlocked to ensure pumps cannot be started unless the valves are fully open.
- *Drain valves*. Consider using automatic drain valves on accumulators to drain the pressurised fluid whenever the machine is switched off. Manual valves should also be included (as a back-up), together with an isolated pressure gauge to verify that the accumulators have been discharged.
- *Components requiring regular attention*. Ensure components requiring regular attention (filters, test points, gauges, etc.) are easily accessible, and that spillage trays are placed where fluid is likely to be spilled during maintenance activities (e.g. changing filter elements).
- *Flush valves*. Consider adding flushing valves around actuators (Fig. 8.15). This will make the flushing process simpler when the system is first installed and will also be useful when actuator connections involve long pipe runs. In long pipe runs the volume of fluid in the pipe may often be greater than the displacement of a cylinder, so the fluid will just shuttle backwards and forwards along the pipe (i.e. it will never get back to the system cooler, filter or reservoir). A bypass valve

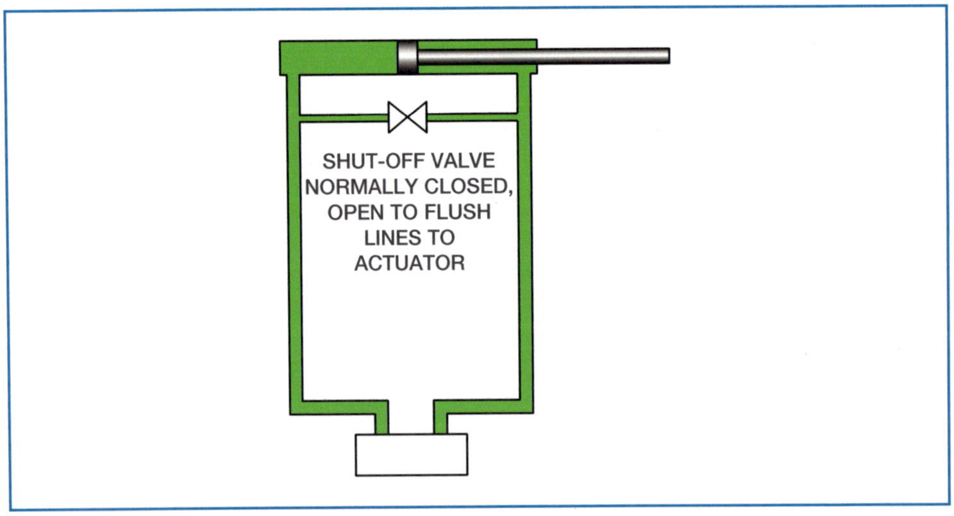

Fig. 8.15 Arrangement for flushing long pipelines

around the cylinder can be opened periodically during maintenance periods to flush new oil into that part of the system.

- *Make the correct way the easiest way*. Adopt the philosophy that the easier it is to do something the more likely it is that it will be done. Also, the correct way of doing something should be made the easiest way of doing it, so that mistakes are less likely to happen.

TOP TIP

When things are easy to do, there is more likelihood that they will be done.

TROUBLESHOOTING

No matter which maintenance methodology is used and how rigorous the process, there will inevitably be occasions when something unexpected occurs to cause a malfunction in a system or of a machine. This is when troubleshooting skills become necessary, to locate the fault as quickly and accurately as possible. So, whereas maintenance deals with the expected, troubleshooting involves handling the unexpected.

TOP TIP

For troubleshooting to be effective, personnel need to be well trained and experienced.

Like maintenance procedures, troubleshooting on hydraulic systems should only be carried out by well-trained, experienced people with a good knowledge of both the machine and the hydraulic system itself. Troubleshooting activities may often involve running a machine in a different way from its normal operation. For example, the machine may have to be operated in manual rather than automatic mode, interlocks or guards may have to be temporarily removed, or directional valve manual overrides may be used instead of normal solenoid operation. During such activities the possibility of dangerous situations arising is much greater than in normal operation, hence the need for well-trained, experienced personnel who are aware of the risks involved and know how to take appropriate measures to minimise them.

Hydraulic specialists also need to have a good working knowledge of electronic control and communication systems, which are increasingly used in today's machinery. While the integration of electronics in mechanical and hydraulic components has added an extra layer of complexity, the diagnostic capabilities of modern electronics have the potential to simplify troubleshooting activities.

Logical troubleshooting approach

As mentioned above, troubleshooting can be defined as the diagnostic and remedial actions taken following an unexpected breakdown of a machine. In some cases, such as a catastrophic pump failure or a hose burst, the reason for the breakdown may be immediately obvious. In other cases the fault may be less obvious and a certain amount of 'detective' work may be required before the failed component or the reason for the fault can be pinpointed.

In both cases the remedy involves not just rectifying the immediate cause of the breakdown (such as replacing the pump or hose) but should also involve an analysis of the potential consequences of the failure and its root cause. For example, if there is a catastrophic pump failure in a system that contains just a return-line filter, contamination from the pump failure is likely to be distributed around the system, and, at least, flushing of the system will be required before a re-start. The same may also be the case for something as apparently innocuous as a seal failure, where particles of rubber can easily jam up valves or block orifices elsewhere in the system. Similarly, a passing relief valve, caused by an incorrect setting, may have overheated the fluid to such an extent that it is no longer fit for purpose and will have to be changed.

A **root-cause analysis** of a failure is not always easy to do but is obviously necessary to prevent the same failure from happening again soon afterwards. For example, a blocked inlet filter or strainer may have resulted in a catastrophic pump failure, caused by the pump cavitating. Simply replacing the pump without checking the condition of the inlet filters would be a recipe for a further disaster in a short space of time.

Root-cause analysis does require a relatively high level of knowledge and experience, but the fact that something may be difficult to do is no excuse for not doing it if it is necessary. The process of locating an unobvious fault is something that can be learned and practised relatively easily, provided the person involved has a good knowledge of hydraulic components and systems.

Step-by-step approach

Troubleshooting should be a logical step-by-step approach based on good knowledge and experience. It also demands good interpersonal skills in order to ascertain facts and be able to resist a request to 'just get on and fix the machine'. Therefore, it is not a task that should be undertaken by inexperienced or undertrained personnel.

Step 1. Gather as much information as possible about the circumstances of the failure

For example:

- What exactly has happened? Describe all the symptoms of the failure in simple terms (slow speed, overheating, erratic movement, creep, etc.).
- Was the failure sudden or has the problem been developing gradually over a period of time?
- What operation was the machine performing at the time when the fault occurred? Was this normal operation or something out of the ordinary?
- Has any work been carried out on the machine or hydraulic system recently?

It is also a good idea to consult any maintenance logs or records at this stage to discover if something similar has happened before.

Step 2. Understand how the machine and hydraulic system function

Depending on how familiar the troubleshooter is with the machine, the time required for this step will vary. Nevertheless, it is a vital step because unless it is understood how something works it is not possible to think about what could go wrong with it. Machine manuals are a good source of information here, as is simply asking the machine operator, but the single most important piece of information on the hydraulic system is the circuit diagram. Normally the functioning of the machine will need to be understood before the circuit diagram will make sense, so these two activities go hand in hand. The circuit diagram will provide information on all the hydraulic components, their function, settings, interconnections, etc., and will indicate where test points or access points are located. The ability to read and interpret circuit diagrams is, however, something that requires both knowledge and experience.

Step 3. Make a list of all the possible causes of the failure and prioritise them

Prioritise the possible causes in order of:

- those which are the most likely
- those which will be the easiest to check.

For example, it may be decided that a lack of flow to a part of the system could be caused by a check valve sticking closed. However, experience indicates that, while this valve becoming stuck is possible, it is a relatively unlikely situation. Therefore, although the check valve should be on the list of items to check, it will probably come well down the list based on it being a relatively less likely cause. Similarly, if the pump is on the list of possible causes but it is mounted in a very inaccessible position, requiring the removal of several pieces of equipment to check it, it will again be placed further down the priority list, as it makes sense to undertake the easiest checks first of all. The list is prioritised, therefore, on the basis of checking the easiest or most likely things first and the hardest or least likely causes last.

POINT OF INTEREST

A large part of the logical troubleshooting approach involves simply studying the available documentation to understand how the system works and thinking about what could be causing the problem.

Step 4. Carry out a preliminary check of components on the 'suspect' list

This step may involve inspecting the hydraulic system while it is in operation, so again the overriding consideration must be safety. The troubleshooter must bear in mind that many components are likely to be hot, there may be fluid leakage from the system under pressure and, especially on mobile equipment, parts of the machine may be moving close by. No investigations should be carried out, therefore, unless the troubleshooter is fully aware of the potential risks and is 100% sure that it is safe to proceed.

Exactly what is included in this preliminary check will depend on the component and its possible mode of failure. For example, it might be difficult to determine if a check valve poppet has stuck open without stripping it down, but it might be possible to discover that a relief valve is passing fluid, either by listening to it or by checking its temperature. This step is sometimes referred to as the 'eyes, ears and hands' stage, as it involves carrying out very simple checks that do not involve adjusting anything, stripping anything down or adding additional instrumentation. It cannot be overemphasised, however, that the 'eyes, ears and hands' must be well protected.

TOP TIP

Remember that instruments can be replaced but eyes, ears and hands cannot.

Unexpected breakdowns in manufacturing or process-industry environments tend to be very expensive in terms of downtime and lost production. As a result there is inevitably a great deal of pressure to resolve the problem as quickly as possible so that production can resume and downtime costs are minimised. Service engineers or troubleshooters can therefore come under a lot of pressure to 'go out and fix the machine' rather than to methodically go through steps 2 and 3. How long step 2 will take depends very much on how familiar the troubleshooter is with the machine and its hydraulic system, but step 3 should always be carried out regardless. Huge amounts of time and money can be (and have been) wasted by jumping to a conclusion too early before considering all the possibilities.

For example, consider the circuitry arrangement shown in Fig. 8.16. The two cylinders are each controlled by a three-position, closed-centre, solenoid directional valve. During a night shift a problem arose with valve 2, and so it was replaced.

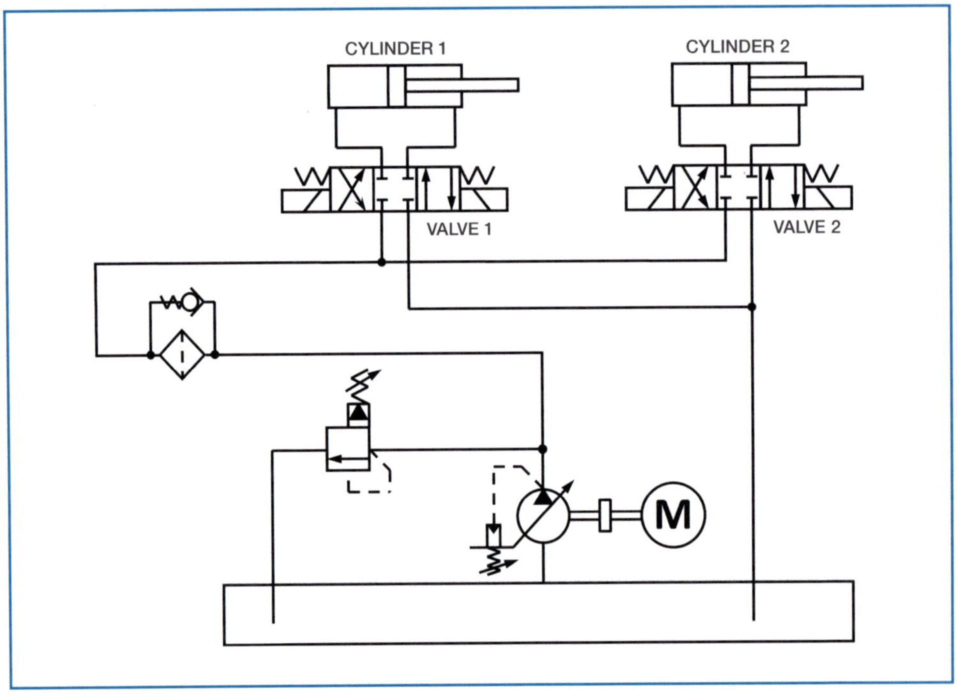

Fig. 8.16 Two-cylinder system

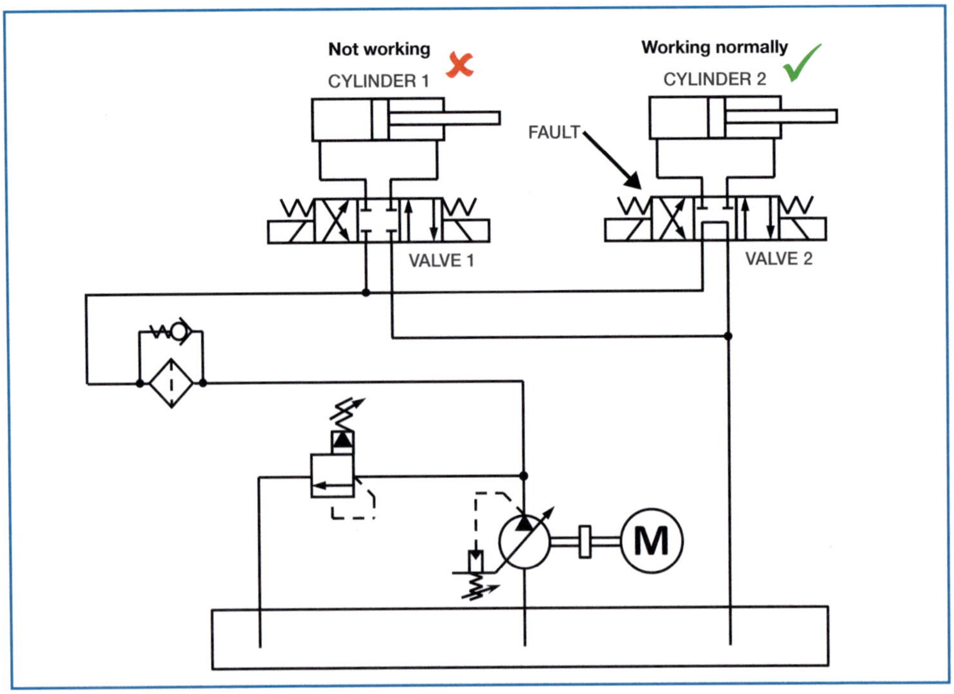

Fig. 8.17 Two-cylinder system fault

Unfortunately, the new valve was not quite the correct one, and had a centre condition P to T instead of P and T blocked (Fig. 8.17).

When the day shift arrived and tried to operate the machine they found that when they operated valve 1, cylinder 1 did not move, as all the pump flow was being unloaded back to tank via valve 2. When they operated valve 2 to move cylinder 2, however, that worked as normal. They therefore assumed that the problem must be with valve 1 or cylinder 1, as that was the part of the circuit that had the fault. They then wasted a lot of time changing or stripping down valve 1 and cylinder 1 when all the time the fault lay elsewhere.

TOP TIP

Consider *all* possibilities of a failure not just the most likely ones.

If they had carried out step 3 thoroughly, valve 1 and cylinder 1 would have been on the list of possible causes, and may well have been at the top of the list in terms of likelihood. But valve 2 should also have been on the list, and had a simple visual check of its nameplate or symbol been carried out (step 4), then the cause of the problem would have then been apparent and a great deal of time and money would have been saved.

The simple things that can be checked in step 4 therefore include:

- Is the component the correct one? Does the nameplate and symbol correspond to the circuit diagram? If not, has it been changed or modified recently?
- If there is the possibility that the component has been replaced recently, is it installed correctly, with all the drain and pilot lines connected and any electrical contacts made?
- Does the component appear to be adjusted correctly? If instrumentation is included in the system to allow this to be checked, then all well and good. If not, are there any signs of it having been recently adjusted (a clean section of screw thread adjuster or a loose lock-nut, for example)?
- If the component is controlled by an electrical signal, is the signal correct and is the component responding to the signal? On a solenoid valve, for example, this check could be carried out by looking at the solenoid indicator light and testing the manual override button. As mentioned previously, great care should always be exercised when using manual override buttons.
- Does the component exhibit any abnormal symptoms such as noise, heat, vibration or external leakage? Experience will again help here, as some components may naturally be hotter than others (flow controls and counterbalance valves, for example), so the key word is 'abnormal'. Also, an experienced ear may often be able to distinguish between a healthy pump and one with problems simply by listening to it, or pick up the 'hissing' noise produce by a high-pressure leak inside a component.

Step 5. Carry out additional instrumentation checks when necessary

In the majority of cases, if steps 1–4 are carried out diligently and correctly the fault can be located. There will, however, be some faults that are more difficult to trace, and will therefore require a deeper investigation involving additional instrumentation

FURTHER READING

See Chapter 9 for more information on instrumentation.

or the stripping down and examination of components. This stage must also be carried out in a logical manner, and the list of potential causes may have to be reprioritised based on their accessibility or ease of checking.

As mentioned earlier, more and more systems are being fitted with relatively low-cost sensors and software, which enable faults to be diagnosed simply by plugging in a handset and reading the appropriate result. In fact, some systems (especially those installed in remote locations such as offshore wind turbines) will flag up faults proactively, without waiting to be interrogated.

While such 'smart' technology inevitably makes life easier for engineers and technicians involved in troubleshooting, there will still be occasions where human knowledge, skill and experience are required. In these cases there is a choice of diagnostic equipment available, for example:

- **Pressure gauges**, either permanently installed or connected via quick-release test points, are useful for checking pressure levels and component adjustments.
- Where it is expected that short-duration pressure peaks are causing problems, a **pressure transducer** connected to an appropriate output device may be necessary.
- Flow rates can be measured using a **flow meter**, which is normally only temporarily connected into the system, as explained in Chapter 9.
- Fluid-condition monitoring (for solid and water contamination) can be carried out online, by using **portable sampling equipment** or by sending a fluid sample to a laboratory. The fluid-additive condition and the acidity level of the fluid are normally determined by laboratory tests.
- **Temperature probes** or **thermal imaging devices** can be useful for identifying local hotspots, which are often associated with internal leakages inside components.
- A simple **multi-meter** can be used to check control valve voltage levels, etc., although an **oscilloscope** or **data logging equipment** may be required to check such things as voltage ripple or short-duration spikes, which can cause problems with proportional-valve electronics, for example.

Step 6. Rectify the fault

Depending on the actual fault, this step may be as simple as re-setting an adjustment or replacing a complete component.

Step 7. Consider, and rectify if necessary, any consequences of the fault

As mentioned earlier, a failed component may have affected other components in the system, in particular the fluid itself if this has become overheated or contaminated. Further tests will therefore often be required to ensure that any damage likely to cause additional failures has been rectified. Again, the nature of the fault will determine how exhaustive this step needs to be, but for catastrophic failures of components or parts within components the process is likely to be lengthy. This is all the more reason to do everything possible to prevent catastrophic failures.

Step 8. Determine, wherever possible, the root cause of the failure and modify the system or maintenance procedures to prevent it happening again

Determining the root cause of a failure is seldom an easy process, but nevertheless as much time and effort as possible should go into this step. Things to consider at this stage might include:

- Why was the component incorrect or incorrectly installed?
  - Modify the documentation if necessary.
  - Error-proof the component-replacement process if possible – colour coding, locating pins, etc.
  - Insufficient knowledge – train maintenance personnel.
- Why was the component not adjusted correctly?
  - Adjustment had been tampered with – fit lockable adjustment.
  - Vibration – fit vibration-proof adjustment/tighten lock-nut.
  - Incorrect setting specification – correct the documentation.
- What caused the component to fail?
  - Lack of fluid – check fluid levels and restrictions, fit limit switches to shut-off valves, etc.
  - Contaminated fluid – check the fluid condition.
  - Wrong fluid temperature – check heaters and coolers.
  - Fluid condition – analyse a fluid sample.
  - Operating beyond its performance envelope – check operating speeds, pressures, etc.
  - Reached the end of its expected life – modify the maintenance schedule to replace it earlier.

Step 9. Put the machine back into operation

Once the fault has been remedied and the consequences and causes of the failure identified and acted upon, the machine can be put back into service. Before restarting the machine, however, safety procedures must be observed to ensure that it is safe to do so. All personnel must be made aware of the restart, temporary alterations should be reinstated, machine guards replaced, etc. The machine performance should then be checked and agreed with the necessary person or people responsible.

TOP TIP

Employ proactive maintenance techniques to prevent the same failure happening again.

FURTHER READING

Fault, cause and remedy checklists and other troubleshooting information can be found at www.webtec.com/education

Step 10. Record all activities and findings in the appropriate manner

A proper maintenance log for a machine will often speed up the troubleshooting process the next time it becomes necessary, and can also be used to modify maintenance procedures to prevent similar faults occurring in the future. Although everything possible should be done to avoid unexpected breakdowns, when they do occur, lessons should be learned from them in a climate of continuous improvement.

CHAPTER

9 SELECTING AND USING MONITORING EQUIPMENT

INTRODUCTION

Selecting which monitoring equipment should be installed, either permanently or temporarily, in a hydraulic system will depend mainly on the machine application and the type of maintenance approach the user is following. The monitoring requirements for a cardboard box baler in a supermarket are likely to be totally different from those for a continuous casting machine in a steelworks. As detailed in Chapter 8, following a 'breakdown maintenance' approach might mean that nothing is measured until a system has failed, at which time a great deal of troubleshooting may be required to diagnose exactly what has caused the failure.

Predictive or proactive maintenance, however, normally requires condition monitoring equipment to be permanently installed or easily fitted into the hydraulic system, and the data this provides can then be used to help eliminate machine breakdowns by repairing or replacing components before they fail completely. This may sound an expensive solution, but when compared with the cost of an unexpected breakdown – which could involve loss of revenue, late-delivery penalties, call-out charges for a service engineer, spares and expedited delivery costs, together with wages for operators left idle, and so on – condition monitoring equipment starts to look very affordable. This can be clearly seen in Fig. 9.1. Also, in recent times, low-cost sensors and communication arrangements (such as **bus systems**) have significantly reduced the cost of installing such equipment.

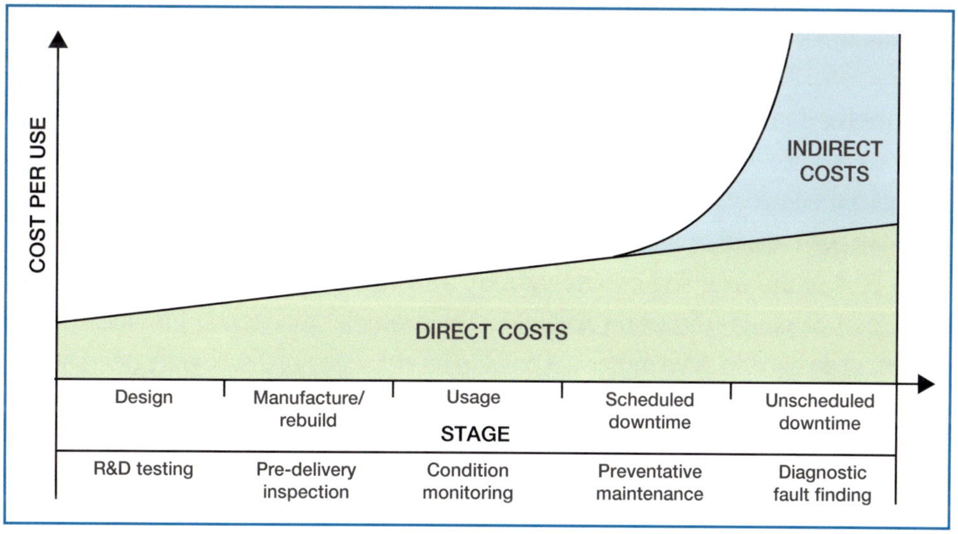

Fig. 9.1 Cost per use of hydraulic monitoring equipment during the life of a hydraulic system

The list given in Table 9.1 is a good example of the properties of a hydraulic system that may need to be checked regularly and the sensors required for each measurement (see also Chapter 8).

Table 9.1 Example of properties of a hydraulic system that may need to be checked regularly and the sensors required for measurement

	Property to be measured	Method of measurement
1.	Fluid level	Visual or reservoir fluid level sensors
2.	Fluid temperature	Temperature sensors
3.	Fluid cleanliness level	Fluid condition monitors
4.	Fluid chemical analysis	Fluid sample testing
5.	Pump temperature	Temperature sensors
6.	Pump noise	Audible or acoustic/vibration monitors
7.	Pump flow rate	Flow sensors
8.	Pump case drain flow rate	Flow sensors
9.	Pump outlet pressure	Pressure gauges or sensors
10.	External leakage	Visual or visual leakage indicators
11.	Actuator performance (speed, force, leakage, temperature, etc.)	Flow, pressure, temperature and speed sensors, and load cells if required
12.	Flexible hose condition	Visual or hose condition sensors
13.	Any abnormal symptoms	Visual, audible or acoustic/vibration monitors

Some measuring equipment is likely to be used frequently, while other equipment may only be required on rare occasions. Experience has shown that, in many cases, temperature, pressure, hose condition and fluid condition are the properties of a hydraulic system most likely to require monitoring on a regular basis, with vibration monitoring being a useful addition on larger or more critical installations. When sensors are not permanently installed, such properties can often be measured without breaking into the system or interrupting the system flow by using contactless sensors, test points or simple T-pieces.

Measurement of flow in a system may also be a maintenance or troubleshooting requirement. If this is the case, it will normally require the installation of an **in-line flow meter** if one is not already permanently installed. Incorporating in the design the means to do this easily (e.g. quick-release couplings, shut-off valves or T-pieces) will make the task much simpler. Other sensors, such as for vibration, load or torque, are often permanently installed in a system, either for condition-monitoring purposes or as part of the machine-control system. In some cases, measurements can be combined or processed using software. For example, monitoring the voltage and current of an electric drive motor will provide an indication of the power consumption of a pump, and monitoring the pump's output pressure and flow will provide an indication of the power output. By comparing input power with output power, the pump efficiency can be determined.

General considerations when selecting hydraulic monitoring equipment are discussed below.

TEMPERATURE MEASUREMENT

The main factors to consider when selecting temperature-measuring devices include:

- operating temperature range
- accuracy and repeatability
- response time
- pressure rating (if the sensor is connected into the system)
- fluid compatibility (if it comes into contact with the system fluid)
- signal type (analogue voltage or current, digital, etc.)
- connector type.

Temperature probes are used to measure the fluid temperature inside a hydraulic system. The three main types of probe are resistance temperature detectors, thermistors and thermocouples.

Resistance temperature detectors (RTDs) function by sensing the change in electrical resistance of a coiled wire, of known resistance versus temperature characteristics, with temperature variation. The wire is normally wound onto a glass or ceramic core inside a protective housing. Typically, RTDs have good accuracy, repeatability and stability, but a relatively slow response time. The most common RTD sensors used in hydraulic systems are designated Pt100, where Pt denotes the wire material (platinum) and 100 represents the wire resistance (100 ohms (Ω)). If greater accuracy is required a Pt1000 sensor could be used, where the resistance change per degree is 10 times greater.

Thermistors work on a similar principle to RTDs except that they use a semi-conductor material rather than a wire coil to detect temperature variation. Depending on the type of thermistor, the current flow will either decrease (**positive temperature coefficient (PTC)**) or increase (**negative temperature coefficient (NTC)**) with temperature. Thermistors are cheaper than RTDs and their response time is faster. While they are not as accurate as RTDs, their accuracy is usually sufficient for temperature sensing in hydraulic systems.

Thermocouples consist of strips of two dissimilar metals joined at one end. A temperature difference along the strips creates a slightly different voltage gradient along each one, and this can be measured and related to the temperature difference. They are the cheapest of all temperature sensors and, although not as accurate or sensitive as RTDs, their performance is often adequate for use in hydraulic systems.

Contactless temperature sensors can be of great use for locating hotspots in a hydraulic system without the need to install temperature probes. As mentioned previously, whenever fluid leaks from high to low pressure, heat is generated locally. Therefore, by identifying areas of higher temperature, problems such as leaking seals or bypassing valves can often be diagnosed. The use of contactless sensors is also considerably safer than using hands as a temperature sensor!

TOP TIP

Whenever fluid leaks from high pressure to low pressure, heat is generated. Localised hotspots can often be identified using an appropriate temperature sensor (not the hands).

Thermal imaging or infra-red (IR) devices (sometimes known as **pyrometers**) use an optical system to measure the electromagnetic waves that are emitted by thermal energy. These contactless devices are now relatively inexpensive (some are available to attach to a mobile phone or tablet), and are available in a range of accuracies and temperature ranges. When using such devices to detect the source of heat generation it is best to do this from a cold start so that the source of the heat will be more obvious (Fig. 9.2).

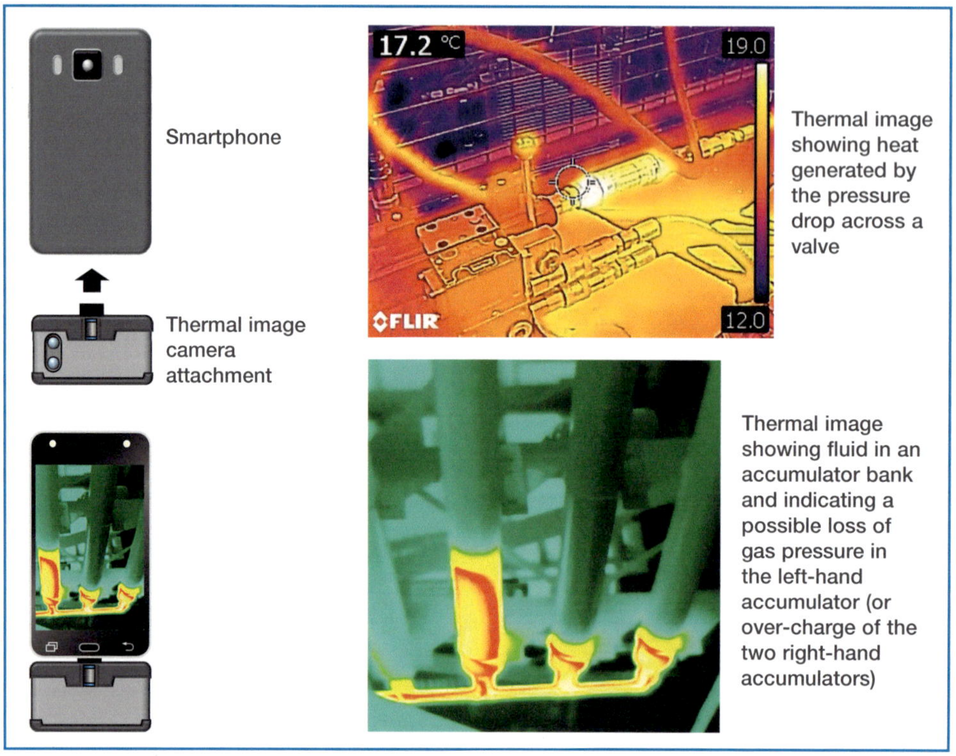

Fig. 9.2 Thermal imaging (Images courtesy of Torishima Service Solutions Europe Ltd and GPM Consulting Inc.)

From a maintenance point of view, it is worth remembering that if an over-temperature situation is identified in a hydraulic system, then damage may already have occurred, either to the fluid or to component seals. Therefore, although it may not accurately predict an impending failure, thermal imaging equipment may help to identify the area in which the problem lies if this is not obvious otherwise.

WARNING

It should be noted that a correct pressure reading does not necessarily mean a system is working correctly unless flow is also measured. A worn pump, for example, which is only producing 10% of its correct flow rate will still be able to move a load (albeit slowly) and would therefore generate normal working pressure. Similarly, if a seal inside a cylinder is leaking and passing 90% of the inlet flow rate, the load will still be moved (slowly), and a pressure gauge on the inlet to the cylinder will read as normal. For this reason, flow and pressure are often measured together where possible.

PRESSURE MEASUREMENT

After temperature, pressure is probably the easiest parameter to monitor in a hydraulic system. When taking a pressure reading it is common practice to connect into the hydraulic system using a pressure test point (Fig. 9.3). These relatively low-cost connection points are often installed by the manufacturer in various places around the system to be used both for commissioning and diagnostic testing. Typically, these have an M16 (7/16 UNF) male threaded connector, and pressure gauges or transducers can be fitted with the mating female threaded connector. The connection can normally be made by hand tightening the two halves together.

If there is no test point fitted, a male–female straight or elbow fitting with a test point fitted can be used. These can be installed at the end of a hose, for example. A wide range of such fittings is available.

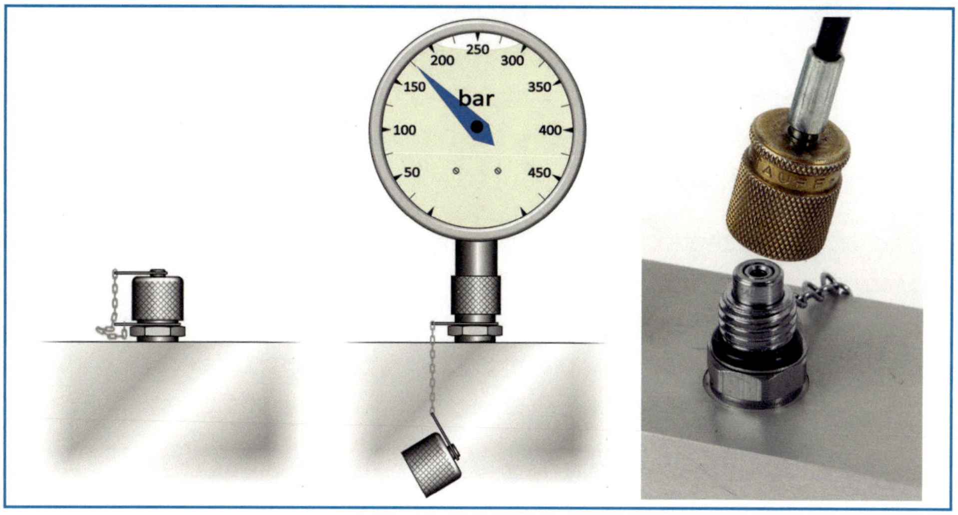

Fig. 9.3 Pressure test point

If a test point is awkward to access, it can be useful to fit a small **micro-bore hose** between the test point and the gauge. However, care should be taken to specify a hose with the correct pressure rating, typically 400 or 600 bar (6000 or 10,000 psi). It is also best practice to keep the hose as short as possible, otherwise it can act as a 'capacitor' by expanding slightly over its length, and increasing the response time and reducing the maximum pressure measured.

Exactly what sensor technology should be used to measure pressure will depend on the application and budget. Mechanical pressure gauges are inexpensive and have been widely used throughout the industry for many years. However, the readings they provide can sometimes mislead, they are regularly misused and misread, and sometimes even blow apart! So, as digital technology becomes less expensive, there is a growing trend towards using pressure transducers instead of simple pressure gauges. If it is necessary to detect pressure spikes, measure differential pressure, capture a small change in a high pressure or measure a low pressure very accurately, a pressure sensor should be used rather than a simple mechanical gauge.

Mechanical pressure gauges

Glycerine-filled **Bourdon tube gauges** (63 mm (2¼ in)) (Fig. 9.4) have been widely used for many years, but their response time is a major limitation. The glycerine

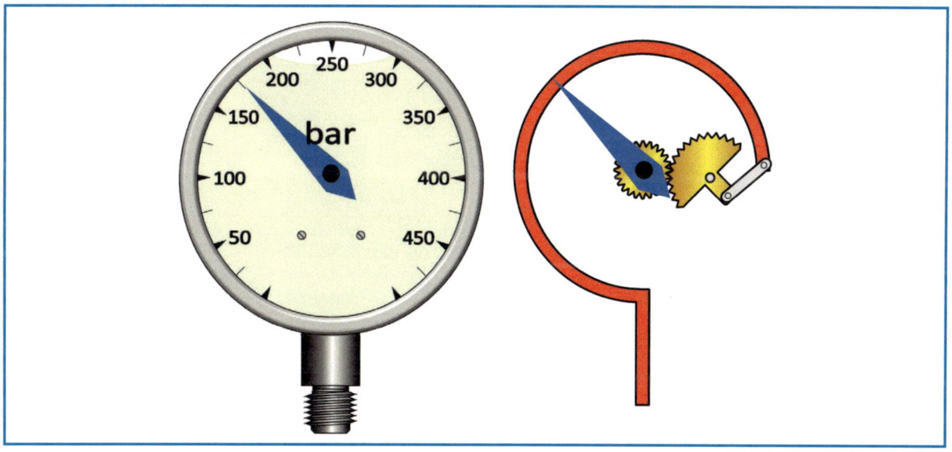

Fig. 9.4 Bourdon tube pressure gauge

is there to help damp the 'flutter' of the needle and prolong the life of the gauge. However, depending on the volume of glycerine in the gauge (which is also affected by temperature) and the angle the gauge is held at, there can be a 10–15 second delay in reading the pressure correctly. This can obviously be a large source of error if the gauge is used incorrectly.

Another area to be aware of with smaller Bourdon tube gauges is accuracy. Typically, accuracy might be stated at around ±1.6% of full scale on a 400 bar (6000 psi) gauge, which means a possible error of ±6.4 bar (96 psi). This could be highly significant if the pressure being measured is only, say, 50 bar (725 psi), so care should be taken not to use a high-pressure gauge to measure low pressures, otherwise errors can be introduced.

Lastly, many hydraulic engineers have at some time blown up a mechanical pressure gauge by accidentally pressurising it above its calibrated range. There are pressure cut-off valves and **snubbers** available to help prevent this from happening, but they are not foolproof. The best rule of thumb is to select a gauge with roughly twice, or at least 1.5 times, the range of the pressure to be measured. So to measure a pressure of 400 bar (6000 psi), for example, select a 600 bar (10,000 psi) gauge.

In summary, the mechanical gauge selected should have a maximum range ideally twice but no more than five times the pressure to be measured, and it should be held vertically and used only as a pressure indicator. The mechanical pressure gauge can be specified 'dry' (i.e. without glycerine), which can be useful in some applications where there is little vibration or pressure ripple. Very large and much more expensive gauges are available for calibration purposes, with a certified accuracy as high as ±0.1% of full scale.

Pressure sensors

Pressure sensor is the generic term for a device that converts a fluid pressure into an electronic signal. Such devices can be either **transducers**, which typically provide a variable voltage signal for transmission over short distances, or **transmitters**, which provide a variable current signal (typically 4–20 mA) for transmission over long distances (Fig. 9.5).

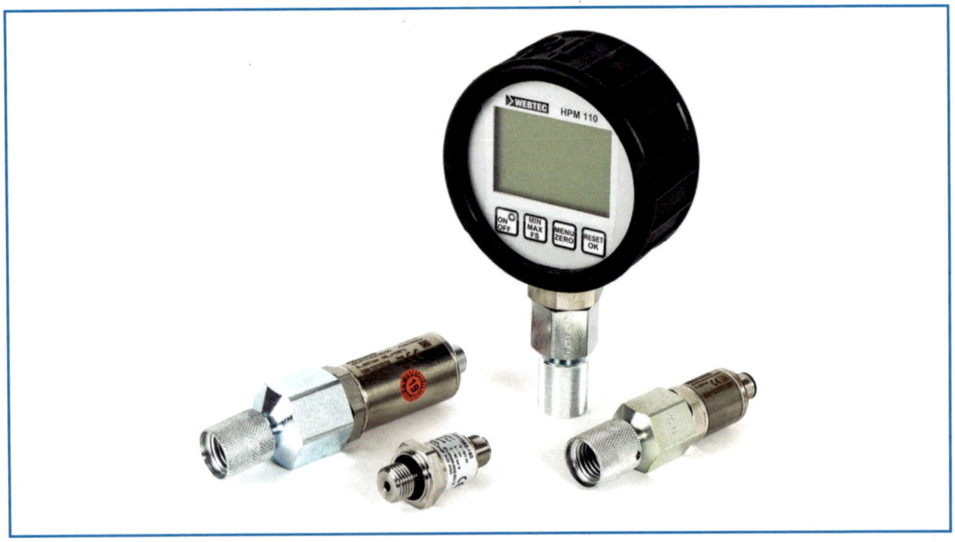

Fig. 9.5 Electronic pressure sensors

Traditionally, pressure sensors have provided an analogue voltage or current output proportional to the pressure they are measuring. However, sensors with a digital output suitable for transmitting via a **serial bus system** (such as **CAN bus**) or **wireless technology** (such as **Bluetooth**) are now becoming more popular.

The factors that should be considered when selecting a pressure sensor include:

- *The pressure range to be measured*. Typically, sensors are selected so that the system operating pressure is 50–60% of the sensor's maximum rated pressure. For example, for a hydraulic system normally operating up to 140 bar (2000 psi), a 300 bar (4500 psi) sensor would typically be used. In addition to providing a safety margin, this practice also provides a good compromise with regard to the performance characteristics of the sensor. Most sensors will have an **over-range capability**, which is the maximum pressure that can be applied to a transducer without causing a change in its performance beyond the specified tolerance. The over-range capability should not be confused with the **burst pressure**, the fluid pressure at which mechanical failure and/or fluid leakage from the transducer is expected. Exceeding the over-range capability can affect a transducer's ability to function, while exceeding the burst pressure can destroy it.
- *Resolution*. The resolution is the smallest change in pressure that can be detected in the sensor's output. It is usually expressed as a percentage of the sensor's maximum output or **full-scale output (FSO)**. For example, if two transducers each have a resolution of 0.1% of FSO, a 10 bar (150 psi) transducer can detect a pressure increase or decrease of 0.01 bar (0.15 psi), whereas a 350 bar (5000 psi) transducer can detect a pressure change of 0.35 bar (5 psi). Resolution generally is not constant across a transducer's range. Manufacturers may publish values for maximum resolution or average resolution. Users should be aware of the difference between them when comparing one transducer's performance with another.
- *Pressure spikes*. Pressure spikes are microsecond to millisecond bursts of pressure that can reach many times the normal system operating pressure. For example, if a valve shifts abruptly to block flow, a shock wave can be generated within the system. Likewise, if a hydraulic system is moving a load and the load suddenly stops, the system may react with a brief surge of pressure. Pressure sensors, which have a much quicker response than mechanical gauges, react to spikes and can show signs of having been over-pressurised. Pressure sensors are the components most vulnerable to damage from pressure spikes, but not because they are less durable than the mechanical gauges. Transducers designed for severe service are available and should be specified where such conditions occur. Spikes also damage the machines that generate them. The erratic flow of fluid, common in systems that generate spikes, reduces efficiency and accelerates wear on valve ports and seals.
- *Other factors*. These include the operating temperature range of the sensor (remembering that hydraulic fluid temperatures may be significantly higher than the ambient atmospheric temperature), the type of output required (voltage, current, digital, etc.) and the physical characteristics of the sensor (size, connections, etc.).

FURTHER READING

For further information and a free guide to selecting a pressure transducer, go to www.webtec.com/education

DEFINITION

Multiply US gpm by 3.785 to obtain L/min.

Multiply imp gpm by 4.546 to obtain L/min.

FLOW MEASUREMENT

Flow is the volume of a fluid that passes a fixed point in a unit of time. For most hydraulic applications, flow is measured in litres per minute (L/min), US gallons per minute (gpm) or, occasionally, UK (imperial) gallons per minutes (imp gpm).

Flow is to the hydraulic engineer is what current is to the electrician, while pressure is the hydraulic equivalent of voltage. Measuring one without the other can lead to a completely incorrect diagnosis of why a system is not performing correctly.

For example, if a hydraulic cylinder is moving slowly, the pressure at the inlet can be checked and found to be correct. Looking no further, it might be concluded that the hydraulic power pack is operating correctly, and so the fault must lie with the cylinder. However, if the flow into the cylinder is measured and found to be too low, this might indicate that perhaps the problem is not with the cylinder, and focus could be switched to the components that can affect flow, such as the pump, flow controls or an incorrectly set relief valve.

Traditionally, flow meters have provided a frequency output or an analogue output proportional to the flow they are measuring. However, sensors with a digital output suitable for transmitting via a serial bus system (such as CAN bus) or wireless technology (such as Bluetooth) are now becoming more popular.

The questions that should be asked when selecting a flow meter are:

- What are the fluid properties?
- What are the hydraulic system operating conditions?
- Why is flow being measured and how accurate does the measurement need to be?
- What effect might the flow meter have on the fluid and/or system, and vice versa?
- How important is it to measure flow and what is the available budget?

If all these questions can be answered, it becomes a lot easier to specify the right flow meter for the application and at the right price to ensure that the right results are obtained.

What are the fluid properties?

Is the flow meter going to be used on the same fluid all of the time? This is typical when a flow meter is permanently installed. However, if it is a portable device and taken to different machines in the field, it might be used with a range of different fluids.

It is important to know about the fluid(s) to be measured, as the characteristics of the fluid can greatly influence the choice of flow meter. Of particular interest are the fluid properties. For example, is it corrosive or a natural lubricant, and what are its material compatibility and viscosity characteristics? This information can be found on the fluid's datasheet under such headings as 'physical properties', 'lubricity' and 'anti-corrosion'.

What are the hydraulic system operating conditions?

Knowing the minimum and maximum flows that need to be measured will directly influence the type of flow meter required and the price of that meter. The ratio of the highest flow to the lowest flow is called the **turndown ratio**. For example, if the minimum and maximum flows are 1000 and 1100 L/min (260 and 290 gpm), then the turndown ratio is 1.1:1, but if the minimum to maximum flow range is 1–400 L/min (0.25–100 gpm), the ratio is 400:1. The higher this ratio is the harder it will be to find a single flow meter that will cover the whole range with consistent accuracy. This could mean that two or more meters will be required to cover the range, or it may be possible to reconsider the minimum operating range and decide that 10–400 L/min (2.5–100 gpm) is acceptable, which reduces the ratio to 40:1.

The other information needed will include the system's typical cleanliness level, the maximum operating pressure, and the ambient and system temperature ranges. Once the fluid properties and the system's operating temperature range are known, it is possible to look up or calculate the **kinematic viscosity** for the fluid over that temperature range. For example, in a typical hydraulic system with an ISO VG32 fluid used between 40 and 60°C, the kinematic viscosity will vary between about 34 and 15 cSt.

Possibly the most difficult situation is when the fluid is subject to wide temperature swings, resulting in large viscosity changes, as this can directly affect the accuracy of flow measurement. In this case, the chosen meter will need to be relatively insensitive to the effect of viscosity, have built-in viscosity compensation or be calibrated for an average viscosity, with allowances being made when the temperature or viscosity is 'out of range'.

FURTHER READING

For further information on the viscosity of hydraulic oils, go to www.webtec.com/education

Why is flow being measured and how accurate does the measurement need to be?

For some applications flow measurement is required simply to monitor trends, such as answering the question 'Is the flow more or less than last week?' At other times, flow measurement is required to compare performance with other systems (in which case repeatability of measurement is more important than accuracy) or against a manufacturer's specification (in which case repeatability is required but accuracy of measurement is more important).

Output signals

Depending on the application, it may be necessary to transmit the signal from the flow meter to a programmable logic controller (PLC) or data logger in order to record the flow going through the system at a particular time. Some basic flow meters, which are often used to monitor trends, have only an analogue dial, and these meters can rarely be upgraded. However, many flow meters offer an electronic output signal as standard, and there is a wide range of different signal options available. These can be broadly split into analogue and digital and linear and non-linear. Many flow meters will have on-board electronics that convert the raw measurement, for example a non-linear frequency, to ensure the output signal is linear.

For optimal accuracy a linear digital signal such as CAN will give the least error and most reliable results, regardless of cable length. Many engineers prefer to use meters that give traditional linear analogue output signals, such as 0–5 V, 0–10 V or 4–20 mA, often because they have existing readout data in this form and the meters are easy to wire up and fault find. Finally, care must be taken when using longer cables (over 5 m (16 ft)) with voltage devices, as it is possible to get a significant voltage drop along the cable, which will in turn increase the error in the displayed values.

Flow meter accuracy

Accuracy is normally quoted by the flow meter manufacturer as a percentage value to indicate the acceptable error band. If requested, most manufacturers can provide a set of flow test results that are traceable to a known standard to demonstrate the flow meter's accuracy when it was calibrated. Typically, accuracies are quoted as a percentage either side of the 'maximum' or **full-scale value**, or as a percentage either side of the 'measured' or 'indicated' reading (Fig. 9.6).

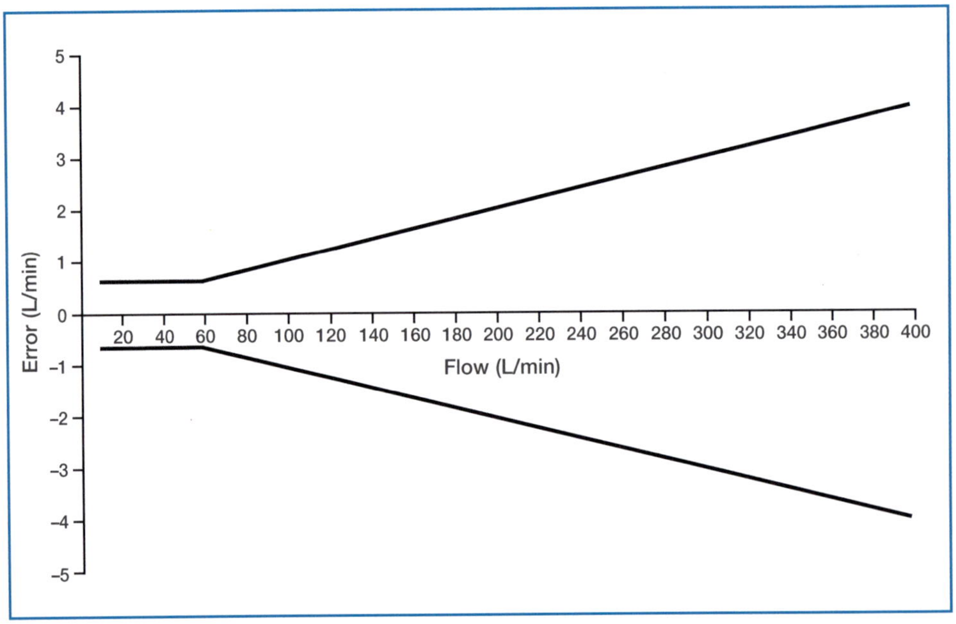

Fig. 9.6 Flow meter accuracy – 1% of indicated reading for a flow meter rated up to 400 L/min

Usually, a meter that has a higher accuracy for the same flow range as another is more expensive. That is, it costs more to have a device that gives traceable readings that are accurate as a percentage of the 'measured' value than a device that only quotes readings as a percentage of the full-scale value.

For example, an accuracy of ±1% of the full scale value (1% FS) means that a flow meter rated to 400 L/min (106 gpm) should measure flow to within ±4 L/min (±1.06 gpm) (0.01 × 400 = 4 L/min). If the same meter is being used to measure a flow of 40 L/min (10.6 gpm), the possible error is still ±4 L/min (±1.06 gpm), although this is equivalent to an error of 10% of the measured value. Hence, a meter that will measure to within 1% of the measured or indicated reading (1% IR) will generally cost more, as this is much more difficult to achieve than 1% FS.

FURTHER READING

For further information on the wide range of flow meters available, not just for hydraulics, including the theory behind how different flow meters work, see *An Introductory Guide to Flow Measurement*, by Roger C. Baker (ISBN 978-1-860-58348-3).

What effect might the flow meter have on the fluid and/or system, and vice versa?

This might seem like a strange question, as the 'effect on the fluid' will depend on whether or not the flow meter is intrusive and also on the flow meter technology. This 'effect' can be measured by the energy loss due to the presence of the meter, which is better known as the pressure drop (ΔP) across the device. This pressure drop has two effects: increasing the upstream pressure and generating heat.

For example, in a case drain line of a piston pump the flow may be quite low, say under 10 L/min (2.6 gpm), and the pressure may not be allowed to exceed 2 bar (30 psi) without risking damage to the seals or affecting operation of the pump. Here, the pressure must be kept as low as possible by using the correct type of flow meter and by ensuring the ports are large enough to minimise the pressure drop, especially in the event of a sudden surge in flow, as can occur with variable-displacement pumps. In a high-flow, high-pressure system, if the flow meter has a high pressure drop, the heat generated might be more important, especially because much of that heat will go into the fluid. This wasted energy might have been avoided by specifying a different type or size of meter.

Regarding the 'effect' of the fluid on the meter, this is not normally a problem if the properties of the fluid are known before selecting the flow meter. The two most common problems encountered are excessive contamination of the fluid and incompatible fluids. Both can severely reduce the life of the flow meter and cause it to malfunction. In the case of the wrong fluid being used, this can result in corrosion, and the flow meter 'sticking' or damage being done to the seals, resulting in leakage.

Maintenance and recalibration

It is worth considering the longer-term effects of the fluid on the flow meter, especially if the fluid is known to be carrying high levels of contaminant or if the fluid has very low lubricity. In any case, it is important not only to identify the correct flow meter in the first place but also to consider how frequently it should be removed for maintenance and recalibration. The quoted accuracy can only be guaranteed to be true on the day the meter was last calibrated. If the accuracy is important, the quality department will add the flow meter to the list of instruments that need recalibration. The exact frequency of recalibration will depend on the meter type, the duty cycle and the manufacturer's recommendations. The typical period for recalibration is every 12 months.

How important is it to measure flow and what is the available budget?

A better question might be 'What happens if flow isn't measured?' If the answer is 'Nothing; it's just one of many indicators to look at', then the budget available is likely to be small. However, if the answer is 'The PLC will think that the hydraulic system has failed and will shut down the machine', then it is possible to visualise how much the production department will be concerned, how important it is to measure the flow and how much budget might be available to stop that from happening.

Just because a budget is available does not mean that it has to be spent. Answering the four questions discussed above will determine what is required of a flow meter, so that the right model can be specified and an explanation provided as to why certain features are important. In addition, by answering the above questions it is often possible to save money because it will be less likely that an over-specified flow meter will be selected just to be 'safe'.

Types of flow meter

Variable-orifice flow meters

The idea of flow displacing an object, usually a piston or a ring, forms the basis of simple variable-orifice flow meters (Fig. 9.7). The momentum of the fluid exerts a force on a piston that is held in place by a spring. As the flow increases, the piston moves and the orifice size increases, along with the spring force on the piston. The piston is linked to the analogue readout via a magnet.

Fig. 9.7 Variable-orifice type flow meters

The flow indicator is purely mechanical and is ideal for situations where trends in data are important rather than exact measurements of flow. Such meters typically have an accuracy of 2–5% of full scale. The flow indicator comes in a range of sizes, any of which will typically cover a 15:1 range. The flow indicator offers a low-cost solution, with a typical pressure drop of about 1.5–2 bar (22–29 psi) at 400 L/min (106 gpm).

Gear-type flow meters

Gear-type flow meters (Fig. 9.8) are positive-displacement flow meters. On the inside they look similar to a gear-type motor. Fluid passes around the outside of a pair of intermeshed gears, rotating the gears on their shafts. A transducer mounted above one of the gears generates a pulse each time a gear tooth passes under it. The rotation of the gears is proportional to the flow rate. Sometimes, two transducers are used in order to measure direction and improve resolution.

The relationship between the frequency measured by the transducer and the flow rate is shown using the K factor (K = frequency/flow). Given a constant K factor, the flow can be easily calculated from the frequency.

Fig. 9.8 Gear-type flow meter

Gear-type flow meters give a very precise measure of flow. Typically, such meters have a large turndown ratio of at least 30:1, and sometimes as high as 200:1. They will measure the flow to an accuracy of better than 1% of full scale without additional **linearisation**. Depending on the flow meter size and sensor linearisation technology, accuracies of ±0.3% to ±0.5% of the indicated reading are achievable.

These flow meters are relatively insensitive to changes in fluid viscosity, and work best with higher-viscosity fluids (greater than 10 cSt to as high as 125,000 cSt). With low-viscosity fluids (less than 10 cSt), leakage can occur across the tips of the gears, which will reduce the flow meter accuracy.

Disadvantages of gear-type flow meters are that they usually have a very high pressure drop, such as 9 bar (130 psi) at 10 L/min (2.6 gpm), and can be quite noisy.

Turbine-type flow meters

In a turbine-type flow meter, a turbine rotor is mounted on a shaft between two sets of flow straighteners (Fig. 9.9). The fluid passes through the flow meter and rotates

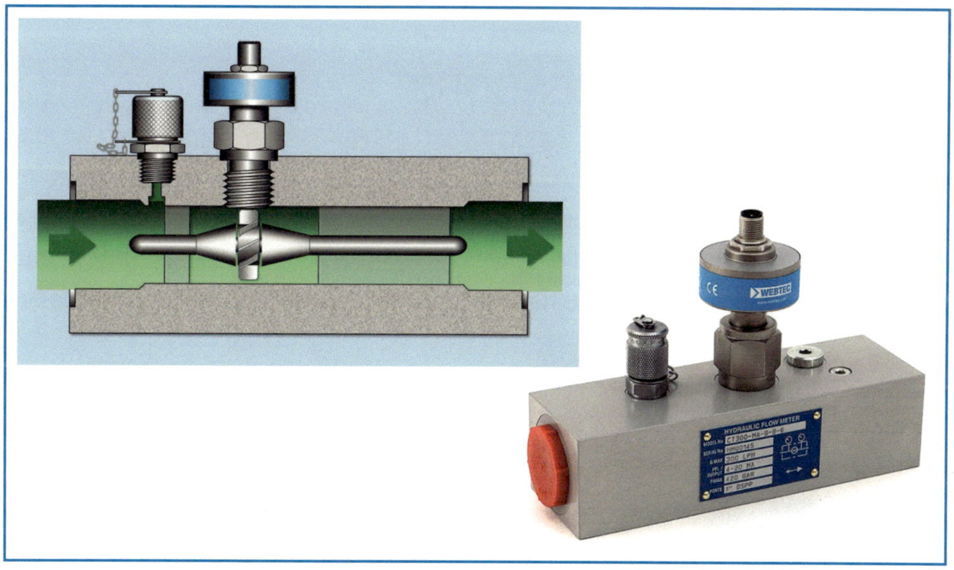

Fig. 9.9 Turbine-type flow meter

the turbine blade. As with the gear-type flow meter, a transducer is mounted above the turbine and generates a pulse each time a blade passes under it.

The frequency from the transducer is proportional to the flow over a limited range. For example, a 25 mm (1 in) turbine flow meter might typically have an accuracy of ±1% of full scale without linearisation. The same 25 mm (1 in) turbine flow meter when used with a look-up linearisation table will operate over a wider range, with an accuracy of 1% of the indicated reading. A turndown ratio of 30:1 is common.

The use of a turbine instead of gears means that the meter requires less energy to operate and has a very low pressure drop, 3 bar (44 psi) at 400 L/min (106 gsm) or lower, depending on the bore size.

The disadvantage of this type of meter is that it is quite susceptible to changes in viscosity. Thus, it is usually used for fluids with a viscosity under 100 cSt.

Other meter types

Oval gear meters are similar to conventional gear meters but use two elliptical gears that rotate together at 90° to one another inside a housing. The fluid is swept around the chamber by the gears, and the frequency of rotation is directly related to the volume of fluid passing through the meter. This type of meter generally works best with higher-viscosity fluids, but has a lower pressure drop than an equivalent conventional gear-type meter. The teeth on the gears tend to be very fine, resulting in the flow meter being more susceptible to fluid contamination than other meter types.

In theory, **non-intrusive meters**, such as **ultrasonic meters**, are very attractive in that they do not require 'breaking into the system' and have virtually no pressure drop. However, they are not yet widely used in hydraulics for two reasons. First, hydraulic pipe sizes are quite small, often less than 25 mm (1 in), making transit times very short and difficult to measure. Second, hydraulic connections often consist of flexible hoses manufactured from two or more materials, making it difficult to transmit signals through them. As a result, non-intrusive meters tend to be used more for permanent installations on industrial hydraulic systems using rigid pipes with larger diameters.

OTHER TYPES OF SENSORS

There are several other types of sensors used in hydraulic systems, such as load cells, torque transducers and speed sensors, which are often permanently installed and may form part of the machine control system. However, their outputs can also be monitored for diagnostic troubleshooting purposes, so a brief description of their function and operation is included here.

Load cells

Load cells measure linear force, either in tension or compression (or in some cases both). There are several different types but the most common for use in hydraulic systems is the **strain gauge** type, although **piezo-electric** types are used in some applications. Strain gauges work on the principle that the electrical resistance of a thin strip of metal will change as it is deformed by a force. Sensing the amount of change in the resistance can, therefore, provide an output proportional to the force applied. Temperature will also affect the resistance, but this is normally compensated for in the construction of the load cell.

Piezo-electric devices use a material that produces a small electric charge when a force is applied to it, which again can be amplified and used to provide an electrical signal proportional to the applied force. Compared with strain gauges, piezo-electric devices often have a wider useful working range, which can be useful when a load cell has to operate under high shock loads but also needs to be accurate when measuring light loads.

The correct selection and installation of load cells is important to ensure that they cannot be overloaded and that the force being measured is transmitted correctly through the load cell. Overloading, even momentarily, can cause mechanical or electrical damage to the load cell, as can high induced voltages caused, for example, by nearby welding.

Torque transducers

Torque transducers are basically load cells designed to sense rotary force (torque) rather than linear force. Again, the most common design used with hydraulically powered machinery is the strain-gauge type, although other constructions are used in specialist applications. An added complication with torque transducers, however, is the fact that the component at which torque needs to be sensed (such as a drive shaft or drill bit) will be rotating, sometimes at high speed. This then requires some form of contact or non-contact device to transmit the signal from the rotating sensor to the non-rotating signal conditioning equipment.

Position sensors

Position sensors are also available to measure either linear or rotary position. Many different types are in use, depending on the accuracy required, the distance over which position is to be measured, the working environment, etc. The simplest type is probably the electrical **potentiometer**, in which a wiper is moved along a coiled wire or carbon track. Applying a voltage difference to the ends of the potentiometer means that the voltage sensed at the wiper is proportional to the wiper position. As with all mechanical contact devices, however, wear will eventually cause deterioration of the sensor performance, so for most applications it is preferable to use non-contact devices, which include the following:

- **Linear variable differential transformers (LVDTs)** move an iron core through electrical coils fed with an AC voltage in order to vary the inductance of one coil relative to another (Fig. 9.10). The change in inductance can then be sensed and converted to a positional signal. They have no contacting components and are suitable for measuring relatively small movements, such as the few millimetres of movement of a proportional valve spool up to several centimetres of movement of a cylinder.
- **Magnetostrictive sensors** use a magnet attached to the moving component (Fig. 9.11). The magnet moves along the outside of a waveguide tube and

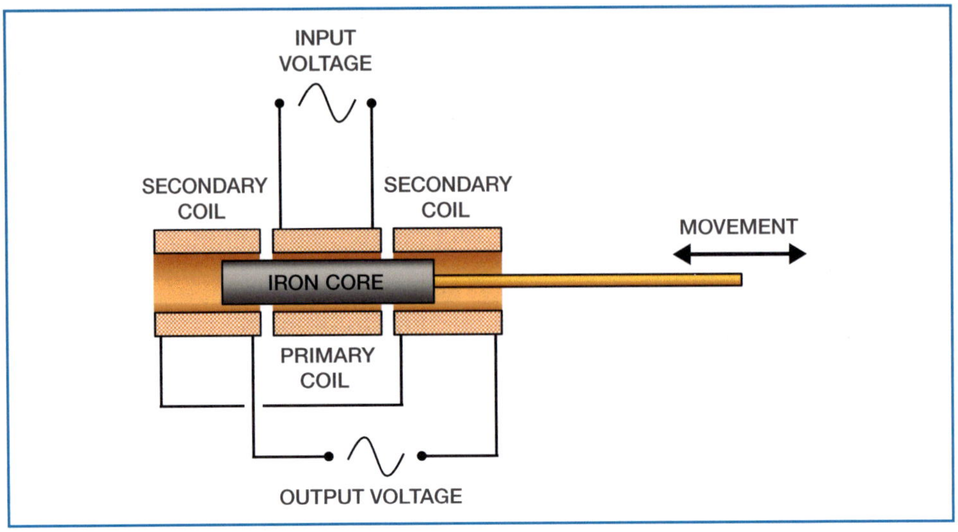

Fig. 9.10 Principle of working of LVDTs

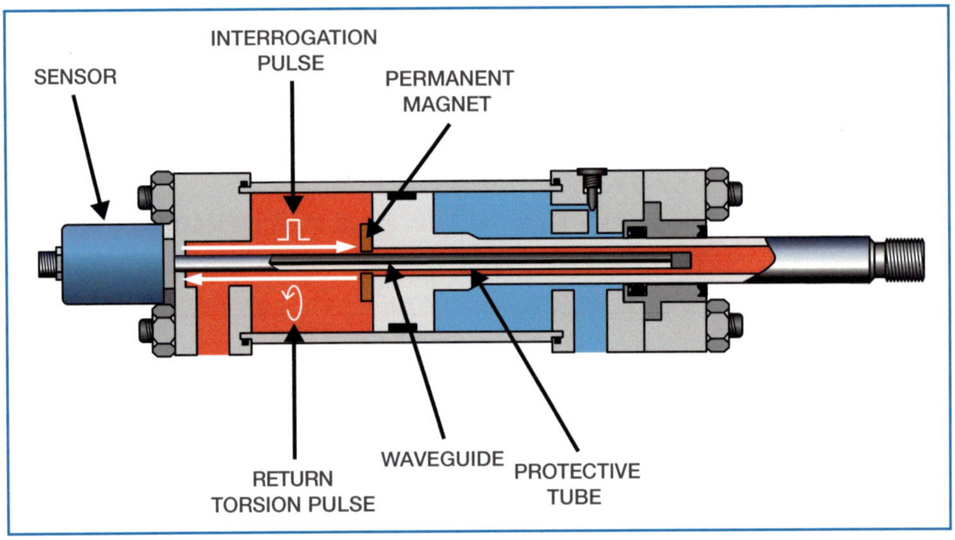

Fig. 9.11 Magnetostrictive position sensor in a hydraulic cylinder

an electrical pulse is transmitted down the waveguide. The magnetic field of the electrical pulse interacts with the magnetic field of the magnet, causing a torsion strain pulse in the waveguide. By sensing the time taken for the pulse to travel out and back, the position of the magnet, and hence the position of the moving component, can be determined. Such devices can be built into hydraulic cylinders, where they are protected from damage and form a very compact arrangement.

Speed sensors

Speed can often be derived electronically from a position signal by relating the change in the position signal to time. **Tachogenerators** have been used for many years to measure rotary speed (rpm). These devices produce an output voltage proportional to their driven speed and a polarity that is dependent on their drive direction.

A more compact arrangement for use with hydraulic pumps and motors is a **pulse-type sensor** (Fig. 9.12), which senses grooves or notches on the rotating shaft and generates a series of electronic pulses (a similar arrangement to that used in

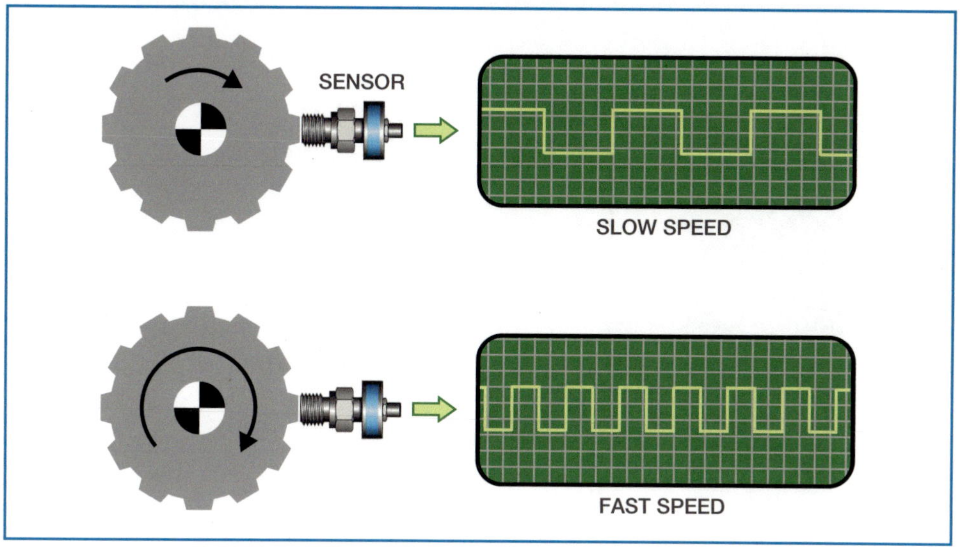

Fig. 9.12 Pulse-type speed sensor (Image courtesy of Eaton Corp.)

gear-type flow meters). The frequency of the pulses indicates the shaft speed, and dual-output sensors also determine the rotational direction.

Where speed sensors are not permanently installed and where a rotating shaft is safely accessible, either contact or non-contact hand-held tachometers (Fig. 9.13) can be used to provide an output-speed reading.

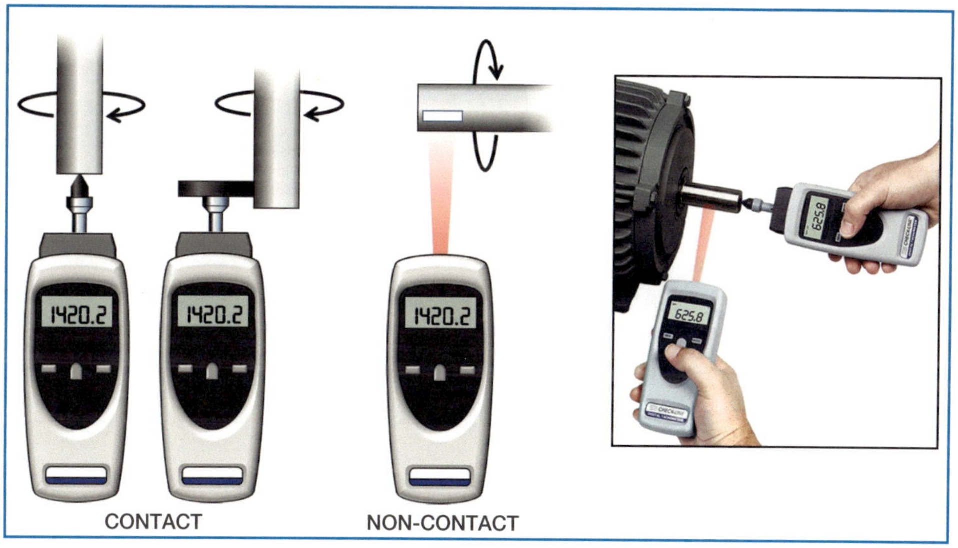

Fig. 9.13 Hand-held tachometers (Photo courtesy of Checkline Europe Ltd)

Vibration sensors

Measuring machine or component vibration can be a very good means of detecting early symptoms of a breakdown, in particular for components such as shaft bearings. Vibration sensors are normally permanently installed (Fig. 9.14) so that vibration can be measured over a period of time and any abnormal trend can be acted upon before complete failure occurs.

The software that accompanies vibration sensors can now identify individual bearings by their distinctive 'signature'. A vibration sensor mounted on a hydraulic pump, for example, is able to identify whether a front or rear bearing is approaching failure or whether some other problem is causing an abnormal operation of the pump.

Fig. 9.14 Portable and permanently installed vibration sensors

Hand-held vibration sensors are also available for routine checking of components, but care is needed to ensure that measurements are taken in exactly the same way on each occasion.

Particle counters

Off-line particle counters have been used for many years to determine the contamination level of a fluid sample taken from a system. They typically use either an infra-red or laser light beam to cast 'shadows' of dirt particles onto a light-sensitive sensor to both count and size the particles passing through in a known volume of fluid.

More recently, particle counters have been developed that can be permanently installed in a system (online) to provide a readout of the contamination level in real time (Fig. 9.15). Where a system has a serious problem and contamination

Fig. 9.15 Online particle counter (Image courtesy of MP Filtri Ltd)

is increasing rapidly, this type of sensor can warn of the issue immediately, thus providing a significant advantage over off-line sensors.

Again, being able to trend the data (recording how it changes over a period of time) will give advance warning of when wear problems are increasing and thus action needs to be taken.

Contamination analysers

More sophisticated analysers can be installed in systems to measure such things as water content and to distinguish between certain types of wear particles (e.g. ferrous or non-ferrous).

SIGNAL TYPES AND WIRING

Often, great care is taken in the selection of a sensor to ensure that it will provide the necessary level of accuracy. However, it is also vital to ensure that the signal generated by the sensor is transmitted to the receiving device without degradation or interference. In some applications the signal transmission has to take place over long distances or through environments with high levels of **electromagnetic interference (EMI)**.

As mentioned in Chapter 5, an analogue voltage is generally the simplest form of signal type. Voltage signals typically operate over the ranges 0–5 V or 0–10 V, or ±10 V for properties that vary either side of a central or neutral condition.

Voltage signals are, however, susceptible to degradation (especially when transmitted over long distances) due to the voltage drop along the cables used to transmit the signal. For this reason, the current flow in the signal cable must be kept as low as possible, so the input connection of the receiving device will need to have a very high resistance (known as **input impedance**). Unfortunately, this requirement makes the signal voltage more susceptible to interference from EMI, which can cause what is known as 'electrical noise'. The situation is similar to trying to listen to a piece of music on the radio with a pneumatic drill operating close by.

EMI can be significantly reduced by using screened cables, where the signal cable itself is surrounded by a **braided metal sheath**, which is connected to an earth or ground connection. This means that any electrical noise is generated in the protective sheath rather than the signal cable. In order to be effective, however, the sheath must have a good (low resistance) ground connection. This is normally connected at one end of the cable only to avoid the possibility of current flows in the sheath, which could also affect the signal voltage. In addition, care must be taken at the cable ends to ensure that the connectors are also shielded from EMI. Using shielded cables, therefore, is similar to listening to the radio via headphones to shield out the noise of the pneumatic drill. However, some drill noise may still be audible, particularly during quieter periods of the music when the signal-to-noise ratio is low.

For transmission over longer distances or through particularly difficult environments, current signals (typically 0–20, 4–20 or ±20 mA) can be used, and in many industries such signals are the standard. Using a 4–20 mA signal means that the working range of the signal is from 4 mA minimum to 20 mA maximum, with the mid-point being 12 mA.

POINT OF INTEREST

Electromagnetic radiation can be created both by equipment that uses high electrical currents (such as welding equipment, arc furnaces and electric motors) and by relatively low-energy devices (such as fluorescent lighting and walkie-talkies).

With a conventional 0–20 mA or 0–10 V signal, if the receiving device detects a signal of 0 mA or 0 V it can mean two things – either the signal being transmitted is zero or the transmitting cable has a fault (such as a broken wire). However, using 4 mA to represent a zero signal means that any signal received that is less than 4 mA indicates there must be a fault with the signal transmission, and appropriate action can be taken.

Although analogue signals have the advantages of simplicity and speed of transmission, in many applications the processing capabilities and potentially reduced wiring requirements of digital signals can be a distinct advantage. Digital signals are conveyed by means of a series of on/off electrical pulses represented by 0 (off) and 1 (on). Each pulse may last only for a fraction of a second, so data can still be transmitted at high speed (although still not as fast as analogue signals). The situation can be likened to listening to a radio message in normal speech (analogue) as opposed to a one transmitted by Morse code (digital).

Digital signals can be transmitted using just two wires, in which case the on/off pulses follow one after another (**serial transmission**), or several wires can be used to transmit pulses simultaneously (**parallel transmission**). As with road networks, a single-lane carriageway is a relatively low-cost solution but can only handle a limited amount of traffic, so progress may be slow. Conversely, a multi-lane highway can speed up the traffic flow but is a higher-cost solution requiring more space and more complex interchanges, etc.

Many different standards exist for coding and decoding digital signals, such as **RS 232**, **RS 485** and **USB (universal serial bus)**, and each has its own particular benefits for different applications. Often, the standard selected is determined by the type of electrical connector used. Due to its common usage on mobile phones and computers, for example, most people can now recognise a USB connected device.

In recent times, digital signal transmission has been developed that requires no wiring at all but instead uses radio waves. Again, the standard that many people will be aware of is Bluetooth, which is often used for connecting computer or mobile phone peripherals (headphones, speakers, etc.) to one another. This wireless approach is generally used only for signal transmission over relatively short distances, although industrial versions can operate over distances of 100 m or more.

In digital signal transmission technology the individual connections to each sensor, switch, actuator, etc., can be replaced by a network (or 'bus' system) that each device is connected into. This can be thought of as being similar to a telephone system, where each user (sensor, switch, etc.) has its own particular identifier (telephone number). Messages can, therefore, be sent backwards and forwards along the network, to and from controllers and every other device connected to it. As with telephone conversations, sometimes the number called may be engaged, so bus systems may not be as fast as other methods, and devices sometimes have to wait until the network is free before a message can be transmitted. However, for applications other than those requiring ultra-fast signal-transmission times, bus systems are now used extensively in both mobile and industrial machinery, including

DEFINITION

An **analogue signal** is a continuously variable signal – typically a voltage or current. Such signals are simple to generate and fast to transmit. However, they are subject to degradation due to cable resistance or electromagnetic interference (EMI), especially when transmitted over long distances or through difficult environments.

A **digital signal** consists of a series of on/off voltage pulses that can be decoded to represent a signal level. They are generally less susceptible to signal degradation and can be used in a serial 'bus' system to significantly reduce the amount of wiring required.

POINT OF INTEREST

The principal advantage of bus systems is the considerable reduction in the amount of physical wiring required and the flexibility they provide for system modification, error-proofing and diagnostics.

passenger cars. Again, several different standards have evolved to suit different types of applications, but for hydraulic systems standards such as **Profibus**, **CAN bus** and **Interbus-S** are typical.

DATA AND SYSTEM PERFORMANCE ANALYSIS

Having provided the means of measuring or sensing various aspects of a system, it is now necessary to record and analyse the data obtained in order to ensure the system is operating to specification or to predict likely failures.

Data logging options

Data loggers record data in one of two main ways. In the first, data is recorded continuously over time, much like an old-fashioned tape recorder: start, record, stop. This is ideal to get a feel for what is happening in a hydraulic system, or where all the data is required for development purposes and it can be analysed afterwards. The only downside to this approach is the large amount of data that may be obtained, which may take a significant amount of time to analyse (Table 9.2). It is usually possible to alter the sampling rate (which is quoted as either the number of readings recorded per second or the time between readings) from milliseconds to minutes or hours, depending on the nature of the property being recorded.

Table 9.2 Typical file size (kB), based on sample rate and duration of test, for an eight-channel data logger

Test duration	Sample rate						
	0.1 ms	1 ms	10 ms	100 ms	1 s	10 s	60 s
1 s	1,800	200	20	0	0	0	0
2 s	3,600	400	40	0	0	0	0
5 s	8,900	900	80	0	0	0	0
1 min	107,200	10,700	1,100	100	20	0	0
5 min	536,100	53,600	5,400	500	60	0	0
1 h	6,433,600	643,400	64,300	6,400	600	60	20
24 h	154,406,300	15,440,600	1,544,100	154,400	15,400	1,500	300

The data can typically be analysed on a data-logging device (Fig. 9.16) by using PC software supplied by the manufacturer, or by exporting it to an analysis package such as Microsoft Excel. A data logger is typically a stand-alone electronic device (normally battery powered) that either interfaces with or incorporates the relevant sensor and stores the data obtained in its memory.

The second method of recording data uses a trigger. A snapshot of data is recorded every time a button is pressed, so a key press could be the trigger. This is ideal when carrying out a pump test, for example, to record the flow and pressure at various points in order to build a power curve. Alternatively, a data logger can be used to record a set of data over a pre-set period every time a measured value hits a certain target value. For example, if the system pressure is normally 210 bar (3000 psi) the data logger can be set to trigger when the pressure exceeds 220 bar (3300 psi) and record all channels for 2 seconds. In this way, much less but more relevant data is captured. This approach is an ideal means of checking for occasional pressure spikes in the system.

Fig. 9.16 A typical data logger (Webtec)

Data analysis

Armed with measurements of what is really happening within the hydraulic system the user will typically want to produce a graph, a table of results or a combination of the two. Using the software supplied, these results can easily be turned into a report format by adding titles, comments, arrows and highlights. This report, or the complete dataset, can then be shared with other engineers at the user's company or at the customer's company to confirm the situation or seek help and direction.

How to select a data logger

While not exhaustive, the following checklist covers the key areas to consider.

- *What is your application?* For example: fault-finding, research and development, condition monitoring, pre-dispatch inspection.
- *Is it fixed or mobile?* Will the data logger be moved from job to job or mounted permanently in a system or test stand?
- *What type and duration of tests need to be run?* See above for examples.
- *What is the distance from the sensors to the readout?* It is better to use digital (CAN) rather than analogue sensors for distances over 5–10 m (15–30 ft).
- *What environment will the data logger be used in?* How much electrical noise, dust and water (rain) will it be exposed to? Check the signal types and ingress protection (IP) rating of the device.
- *What needs to be measured?* For example: flow, pressure, temperature, contamination, any custom sensors.
- *How many channels/sensors are required simultaneously?* Fewer channels generally mean lower cost. Some sensors can transmit two values, such as flow and temperature.

DEFINITION

The **ingress protection (IP)** rating of a component provides an indication of how resistant the component is to environmental dust and moisture. The higher the number the better the protection.

- *What is the fastest sampling rate required?* Most data loggers will sample at 1 ms, some sample at 0.1 ms. The sampling rate may have an impact on the amount of memory required if long tests are carried out at high speed.
- *What is the skill level of the operator?* Consider ease of use from the operator's perspective. A simpler data logger may get used much more than a more complex and expensive model.
- *What is the available budget?* There are lots of solutions available at all price levels. If it is not possible to find a device within budget, consider which of the points above may be flexible. Check that the data logger is not being over-specified for the job in hand.

FURTHER READING

For further information and white papers on how to select flow or pressure test equipment, go to www.webtec.com/education

RESOURCES

CHAPTER

10 TOP TIPS FROM THE EXPERTS

SEVEN COSTLY MISTAKES MOST HYDRAULICS USERS MAKE ... AND HOW YOU CAN AVOID THEM

Brendan Casey is the founder of HydraulicSupermarket.com and the author of *The Hydraulic Maintenance Handbook*, *The Hydraulic Troubleshooting Handbook*, *Insider Secrets to Hydraulics*, *Preventing Hydraulic Failures*, *The Mobile Hydraulics Handbook*, *Hydraulics Made Easy*, *Advanced Hydraulic Control*, *The Definitive Guide to Hydraulic Troubleshooting* and *The Definitive Guide to Mobile Hydraulics*.

A hydraulics specialist with an MBA, Brendan has more than 25 years' experience in the design, maintenance and repair of mobile and industrial hydraulic equipment.

Visit his website: www.HydraulicSupermarket.com

The following seven mistakes are not only common, together they cost hydraulic equipment users around the globe millions every year. The good news is, if you're aware of them, you can avoid them. And if you avoid them, you will save the business you work for a lot of money. It's as simple as that.

Mistake 1 – Changing the oil

There are only two conditions that necessitate a hydraulic oil change – and they are degradation of the base oil and depletion of the additive package. Because there are so many variables that determine the rate at which oil degrades and additives get used up, changing hydraulic oil on hours in service, without any reference to the actual condition of the oil, is like shooting in the dark.

Given the current high price of oil, dumping oil which doesn't need to be changed is the last thing you want to do. And the larger the reservoir the more expensive this mistake. On the other hand, if you continue to operate a hydraulic machine with the base oil degraded or additives depleted, you compromise the service life of every other component in the hydraulic system. And that's the last thing you want.

As you can see, changing hydraulic oil on a fixed number of hours in service is a bad idea – for all but the smallest of hydraulic systems. And the only way to know when the oil *does* need to be changed is through oil analysis.

Oh, and contamination by particles or water doesn't mandate an oil change either. Both of these contaminants can usually be economically removed using offline filtration.

So to avoid mistake number 1, don't change hydraulic oil on hours in service. Change the oil when its additives are used up or the base oil is shot. And the only way to know when this point is reached is to do regular oil analysis.

Mistake 2 – Changing the filters

A similar situation applies to hydraulic filters. If you change them on hours, you're either changing them too early or too late. If you change them early – before all their dirt-holding capacity is used up, you're wasting money on unnecessary filter changes. If you change them late, after the filter has gone on bypass, the increase in particles in the oil quietly reduces the service life of every other component in the hydraulic system, costing a lot more in the long run.

The solution is to change your filters when all their dirt-holding capacity is used up but before the bypass valve opens. This requires a mechanism to monitor the restriction to flow (pressure drop) across the filter element and alert you when this point is reached. A clogging indicator is the crudest form of this device, and although better than nothing, continuous monitoring of pressure drop across the filter using a differential pressure gauge or transducer is the better solution.

So to avoid mistake number 2, don't change hydraulic filters on hours in service. Change the filters when all their dirt-holding capacity is used up, but *before* they go on bypass. And the only way to know when this point is reached is to monitor the pressure drop across the filter element using a clogging indicator (that works!), or better still, a differential pressure gauge.

Mistake 3 – Running too hot

There are not too many equipment owners or operators out there who would continue to operate an engine that was overheating. Unfortunately, the same can't be said when the hydraulic system gets too hot. But, like an engine, the fastest way to destroy hydraulic components, seals, hoses and the oil itself is high-temperature operation.

But how hot is too hot for a hydraulic system? Well, it mainly depends on the viscosity and viscosity index (rate of change in viscosity with temperature) of the oil used, and the type of hydraulic components in the system.

As the oil's temperature increases, its viscosity decreases. And so a hydraulic system is operating too hot when it reaches the temperature at which oil viscosity falls below that required for adequate lubrication.

A vane pump requires a higher minimum viscosity than a piston pump, for example. And this is why the type of hydraulic components used also influences the system's safe maximum operating temperature.

If your hydraulic system contains a vane pump, the minimum viscosity you should be looking to maintain is 25 centistokes (cSt or mm^2/s). For mineral oils with a viscosity index of around 100, this equates to a maximum allowable operating temperature of 35°C (95°F) if you're using ISO VG22 oil or 65°C (150°F) for ISO VG68.

Apart from the lubrication issue, the importance of which can't be overstated, operating temperatures above 82°C (180°F) damage most seal and hose compounds and accelerate degradation of the oil. But, for the reasons explained above, a hydraulic system can be running too hot well below this temperature.

So to avoid mistake number 3, do the exercise at the end of mistake number 4, below, and never let your hydraulic equipment operate above 82°C (180°F) or the temperature at which viscosity falls to 10 cSt, whichever is the lower of the two (you'll see what I mean in a moment).

Mistake 4 – Using the wrong oil

The oil is *the* most important component of any hydraulic system. Not only is hydraulic oil a lubricant, it's also the means by which power is transferred throughout the hydraulic system. It's this dual role which makes viscosity the most important property of the oil – because it affects both machine performance and service life.

Expanding on what I said about mistake number 3 (running too hot), the viscosity of the oil largely determines the maximum and minimum oil temperatures within which the hydraulic system can safely operate. This is sometimes referred to as the temperature operating window (TOW).

If you use oil with a viscosity that's too high for the climate the machine has to operate in, the oil won't flow properly or lubricate adequately during cold start. If you use oil with a viscosity that's too low for the prevailing climate, it won't maintain the required minimum viscosity, and therefore adequate lubrication, on the hottest days of the year.

But that's not the end of it. Within the allowable extremes of viscosity required for adequate lubrication, there's a narrower viscosity band where power losses are minimised. If the operating oil viscosity is higher than ideal, more power is lost to fluid friction. If the operating viscosity is lower than ideal, more power is lost to mechanical friction and internal leakage.

So using the wrong viscosity oil not only results in lubrication damage and premature failure of major components, it also increases power consumption (diesel or electricity) – two things you don't want.

And despite what you might think, you won't always get this right by blindly following the machine manufacturer's oil recommendation. It's not the machine manufacturer who needs to sweat about this though because they aren't the one who's going to pay dearly if the oil selection is wrong.

The only way to be certain is to check your machine's *actual* TOW lies within the *allowable* TOW and ideally within the *optimum* TOW for the oil you're using.

So to avoid mistake number 4 (and number 3), you must define the TOW for the viscosity grade and the viscosity index of the hydraulic oil you're currently using. And then make sure your machine operates within it at all times! The procedure for doing this is a bit involved and I don't have the space to explain it here. But it is explained in detail in Chapter 5 of my book *The Hydraulic Maintenance Handbook*.

Mistake 5 – Wrong filter locations

Any filter is a good filter, right? Wrong! There are two hydraulic filter locations that do more harm than good, and can rapidly destroy the very components they were installed to protect. These to-be-avoided filter locations are the pump inlet, and the piston pump and motor case drain lines.

At this point, it wouldn't surprise me if you're shaking your head in disagreement. After all, this flies in the face of conventional wisdom doesn't it? That you have to have a strainer on the pump inlet to protect it from 'trash'. Well, firstly, the pump draws its oil from a dedicated reservoir not a rubbish bin. Secondly, if you believe it's normal or acceptable for trash to get into the hydraulic tank, then you're probably wasting your time reading this book.

If getting maximum pump life is the primary concern here, and it should be, then it's far more important for the oil to freely and completely fill the pumping chambers during every intake than it is to protect the pump from nuts, bolts and 9/16 in combination spanners, which pose no danger in a properly designed reservoir where the pump inlet penetration is a least 5 cm (2 in) off the bottom.

Research has shown that a restricted intake can reduce the service life of a gear pump by 56%. And it's worse for vane and piston pumps because these designs are less able to withstand the vacuum-induced forces caused by a restricted intake. Hydraulic pumps are not designed to 'suck'.

A different set of problems arise from filters installed on the drain lines of piston pumps and motors, but the result is the same as suction strainers – they can reduce service life and cause catastrophic failures in these expensive components. If these filters are fitted to any of your hydraulic machines and you don't get rid of them, there's a very good chance they'll end up costing you serious money.

And if you're still not convinced or are nervous about discarding a filter the machine manufacturer thought was wise to install in the first place, there's no need to take my word for it. Ask the pump or motor manufacturer. And if you do manage to find a hydraulic pump or motor manufacturer who recommends the use of suction strainers and/or conventional depth filters on case drain lines, I'd like to hear about it.

So to avoid mistake number 5, check each of your hydraulic machines, and if there's a suction strainer on the pump inlet, or a *depth* filter on any piston pump or motor case drain line, remove and discard them – or at the very least, get a second opinion from the pump or machine manufacturer.

Mistake 6 – Believing hydraulic components are self-priming and self-lubricating

You wouldn't start an engine with no oil in the sump – not knowingly anyway. And yet I've seen what amounts to the same thing happen to a *lot* of pricey hydraulic components. Fact is, if the right steps aren't followed during initial start-up, hydraulic components can be seriously damaged. In some cases they may work OK for a while, but the harm done at start-up dooms them to premature failure. You'd be amazed at the number of these types of failures which wrongly end up as warranty claims by hydraulic equipment owners. And it's frustrating for everyone concerned because they're totally preventable.

There's two parts to getting this right: knowing what to do; and remembering to do it. If you don't know what to do, that's one thing. However, if you do know but forget to do it, well that's soul destroying. You can't pat yourself on the back for filling the pump housing with clean oil when you forgot to open the intake isolation valve before starting the engine!

This sort of mistake is easily prevented by using a start-up procedure and checklist. I don't know about you, but these days I don't like relying too much on memory, not for important stuff anyway. So even after more than 25 years of working on hydraulic equipment, I would never attempt to commission or re-commission a hydraulic system without having a piece of paper to remind me of what I need to do and the order I need to do it in. This simple technique eliminates all possibility of error.

So to avoid the consequences of mistake number 6, never attempt (or allow anyone else to attempt) to restart a hydraulic machine after changing components without a written checklist which tells you exactly what to do, and the order in which to do it. And I can't tell you precisely what that is here. Because, to be effective, it must be machine specific. The pre-flight check list for a Boeing 767 is no use at all to a pilot flying an Airbus A380!

Mistake 7 – Assuming the machine manufacturer knows best

This can be a fatal mistake. After all, if the machine manufacturer always knew best, and assuming you're following the machine's service manual to the letter, you wouldn't be making any of the above mistakes, would you?

Bonus Mistake 8 – Not getting an education in hydraulics

As I hope the preceding discussion have shown, if you own, operate, repair or maintain hydraulic equipment, and you aren't 'clued up' on hydraulics, a lot of money – yours or somebody else's – can slip through your fingers. It also means that reading this book is not the end. It's the beginning of something. The beginning, I hope, of your commitment to an ongoing education in hydraulics. Make this your mission.

A FAULT IS ALWAYS A DEVIATION FROM THE NORM

John Savage is Director of the National Fluid Power Centre in the UK and has been Chairman of the British Fluid Power Association's Education and Training Committee since 1992. His involvement with hydraulics started in 1962 as a mining craft apprentice with the National Coal Board. He worked underground for some 12 years on a range of equipment involving hydraulic systems working in a very harsh environment. This experience set the foundation for his progress from apprentice to fully qualified mining mechanical engineer.

To this day he remains passionate and enthusiastic about hydraulics and the development of knowledge and skills in the UK workforce. In 2017 he was awarded the Institution of Mechanical Engineers' prestigious Bramah Medal for his outstanding services to the fluid power industry.

1. Your greatest guarantee of success when carrying out fault diagnosis, establishing the root cause and preventing a reoccurrence is your knowledge of the system on which the fault has occurred, including:

- the ability to read and interpret the circuit diagram(s)
- knowing the operation of the system (random or sequential)
- recognising specific abnormal symptoms and associated causes
- identifying the nature of the fault, based upon the norm
- knowing the function, operation and control of the individual components and their operating relationship.

2. Fault finding on site can be extremely stressful. Develop a list of key questions to be asked:
   - before arriving on site
   - when you have arrived on site.

   The answers to your specific questions may well save your time, reduce downtime and enable you to focus on the problem much quicker. You may also be able to have the site prepared and made safe before you arrive and have supporting equipment waiting for you to assist in your diagnosis.
3. A fault is always a deviation from the norm. Knowing how a system performs when working correctly enables you to analyse a problem with greater accuracy compared with the direct opposite (hit and miss – 'I don't understand this system but I will have a go'.) Fault finding consists of logical step-by-step thinking, involving an extensive range of knowledge and skills, and correct and effective training is necessary to achieve safe and effective results.

IT IS OFTEN DIFFICULT TO PROVE THAT SYSTEM FAILURES ARE CAUSED BY CONTAMINATION

Adrian Wright is a Chartered Engineer and a member of the Institution of Mechanical Engineers. He has 30 years of experience of fluid power systems.

He started his career as an apprentice working on large hydraulic forging presses, before moving on to become a service engineer and then design engineer. Working on all types of fluid power systems, including industrial, marine, off-highway and aerospace systems. He has many years of experience of systems design and component design, including pumps, valves and cylinders.

Adrian graduated from Birmingham City University in 2012 with a first-class honours degree in mechanical engineering. He was formerly Engineering Manager at Caterpillar, where he had responsibility for the hydraulic systems on Caterpillar's back hoe loaders and compact wheel loaders.

Safety

We take time and care to make sure that the systems we design are quiet, safe, efficient and reliable. The negative side of this effort – and it's the same across engineering – is that some people tend to become unaware or complacent of the danger contained within. Hoses, heat, power, energy and accumulators form the basis of this aspect of the discussion.

Contamination control

The importance of contamination control is probably the most difficult subject of fluid power to convince people of. This is no doubt because, as with most things, what the eye doesn't see, the heart doesn't grieve over. It is often difficult to prove that system failures are caused by contamination, and people don't always seem to understand the importance of filtration. That's been true of pretty much everyone I have ever tried to talk to about contamination control. That is apart from the people who work within the business and do not doubt the damage that all forms of contamination can do to a system.

I believe that what we see here is reflection of the wider issues within engineering in general. Indeed, even professional engineering institutions do not always give fluid power the attention and positive support that it deserves. This is especially true if we consider the vast spread of applications in which fluid power is to be found.

On the whole, it is my opinion that fluid power as an engineering solution has suffered because of loose or vague anecdotal advice that has been passed down from person to person over the years. That is not to say that everyone gives out bad advice, it is more to say that a tank can be filled with water and a hydraulic system will run for at least a few minutes before things start to go wrong. It is these few minutes of running that tend to give the false impression that all is well. Bad designs and bad practice are often forgiven within a fluid power system, and it often takes time for problems to emerge. Hot, leaky and inefficient systems give a very bad impression of fluid power.

All of the above is a summary of my 30 years in fluid power. I've made the mistakes that I mentioned above, I've seen spectacular failures occur and jobs lost because of folks that 'have a go' at fluid power design, in the naive belief that it's 'just plumbing isn't it?'

General tips on troubleshooting

When encountering a system fault, assume nothing and suspect everything. Common sense is a very valuable tool in these circumstances, and based upon the symptoms, one can begin to eliminate or reject different parts of the system in order to narrow the suspects down to a smaller and more manageable number.

Never be afraid to ask questions. Find out what the machine or system should be like and also how it was performing before the fault occurred.

It is often the case that adjustments to a system will have unintended and perhaps unnoticed consequences. Adjustments made to a cold system can cause issues when the system becomes hot. The same applies when making certain adjustments to a hot system. Changes in the density and viscosity of the fluid can change the system characteristics in many ways.

It is safe to assume that nothing is right. There are no guarantees in this game. Design errors can go unnoticed for many months or years. Do not assume that the designer knew what he or she was doing. If there are strong suspicions that the design is wrong, do not be afraid to speak up – just be sure to do in a respectful manner.

Don't rush things – Don't take risks – If in doubt, ask

There is an old Chinese proverb which says:

> *He that does not know, but asks, is a fool for five minutes.*
> *He who does not know, but does not ask, he is a fool forever.*

A fluid power system will run on water, at least for a short time. The basic principles of fluid dynamics always apply. It is good engineering that keeps things moving. If a motor or engine turns a pump, the pump will try to move the fluid. The fluid will move in accordance with the laws of physics and fluid dynamics. When working on fluid power systems, use your ears and eyes to survey the system. Look for signs of problems. Ask yourself if there can be a good explanation of what you can see or hear. When it is *safe* to do so, put your hands on pipes and hoses to feel them. Excessive heat or vibration is a sign of a problem.

ALWAYS ENSURE THAT YOU KNOW WHAT YOU ARE DOING BEFORE YOU DO IT

Steve Skinner retired from full-time employment with the Eaton Corporation in 2013 after a 46-year career in the hydraulics industry. Steve spent much of that time preparing and presenting training courses on hydraulic equipment and systems to audiences around the world. He is now an associate lecturer at the National Fluid Power Centre in the UK.

As with any diagnosis process, fault finding in a hydraulic system should be a logical, step-by-step procedure where all of the possible causes of a fault are considered before narrowing the list down. In the pressure of a breakdown situation it is all too easy to jump to false assumptions just to appear 'to be doing something'. Most failures are caused by relatively simple faults so always look for the obvious first. It's the principle of Occam's razor – the simplest explanation is usually the right one.

Before the diagnosis process can start, it is vital to have a thorough understanding of both how the hydraulic system and the machine it operates are designed to function. Until it is known how something works, it's not really possible to think about what could go wrong with it. It should be remembered, however, that the way things have been designed to function is not always the way they have been functioning!

The next important step in the troubleshooting process is to gather all the facts relating to the breakdown. How did it occur? What was being done differently at the time? Were there any abnormal symptoms prior to the breakdown? Only then should the examination process start, by checking each of the suspect areas in turn, looking for the obvious symptoms of non-conformance.

The following is a list of recommendations based on the author's many years of experience in the hydraulics industry.

1. If a system is fitted with a pressure-compensated variable pump and the system is overheating, check that the pump relief valve setting is higher than the pump compensator setting. There's an old saying that if something is adjustable then sooner or later someone will adjust it, often for no good reason. If the relief valve is not set at *least* 20 bar higher than the pump

compensator setting, then it is quite likely to pass flow and create heat. Relief valves are fitted for protection, not to pass flow continuously and create heat.

2. Pay particular attention to pump case drain lines, especially variable-displacement pumps. Make sure the case is full of clean fluid before the pump is started and ensure that the case drain line is as large and free flowing as possible. Avoid check valves in pump case drain lines or anything else that tends to increase the pump case pressure. As well as causing shaft seal leakage, high case pressures can cause piston shoes to lift off the surface of the swashplate, causing rapid failure of the pump. Ensure that the case drain line terminates below the fluid level in the reservoir to avoid foaming of the fluid.
3. If inlet filters on a pump are unavoidable, then ensure that they are readily accessible (mounted external to the reservoir), and fitted with bypass valves and indicators. They should also only normally be used with positive head reservoir layouts.
4. Trapped air in a system can create strange and often intermittent faults to occur, so always ensure that systems are bled correctly, especially after carrying out maintenance work. Even a simple task such as changing a filter element can introduce air into a system.
5. Modern electronic components (such as proportional valves) require a reliable power supply. If the power supply voltage is too low (or too high) or it contains too much residual ripple (from a rectified AC supply), then electronic controls can be very unpredictable. Always check the power supply voltage at maximum load (i.e. with all components drawing maximum current). It is often the case that the power supply voltage appears correct when nothing is operating, but as soon as valves, etc., start to draw a current the voltage drops.
6. Erratic or unpredictable behaviour of electronic components is often caused by inadequate shielding of signal cables. Be sure to use the correct cables and ensure that they are properly shielded (including connectors).
7. Treat accumulators with a great deal of respect. Check them regularly for gas pre-charge pressure, ensure they are drained of fluid before carrying out work on the system, remember to close off the drain valves before re-starting, and check and comply with all relevant legislation regarding their use and inspection.
8. Carry out a regular inspection of flexible hoses to check for damage, entrapment, chafing, etc., and replace/re-route where necessary.
9. Remember that prevention is always better than cure. If a pump can be replaced before it fails, this will prevent considerable amounts of debris entering the system. If a hose can be replaced before it fails, this will prevent a large clean-up bill. Preventing water and contamination entering a system will be many times less costly than the effort involved in removing them.
10. Always ensure that you know what you are doing before you do it. If you don't understand hydraulic systems, leave them alone and find someone who does. The person most responsible for your safety is you.

The following is based on an article by the author which appeared in the *Fluid Power Journal* (Volume 21, Issue 2).

Perfect reliability

Is it possible, in the 21st century, to build a hydraulic system that doesn't break down? Mechanical devices will eventually wear out, of course, so by 'break down' we really mean 'fail unexpectedly'. To help us achieve an aim of perfect reliability we now have systematic tools available such as proactive maintenance procedures and design for six sigma (DFSS). The objective of DFSS is to design systems that have a target reliability level of 99.99966%, which equates to no more than 3.4 failures in every million opportunities. Proactive maintenance combines many of the techniques of preventive and predictive maintenance into a process also designed to achieve similar levels of reliability. The essence of both tools is to attempt to think of all possible failure mechanisms and then to prevent them happening, either by design or maintenance.

However, safety experts tell us that 80–85% of industrial accidents are caused by human error, so is perfect reliability already a lost cause? Fortunately, human beings are very predictable animals, so it's a fair bet that at some time during the life of a hydraulic system someone will try to start it up with no oil in it. It's a 50:50 chance that when the electricians first wire up the electric motor, the motor (and pump) will run backwards. If there's a shut-off valve on the inlet or drain line of the pump, someone's inevitably going to start the pump up with one or both of the valves closed. If something is adjustable then, as sure as eggs is eggs, at some time or another someone will adjust it. If there is an accumulator on the system, sooner or later someone will forget to drain it down before starting maintenance work on the system. And one thing that's 100% certain is that if there's a pressure compensated pump on the system with a relief valve to protect it, one day the compensator will be wound up higher than the relief valve, and the oil's going to boil.

Anyone who's worked with hydraulics for any length of time could probably come up with a whole page full of such instances. It's not that maintenance people are fools, it's just that sometimes they get tired and lose concentration, sometimes their mind is elsewhere and sometimes they forget things. All perfectly normal human weaknesses. Sometimes they don't really know enough about the job they're doing, so they shouldn't really be doing it ... but, hey, 'someone's got to do it'. So there are all sorts of reasons why people sometimes do stupid things; I've done enough myself to know.

Engineers therefore need to think about all these things that might (or rather, will) go wrong and try to design them out. Automatic drain valves for accumulators, interlock switches on shut-off valves, float switches in tanks, thermal cut-off switches, lockable adjusters, etc. It all sounds very expensive, but probably not as expensive as the first breakdown, if anyone ever stops to reckon up its true cost. Not only are we talking about lost production, consequential repairs, premium labour rates and shipping costs, clear-up costs, etc., but we may also be talking about people's well-being and even their lives.

To illustrate the point, in 1886 it was decided that a new bridge was required across the River Thames in London, but being downstream of what was then the biggest port in the world, it had to allow tall-masted ships to pass freely beneath it. The result was Tower Bridge, one of the best-known landmarks of the City of London, with its two opening sections or 'bascules'. The problem was, each 162 ft long bascule weighed around 1500 tons, and for the bridge to open each had to tilt through almost 90° in just over a minute, then close again once the ship had passed through.

To begin with, this would happen more than 20 times a day. The solution – eight 20 gal/rev hydraulic motors (with built-in brakes) operating via a curved rack and pinion arrangement. But, being aware of the consequences of a failure of the bridge, the Victorian engineers built in numerous parallel systems and devices, and no doubt when the bridge opened in June 1894 they were confident that they had covered every eventuality. Unfortunately, they were wrong, and one sunny afternoon in July the bridge suffered a mechanical failure and failed to close. However, this was in July 1968, and in the interim 74-year period the bridge had hydraulically opened and closed its 1500 ton bascules 352,713 times.

So, is it possible, in the 21st century, to build a hydraulic system that doesn't break down? Maybe not, but it should be possible to get pretty close if they could achieve 99.99972% reliability in the 19th century.

BEWARE OF MAN-MADE FAULTS – THEY ARE HARD TO DETECT

Ian McKenna joined Bachy Soletanche in 1988. He started in the workshops, then worked as a mobile fitter for many years before returning to the workshops in 1999 to take on the role of workshop supervisor. In 2008 he took on his current position of Plant Technical Support Supervisor for the fleet of UK equipment. In 2016 he joined the British Fluid Power Association's Education and Training Committee.

Hydraulics are widely used throughout modern-day industry, and construction is considered one of the big users. With an internal combustion engine as the prime mover, it makes mobile hydraulic plant a versatile and valuable tool. However, mobile plant failures (burst hoses, failed valves, etc.) result in repairs being carried out in less than desirable conditions, often with tight time constraints to get the job done as quickly as possible. A broken-down machine means lost production and crews standing idle, and because of this it's often hard for the fitter attending the breakdown to follow 'best practices'.

Equipment designers and manufacturers should be mindful of these potential conditions, and install equipment that makes it easier for certain 'critical' aspects of the job to be done correctly.

For example, most on-site hydraulic repairs result in new oil having to be added. The oil filling should be done via a dedicated pump and hose assembly installed on the machine that transfers new oil from a drum through the filter before it enters the reservoir. That way there is no need to remove the reservoir filler cap and filter to pour the oil straight in.

If a machine manufacturer values your business they'll work with you at the negotiation stages to help give you what you specify – don't settle for anything less.

Be methodical and keep everything clean throughout the repair

Condition yourself to be methodical in your approach to all repairs. It may only be a hose-replacement task, but photograph the area before you start (so you know how it was before and how it should look on completion). If a hose has failed due to rubbing against something, re-route it to prevent a similar failure. If you have to remove or strip a component, photograph it from all angles before removal and during disassembly so you know how it goes back together – believe me, it's time well spent.

It's far better to clean an area prior to removing any component than to spend time trying to fish out debris later that could contaminate a system and cause a malfunction.

Don't take the operator's word as being correct, question everything, assume nothing, build up the full picture before 'jumping in'

In the construction industry we have a lot of mobile plant with operators who can sometimes be a little bit lean with the truth when a problem has occurred.

The operator may be a good work mate and fun to be with outside of work, but they may be covering up a mistake they've made. So before you start making adjustments, alterations or component replacement, 'witness the reported problem for yourself', providing it's safe to do so.

Have the correct hydraulic schematic to hand and use it

Always have the correct hydraulic schematic for the machine to hand. Spend time familiarising yourself with the schematic. Identify the components that are in the circuit that is giving the problem and understand each component's function in the circuit. It can save hours and make the difference between solving a problem or not.

Beware of man-made faults – man-made faults are hard to detect

If someone has made an adjustment or replaced a component without informing you (or recording it in the machine documentation), you could be chasing a problem for days before it comes to light.

Our company had two identical rigs on the same job: one was working and the other was a standby rig to keep production going in the event of a major breakdown. When the time came that the standby rig was needed, it would not function correctly. So the rig was returned to the workshop with a request for a refund for the hire and transport charges. After a lengthy investigation in the workshop, using flow meters and pressure gauges, the fault was narrowed down to a particular part of the circuit. The schematic showed a 1.2 mm restrictor in the circuit, but on the problematic machine this restrictor was missing. It had been replaced with a fitting that was similar but full bore.

It turned out that at some point the working rig had been damaged and a hose needed to be replaced on the drill head. As per standard company practice, a fitter was called out from a local hose-repair company. The fitter found that, in addition to the damaged hose, there was also damage to a fitting between the hose and a valve on the drill head, and this also needed to be replaced. However, the hose-repair fitter did not have the correct fitting. A site worker suggested that the same fitting be removed from the standby rig and used for the repair, in order to get the working rig back in production. The hose fitter could then return at a later time with a replacement fitting for the standby rig. What wasn't noticed was the restrictor built into the fitting, so when the new fitting (full bore, no restriction) was fitted in the standby rig it would not build up sufficient back pressure to allow the drill head to function correctly.

When the site team was questioned about this, a reference was found in the site diary to a breakdown and to the hose-repair company attending the site and returning the following day to replace a fitting on the standby rig. A new 'correct' fitting was installed in the rig and the drill head then worked correctly.

Check the obvious, observe, be patient, ask questions and educate

Man-made intermittent faults can be even harder to detect.

We had a drilling rig on site that would occasionally overheat on the hydraulics. When it overheated, the machine operation would slow down and production was affected. The operator was fully familiar with the machine and had driven it many times before.

Initial checks carried out showed that the oil level in the reservoir was correct, the oil cooler radiator fins were clear, the cooler-fan speed was correct to the machine information we had, the temperature drop across the radiator was sufficient and the pilot pressure was correct – in fact the machine was working fine.

So I sat in the van and watched the rig working satisfactorily, and waited for the problem to manifest itself, to see what changed. The typical UK autumn weather meant that there was occasionally a spell of rain. Shortly after a sudden downpour, the banksman took his jacket off and hung it over the oil cooler grill to dry off. The fan sucked the jacket against the grill, restricting the air flow and thus reducing the efficiency of the oil cooler. Shortly afterwards the rig started to overheat. On speaking with the banksman, he confessed that he had done this several times, but neither he nor the operator had related it to the machine-overheating problem. After explaining to them both what was happening, they took it on board and the overheating problem went away. Education is a great tool.

FURTHER READING

For details of other useful sources of information on hydraulic maintenance and troubleshooting, go to www.webtec.com/education

CHAPTER

11 HYDRAULIC CALCULATIONS

CYLINDERS

Normally the first requirement for cylinders is to determine the full-bore and annulus areas of the piston (A_1 and A_2) (Fig. 11.1). Assuming the cylinder is a single-rod type, the full-bore piston area can be determined simply from the bore diameter of the cylinder by using the formula for the area of a circle:

$$A = \frac{\pi \times D^2}{4}$$

which is approximately equal to $0.8 \times D^2$. To determine the annulus area (A_2) it is necessary to subtract the rod area from the full piston area. For double-rod cylinders it will be necessary to do this on both sides.

If it is necessary to know the cylinder volume, this can now be obtained by multiplying the piston area by the stroke. The maximum volume on the full-bore side (V_1) will therefore be equal to the full piston area (A_1) multiplied by the maximum piston stroke (S):

$$V_1 = A_1 \times S$$

Note that in the metric system of units, cylinder bore and stroke dimensions are normally quoted in millimetres (mm). Using these units for volume calculations can therefore result in very large numbers, for example a 200 mm bore, 1000 mm stroke cylinder would have a maximum full-bore volume (when fully extended) of 31,416,000 mm^3. Although not generally recognised as an engineering unit of measurement, the centimetre does have advantages for hydraulic engineers, for the following reasons:

- Many hydraulic calculations involve volumes or volume flow rates. The normal unit of volume used in hydraulic systems is the litre, which is one-thousandth of a cubic metre (m^3). One-thousandth of a litre is a millilitre, which is equal to one cubic centimetre (cm^3). Therefore, the three units of millilitre (cm^3), litre and cubic metre are each separated by a factor of 10^3, and together will cover most requirements of hydraulic system calculations.
- Determining cylinder volumes in cm^3 results in more manageable numbers that can easily be converted to litres simply by dividing by 1000.
- Pump and motor displacements are almost always quoted in terms of cm^3/rev.
- When determining system pressures, dividing the load in kilograms by the area in square centimetres will give the pressure in bar (approximately, within 2%).
- Measurements used outside of the engineering industry are most often quoted in centimetres. School children, for example, use a 30 cm ruler, rather than a 300 mm or 0.3 m ruler.

SI UNITS

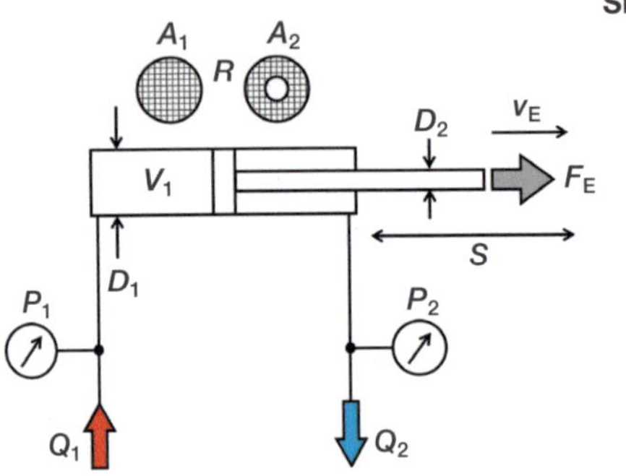

$$A_1 = \frac{\pi \times D_1^2}{4}$$

$$Q_2 = \frac{Q_1}{R}$$

$$A_2 = A_1 - \frac{\pi \times D_2^2}{4}$$

$$v_E = \frac{Q_1 \times 16{,}667}{A_1}$$

$$V_1 = \frac{A_1 \times S}{10^6}$$

$$F_E = \frac{(P_1 \times A_1) - (P_2 \times A_2)}{10}$$

- D_1 Cylinder bore diameter (mm)
- D_2 Cylinder rod diameter (mm)
- A_1 Full-bore area (mm²)
- A_2 Annulus area (mm²)
- R Cylinder area ratio (A_1/A_2)
- S Stroke (mm)
- V_1 Full-bore volume fully extended (L)
- Q_1 Full-bore flow rate (L/min)
- Q_2 Annulus flow rate (L/min)
- v_E Extending velocity (mm/s)
- P_1 Full-bore pressure (bar)
- P_2 Annulus pressure (bar)
- F_E Extending force (N)

IMPERIAL UNITS

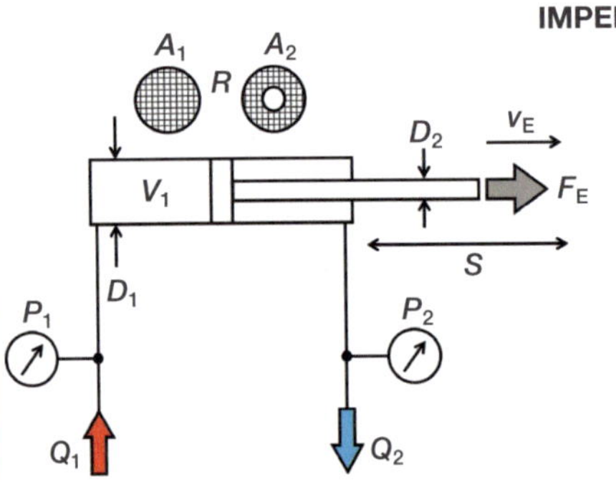

$$A_1 = \frac{\pi \times D_1^2}{4}$$

$$Q_2 = \frac{Q_1}{R}$$

$$A_2 = A_1 - \frac{\pi \times D_2^2}{4}$$

$$v_E = \frac{Q_2 \times 3.85}{A_2}$$

$$V_1 = A_1 \times S$$

$$F_E = (P_1 \times A_1) - (P_2 \times A_2)$$

- D_1 Cylinder bore diameter (in)
- D_2 Cylinder rod diameter (in)
- A_1 Full-bore area (in²)
- A_2 Annulus area (in²)
- R Cylinder area ratio (A_1/A_2)
- S Stroke (in)
- V_1 Full-bore volume fully extended (in³)
- Q_1 Full-bore flow rate (gpm)
- Q_2 Annulus flow rate (gpm)
- v_E Extending velocity (in/s)
- P_1 Full-bore pressure (psi)
- P_2 Annulus pressure (psi)
- F_E Extending force (lbf)

Fig. 11.1 Cylinder extending

Once the flow rate and cylinder volume are known, the piston velocity (in mm/s) can be determined by dividing the volume flow rate by the piston area. However, the units used must be consistent, so the flow rate in L/min must be multiplied by 16,667 to convert it into mm³/s or 16.7 to convert it into cm³/s. This does not take into account any leakage across the piston seals. Leakage rates for cylinders and seals in good condition will normally be very small and are often ignored (and they are rarely published by manufacturers anyway). However, as seals or bore surfaces wear, or where special-purpose seals have been used, leakage rates may be more significant. Flow rates in and out of the cylinder will be in the same ratio (R) as the piston areas.

The most fundamental equation for hydraulic systems is often quoted as

$$\text{Force } (F) = \text{Pressure } (P) \times \text{Area } (A)$$

However, as pressure in a cylinder is created by the resistance to movement of the cylinder piston, the equation should really be expressed as $P = F \div A$. Mathematically the two expressions are the same. When determining the necessary pressure

SI UNITS

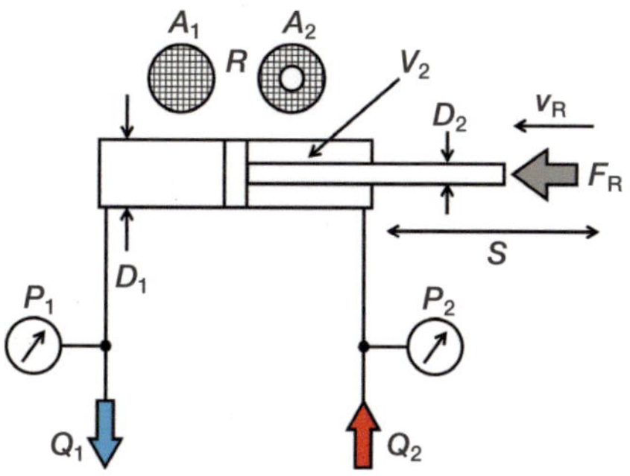

$$A_1 = \frac{\pi \times D_1^2}{4}$$

$$A_2 = A_1 - \frac{\pi \times D_2^2}{4}$$

$$V_2 = \frac{A_2 \times S}{10^6}$$

$$v_R = \frac{Q_2 \times 16{,}667}{A_2}$$

$$Q_1 = Q_2 \times R$$

$$F_R = \frac{(P_2 \times A_2) - (P_1 \times A_1)}{10}$$

D_1	Cylinder bore diameter (mm)
D_2	Cylinder rod diameter (mm)
A_1	Full-bore area (mm²)
A_2	Annulus area (mm²)
R	Cylinder area ratio (A_1/A_2)
S	Stroke (mm)
V_2	Annulus volume fully retracted (L)
Q_1	Full-bore flow rate (L/min)
Q_2	Annulus flow rate (L/min)
v_R	Retracting velocity (mm/s)
P_1	Full-bore pressure (bar)
P_2	Annulus pressure (bar)
F_R	Retracting force (N)

IMPERIAL UNITS

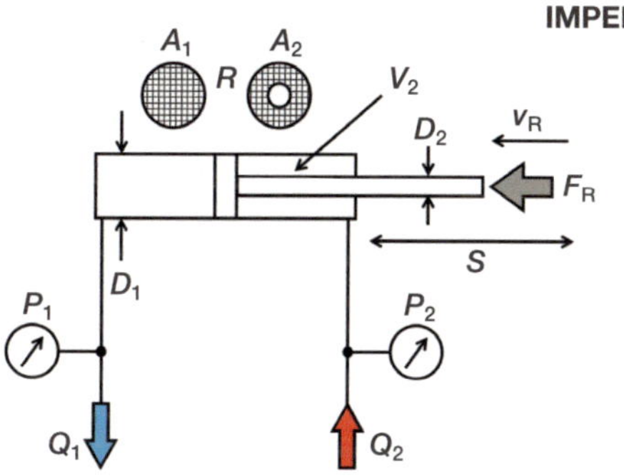

$$A_1 = \frac{\pi \times D_1^2}{4}$$

$$A_2 = A_1 - \frac{\pi \times D_2^2}{4}$$

$$V_2 = A_2 \times S$$

$$v_R = \frac{Q_2 \times 3.85}{A_2}$$

$$Q_1 = Q_2 \times R$$

$$F_R = (P_2 \times A_2) - (P_1 \times A_1)$$

D_1	Cylinder bore diameter (in)
D_2	Cylinder rod diameter (in)
A_1	Full-bore area (in²)
A_2	Annulus area (in²)
R	Cylinder area ratio (A_1/A_2)
S	Stroke (in)
V_2	Annulus volume fully extended (in³)
Q_1	Full-bore flow rate (gpm)
Q_2	Annulus flow rate (gpm)
v_R	Retracting velocity (in/s)
P_1	Full-bore pressure (psi)
P_2	Annulus pressure (psi)
F_R	Retracting force (lbf)

Fig. 11.2 Cylinder retracting

required to extend a cylinder, however, it is necessary to take into account not only the mechanical force to overcome but also:

- The effect of back pressure in the annulus side of the cylinder acting on the annulus area. Such back pressure may be the result of valves in the exhaust line from the cylinder or simply the pressure drop through pipes, hoses, filters, coolers, etc.
- The effect of piston and rod seal friction. Seal friction in cylinders is very difficult to determine accurately because there are several factors which affect it, such as:
  - seal material and construction
  - fluid pressure
  - direction of travel (extending or retracting)
  - seal condition.

In practice, therefore, and in the absence of any measurable data, an assumption is usually made that 2–5% of the theoretical piston force is used to overcome seal friction.

Multiplying a pressure in megapascals by an area in square millimetres gives a force in newtons, so if the more usual unit of bar is used for pressure, this must be divided by 10 to make the units consistent (1 MPa = 10 bar).

Similar formulae are used for the retracting stroke of the cylinder, as can be seen from Fig. 11.2. Two things need to be remembered during the retracting stroke of a cylinder:

- The effect of back pressure in the full-bore side of the cylinder when retracting will now be more significant as it now acts over a larger area.
- The exhaust flow from the full-bore side of the cylinder will now be greater than the inlet flow to the annulus side. If the area ratio of the piston is large and full pump flow is directed to the annulus side, the flow in the system return line will be significantly greater than the pump outlet flow. This means that all components in the return line have to be sized for this increased flow rate.

When cylinders are used regeneratively, as shown in Fig. 11.3, both sides of the cylinder piston are connected to the pressure line simultaneously.

SI UNITS

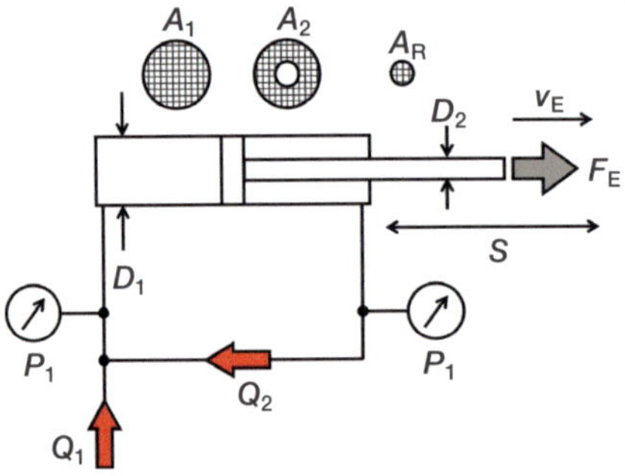

D_1	Cylinder bore diameter (mm)
D_2	Cylinder rod diameter (mm)
A_1	Full-bore area (mm²)
A_2	Annulus area (mm²)
A_R	Piston rod area (mm²)
R	Cylinder area ratio (A_1/A_2)
S	Stroke (mm)
Q_1	Inlet (pump) flow rate (L/min)
Q_2	Annulus flow rate (L/min)
Q_P	Pump flow (L/min)
v_E	Extending velocity (mm/s)
P_1	Supply pressure (bar)
F_E	Extending force (N)

$$A_1 = \frac{\pi \times D_1^2}{4} \qquad Q_1 = \frac{R \times Q_P}{R - 1} \qquad v_E = \frac{Q_P \times 16{,}667}{A_R}$$

$$A_R = \frac{\pi \times D_2^2}{4} \qquad Q_2 = \frac{Q_P}{R - 1} \qquad F_E = \frac{P_1 \times A_R}{10}$$

$$A_2 = A_1 - A_R$$

IMPERIAL UNITS

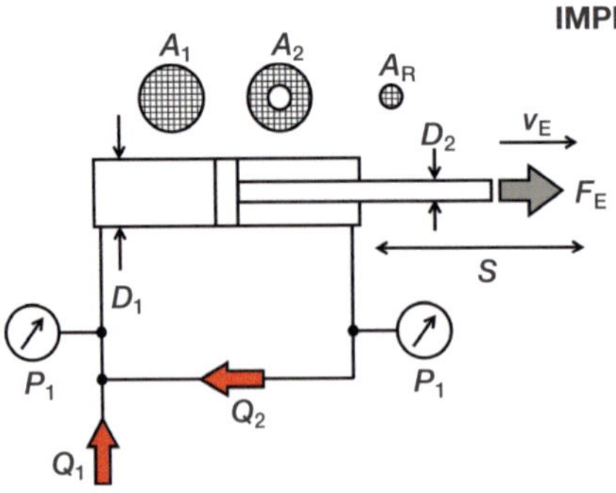

D_1	Cylinder bore diameter (in)
D_2	Cylinder rod diameter (in)
A_1	Full-bore area (in²)
A_2	Annulus area (in²)
A_R	Piston rod area (in²)
R	Cylinder area ratio (A_1/A_2)
S	Stroke (in)
Q_1	Inlet (pump) flow rate (gpm)
Q_2	Annulus flow rate (gpm)
Q_P	Pump flow (L/min)
v_E	Extending velocity (in/s)
P_1	Supply pressure (psi)
F_E	Extending force (lbf)

$$A_1 = \frac{\pi \times D_1^2}{4} \qquad Q_1 = \frac{R \times Q_P}{R - 1} \qquad v_E = \frac{Q_1 \times 3.85}{A_R}$$

$$A_R = \frac{\pi \times D_2^2}{4} \qquad Q_2 = \frac{Q_P}{R - 1} \qquad F_E = P_1 \times A_R$$

$$A_2 = A_1 - A_R$$

Fig. 11.3 Cylinder extending regeneratively

At first sight it may appear that the piston will not move, as the pressure is the same on either side of it. However, the areas are not the same on either side ($A_1 > A_2$), so there will still be a net force pushing the piston and rod outwards. The force will now be less than in a conventional arrangement, and will be equal to $(P_1 \times A_1) - (P_1 \times A_2)$ if seal friction is ignored. The difference between A_1 and A_2 is equal to the cross-sectional area of the rod, so the net cylinder force is now effectively equal to $P_1 \times A_R$. In practice, however, the pressure in the annulus side of the cylinder during regeneration will be slightly higher than that in the full-bore side due to the pressure drop caused by valve and pipework restrictions.

The advantage of a regenerative arrangement is that the exhaust flow from the annulus side of the cylinder is added to the pump flow, thereby increasing the speed of piston movement. The pump flow or inlet flow is now simply displacing the volume of the rod from the cylinder.

Circuits are often arranged so that cylinder operation can be switched between regenerative and conventional flow. This then provides a fast cylinder speed for light-load conditions (regenerative flow) but full force/slower speed operation (conventional flow) when required.

MOTORS

The equivalent property to cylinder piston area for a hydraulic motor is known as its 'displacement' (*d*) (Fig. 11.4). It is determined by the volume of fluid required to rotate the shaft one revolution. This is a theoretical or geometric property, as there will be leakage internally in the motor, which means that slightly more volume will be required in practice for one turn of the motor shaft.

Whereas cylinders normally use elastomeric seals, which are very efficient and reduce internal leakage to very low levels, hydraulic motors rely mainly on small clearances between moving components to reduce internal leakage. Although small, such clearances have to be large enough to allow components to move relative to each other (pistons, gears, etc.), and allow for variations in expansion as the components heat up and cool down. Internal leakage in motors, therefore, will always be significantly higher than that in cylinders and has to be allowed for in speed and flow calculations by use of the property called 'volumetric efficiency' (η_{VOL}). Volumetric efficiency is a measure of the actual shaft rotation compared with the theoretical rotation for a given volume of fluid, and will vary primarily according to the type and condition of the motor, the fluid viscosity, the motor speed and the operating pressure.

Similarly, when determining the torque output of a motor, the internal friction of the moving parts, together with pressure drops along passages within the motor, have to be taken into account. The mechanical efficiency of the motor (η_{MECH}) is a measure of the actual torque output compared with the theoretical output for a given pressure difference across the motor ports. As with volumetric efficiency, its value will vary with factors such as motor type and condition, fluid properties and operating pressure.

The 20π term in the torque equation (see Fig. 11.5) is a combination of 2π, which is mathematically necessary to convert revolutions into radians, and 10, to make the units of pressure consistent.

SI UNITS

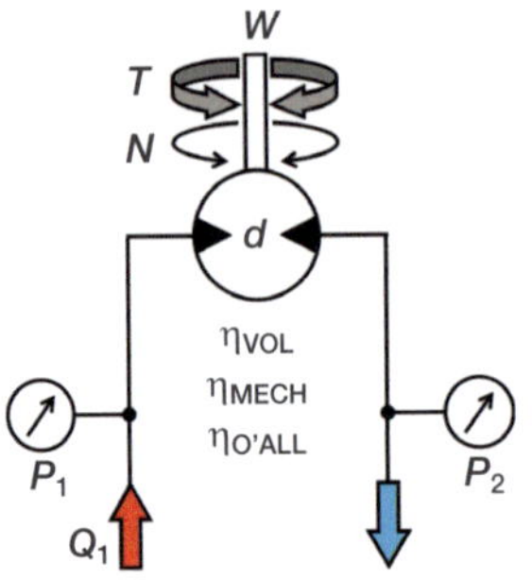

d	Motor displacement (cm³/rev)
η_{VOL}	Volumetric efficiency
η_{MECH}	Mechanical efficiency
$\eta_{O'ALL}$	Overall efficiency
N	Rotational speed (rpm)
Q_1	Inlet flow rate (L/min)
P_1	Inlet pressure (bar)
P_2	Outlet pressure (bar)
T	Shaft torque (N m)
W	Shaft power (kW)

$$N = \frac{Q_1 \times 1000 \times \eta_{VOL}}{d}$$

$$T = \frac{(P_1 - P_2) \times d \times \eta_{MECH}}{20 \times \pi}$$

$$\eta_{O'ALL} = \eta_{VOL} \times \eta_{MECH}$$

$$W = \frac{Q_1 \times (P_1 - P_2) \times \eta_{O'ALL}}{600}$$

IMPERIAL UNITS

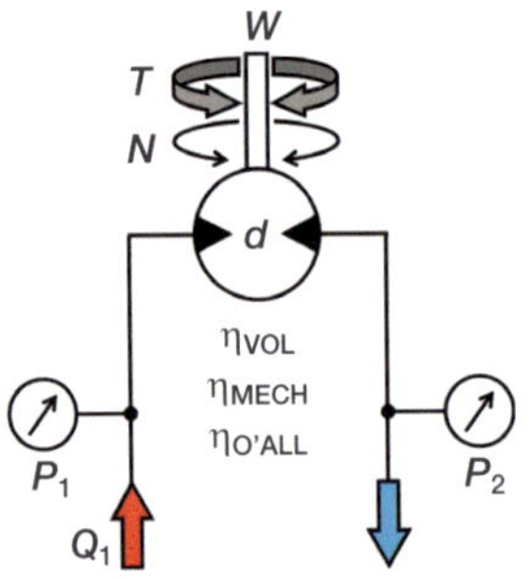

d	Motor displacement (in³/rev)
η_{VOL}	Volumetric efficiency
η_{MECH}	Mechanical efficiency
$\eta_{O'ALL}$	Overall efficiency
N	Rotational speed (rpm)
Q_1	Inlet flow rate (gpm)
P_1	Inlet pressure (psi)
P_2	Outlet pressure (psi)
T	Shaft torque (lbf ft)
W	Shaft power (hp)

$$N = \frac{Q_1 \times 231 \times \eta_{VOL}}{d}$$

$$T = \frac{(P_1 - P_2) \times d \times \eta_{MECH}}{2 \times \pi}$$

$$\eta_{O'ALL} = \eta_{VOL} \times \eta_{MECH}$$

$$W = \frac{Q_1 \times (P_1 - P_2) \times \eta_{O'ALL}}{1714}$$

Fig. 11.4 Hydraulic motor

The overall efficiency of the motor ($\eta_{O'ALL}$), is determined by multiplying the volumetric efficiency by the mechanical efficiency, and must be known in order to determine the actual power output of the motor. Efficiencies are often quoted as percentages, so this number must be divided by 100 for use in formulae (e.g. 90% efficiency must be expressed as 0.9 in the formulae).

PUMPS

The formulae for pumps are very similar to those for motors except for the position of the efficiency terms (Fig. 11.5). It is worth remembering that whether determining the output of a pump or motor, inefficiencies always act to reduce the theoretical output.

For pumps used in open-circuit systems, the inlet pressure (P_1) will be close to atmospheric pressure, so is often ignored in the calculation. Where the pump inlet port is supercharged, however (typically in closed-circuit transmission circuits), it should be taken into account in the relevant formulae.

To determine the power required to drive a pump the theoretical power (output flow multiplied by output pressure) must be divided by the overall pump efficiency.

SI UNITS

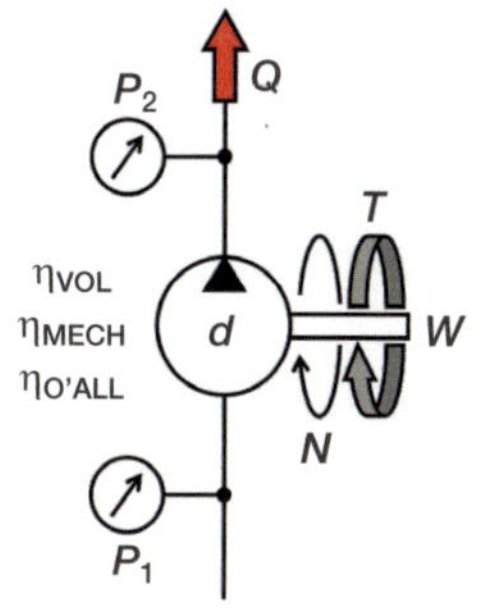

d	Pump displacement (cm³/rev)
η_{VOL}	Volumetric efficiency
η_{MECH}	Mechanical efficiency
$\eta_{O'ALL}$	Overall efficiency
N	Rotational speed (rpm)
Q	Pump flow rate (L/min)
P_1	Pump charge pressure (bar)
P_2	Pump outlet pressure (bar)
T	Shaft torque (N m)
W	Shaft power (kW)

$$Q = \frac{N \times d \times \eta_{VOL}}{1000}$$

$$T = \frac{(P_2 - P_1) \times d}{20 \times \pi \times \eta_{MECH}}$$

$$\eta_{O'ALL} = \eta_{VOL} \times \eta_{MECH}$$

$$W = \frac{Q_1 \times (P_2 - P_1)}{600 \times \eta_{O'ALL}}$$

IMPERIAL UNITS

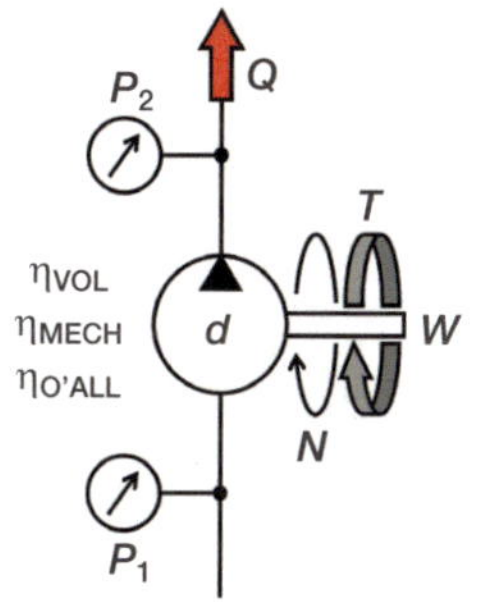

d	Pump displacement (in³/rev)
η_{VOL}	Volumetric efficiency
η_{MECH}	Mechanical efficiency
$\eta_{O'ALL}$	Overall efficiency
N	Rotational speed (rpm)
Q	Pump flow rate (gpm)
P_1	Pump charge pressure (psi)
P_2	Pump outlet pressure (psi)
T	Shaft torque (lbf ft)
W	Shaft power (hp)

$$Q = \frac{N \times d \times \eta_{VOL}}{231}$$

$$T = \frac{(P_2 - P_1) \times d}{20 \times \pi \times \eta_{MECH}}$$

$$\eta_{O'ALL} = \eta_{VOL} \times \eta_{MECH}$$

$$W = \frac{Q_1 \times (P_2 - P_1)}{1714 \times \eta_{O'ALL}}$$

Fig. 11.5 Hydraulic pump

FLOW RESTRICTIONS

Restrictions to flow in a hydraulic system will be created by almost all the components within it (pipes and hoses, fittings, manifold blocks, filters, coolers, etc.). However, control valves, such as throttle valves, pressure-reducing valves and directional valves, will probably be the main cause of pressure losses in the system. Whenever pressure is lost across a component where no work is done (i.e. no mechanical output), heat will be generated. The amount of heat generated will depend on the amount of pressure drop and the flow rate passing through the component. The factor of 600 in the formula (Fig. 11.6) converts the units of pressure and flow into consistent units in order to provide the heat generated in kilowatts.

When a component creates straightforward restriction to the fluid flow (i.e. there are no pressure-compensating devices involved), the pressure drop across the restriction ($P_1 - P_2$) is approximately proportional to the flow rate squared (Q^2). If the restriction is a sharp-edged orifice the relationship is exactly proportional, but as most restrictions also include some frictional losses the relationship is only approximately equal (≈). The value of *k* depends on the shape and area of the restriction and

SI UNITS

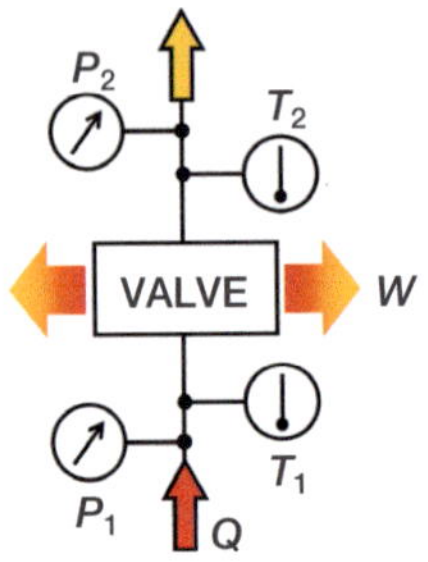

Q	Valve flow rate (L/min)
P_1	Valve inlet pressure (bar)
P_2	Valve outlet pressure (bar)
T_1	Fluid inlet temperature (°C)
T_2	Fluid outlet temperature (°C)
W	Heat generated (kW)
k	A constant dependent on valve characteristics and fluid density

$$W = \frac{(P_1 - P_2) \times Q}{600} \qquad (P_1 - P_2) \approx k \times Q^2 \qquad T_2 - T_1 \approx \frac{P_1 - P_2}{17.5}$$

IMPERIAL UNITS

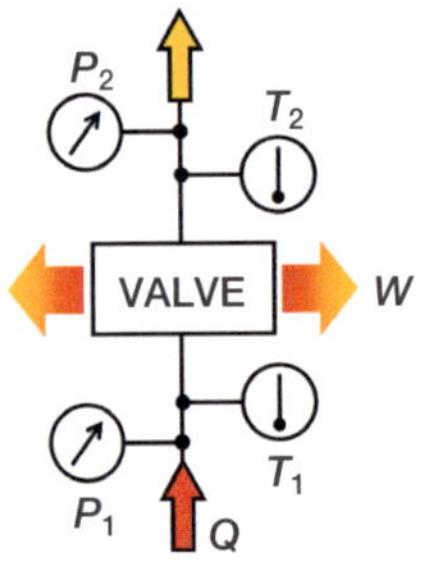

Q	Valve flow rate (gpm)
P_1	Valve inlet pressure (psi)
P_2	Valve outlet pressure (psi)
T_1	Fluid inlet temperature (°F)
T_2	Fluid outlet temperature (°F)
W	Heat generated (hp)
k	A constant dependent on valve characteristics and fluid density

$$W = \frac{(P_1 - P_2) \times Q}{1714} \qquad (P_1 - P_2) \approx k \times Q^2 \qquad T_2 - T_1 \approx \frac{P_1 - P_2}{140}$$

Fig. 11.6 Flow through a restriction

the characteristics of the fluid. Normally, pressure drops through components are determined from data in manufacturers' catalogues, but the formula is useful for estimating pressure drops at flow rates other than those quoted by the manufacturer.

POINT OF INTEREST

The formula for fluid temperature rise is an approximation and assumes that most of the heat generated is transferred to the fluid. In practice, some heat will be dissipated by radiation and conduction also.

As mentioned above, when a pressure drop occurs across a component and no mechanical work is done, heat is generated. This means that the fluid leaving the component will be hotter than the fluid entering. The higher the flow rate the greater will be the heat generated, but the volume of fluid to heat will also be greater. In practice, therefore, the rise in temperature of the fluid depends only on the pressure drop and the type of fluid used (its value of specific heat). The formula provides an approximate value of temperature rise for a typical mineral oil fluid.

EXAMPLE CALCULATIONS

Example 1

Determine the pressure and flow required to extend a hydraulic cylinder based on the following data:

Cylinder bore diameter: 100 mm

Cylinder rod diameter: 56 mm

Cylinder required velocity: 150 mm/s

Output force requirement: 100 kN

Back pressure in cylinder exhaust port: 10 bar

$$A_1 = \frac{\pi \times D_1^2}{4} = \frac{\pi \times 100^2}{4} = 7854\ \text{mm}^2$$

$$A_2 = A_1 - \frac{\pi \times D_2^2}{4} = 7854 - \frac{\pi \times 56^2}{4} = 5390\ \text{mm}^2$$

$$F = \frac{(P_1 \times A_1) - (P_2 \times A_2)}{10}$$

$$100{,}000 = \frac{(P_1 \times 7854) - (10 \times 5390)}{10}$$

$$P_1 = \frac{1{,}000{,}000 + 53{,}900}{7854}$$

$$P_1 = 134\,\text{bar}$$

Adding an additional estimated 3% to account for cylinder seal friction gives a required input pressure of:

$$P_1 = 134 \times 1.03$$

$$\mathbf{P_1 = 138\,bar}$$

$$V_E = \frac{Q_1 \times 16{,}667}{A_1}$$

$$150 = \frac{Q_1 \times 16{,}667}{7854}$$

$$Q_1 = \frac{150 \times 7854}{16{,}667}$$

$$\mathbf{Q_1 = 70.7\,L/min}$$

Example 2

Determine the maximum extending force and velocity if the cylinder is connected for regenerative flow using the same values of supply pressure and flow.

$$A_R = \frac{\pi \times D_2^2}{4} = \frac{\pi \times 56^2}{4} = 2463\ \text{mm}^2$$

$$F = \frac{P_1 \times A_R}{10}$$

$$F = \frac{138 \times 2463}{10}$$

$$F = 34\,\text{kN}$$

Again, assuming 3% of the pressure force is used to overcome seal friction, the net output force would be:

$$F = 34 \times 0.97$$

$$\boldsymbol{F = 33\,\text{kN}}$$

$$V_E = \frac{Q_1 \times 16{,}667}{A_R}$$

$$V_E = \frac{70.7 \times 16{,}667}{2463}$$

$$\boldsymbol{V_E = 478\,\text{mm/s}}$$

Example 3

Determine the pressure and flow required to drive a hydraulic motor based on the following data:

Motor displacement: 100 cm^3/rev

Motor mechanical efficiency: 94%

Motor volumetric efficiency: 92%

Motor outlet port pressure: 5 bar

Required torque output: 200 N m

Required speed output: 850 rpm

$$T = \frac{(P_1 - P_2) \times D \times \eta_{MECH}}{20 \times \pi}$$

$$200 = \frac{(P_1 - 5) \times 100 \times 0.94}{20 \times \pi}$$

$$P_1 = \frac{200 \times 20 \times \pi}{100 \times 0.94} + 5$$

$$\boldsymbol{P_1 = 138.7\,\text{bar}}$$

$$N = \frac{Q_1 \times 1000 \times \eta_{VOL}}{d}$$

$$850 = \frac{Q_1 \times 1000 \times 0.92}{100}$$

$$Q_1 = \frac{850 \times 100}{1000 \times 0.92}$$

$$\boldsymbol{Q_1 = 92.4\,\text{L/min}}$$

Example 4

Based on the following data, determine the minimum displacement of a pump required to deliver a flow of 100 L/min, and the drive power required if the pump outlet pressure is 150 bar

Pump drive speed: 1420 rpm

Pump mechanical efficiency: 95%

Pump volumetric efficiency: 90%

Pump inlet pressure: atmospheric pressure

Required pump flow: 100 L/min

Required outlet pressure: 150 bar

$$Q = \frac{N \times d \times \eta_{\mathrm{VOL}}}{1000}$$

$$100 = \frac{1420 \times d \times 0.9}{1000}$$

$$d = \frac{100 \times 1000}{1420 \times 0.9}$$

$$\boldsymbol{d = 78.2\,\mathrm{cm^3/rev}}$$

$$\eta_{\mathrm{O'ALL}} = \eta_{\mathrm{VOL}} \times \eta_{\mathrm{MECH}}$$

$$\eta_{\mathrm{O'ALL}} = 0.9 \times 0.95 = 0.855$$

$$W = \frac{Q \times P}{600 \times \eta_{\mathrm{O'ALL}}}$$

$$W = \frac{100 \times 150}{600 \times 0.855}$$

$$\boldsymbol{W = 29.2\,\mathrm{kW}}$$

Example 5

A proportional throttle valve has a maximum rated flow of 150 L/min at 10 bar pressure drop. Determine the pressure drop across the valve when passing 100 L/min.

Assume ΔP ($P_1 - P_2$) is approximately proportional to the flow rate squared, then:

$$\Delta P \approx 10 \times \frac{100^2}{150^2}$$

$$\boldsymbol{\Delta P \approx 4.44\,\mathrm{bar}}$$

Example 6

If a relief valve is set to a pressure of 150 bar and is passing a flow of 100 L/min, determine how much heat is being created in the valve and the approximate difference in fluid temperature (ΔT) between inlet and outlet.

$$W = \frac{Q \times P}{600}$$

$$W = \frac{100 \times 150}{600}$$

$$\boldsymbol{W = 25\,\mathrm{kW}}$$

$$\Delta T = \frac{P_1 - P_2}{17.5}$$

$$\Delta T = \frac{150}{17.5}$$

$$\boldsymbol{\Delta T = 8.6^\circ\mathrm{C}}$$

FURTHER READING

For further hydraulic formulae and calculation procedures, go to www.webtec.com/education

CHAPTER

12 FURTHER READING AND OTHER RESOURCES

FLUID POWER BOOKS AND JOURNALS

Fluid Power Journal – official journal of the US International Fluid Power Society: http://fluidpowerjournal.com

Fluid Power World – US-based publication on fluid power: https://www.fluidpowerworld.com

Hydraulics and Pneumatics – US-based publication on fluid power: http://www.hydraulicspneumatics.com

Hydraulics & Pneumatics – UK-based publication on fluid power: http://hpmag.co.uk

Machinery Lubrication – a well-respected publication, written by training organisation Noria, about lubrication and oil analysis to aid machine reliability: http://www.machinerylubrication.com

Hydraulic Supermarket – founded by Brendan Casey and based in Australia. Brendan writes a very popular newsletter, numerous books and runs training seminars followed by fluid power engineers all over the world: https://www.hydraulicsupermarket.com

FLUID POWER TRADE ASSOCIATIONS, RESEARCH CENTRES AND TRAINING SCHOOLS

Trade associations are generally an excellent resource for finding local training schools, information on industry standards and lists of members grouped by products they supply. Many training schools offer fluid power qualifications accredited by their local trade association, ranging from an awareness level right up to complex hydraulic system design and maintenance.

CETOP (Comité Européen des Transmissions Oléohydrauliques et Pneumatiques) – the European Fluid Power Committee. An excellent starting place to find all other fluid-power related trade associations in Europe: https://www.cetop.org

British Fluid Power Association (BFPA) (http://bfpa.co.uk) – publishes a list of British training suppliers: https://bfpa.co.uk/training/suppliers

National Fluid Power Training Centre (NFPC) – the largest training supplier in the UK. It offers a very wide range of short courses on 'integrated systems engineering', including CETOP-certified courses. It assists more than 25 sectors of industry for those involved in the maintenance and management of fluid power systems. It is probably the best-equipped hands-on fluid power training facility in Europe: http://www.nfpc.co.uk

National Fluid Power Association (NFPA) – the principal fluid power association in the USA (http://www.nfpa.com). It publishes a comprehensive list of fluid power training establishments in the USA: http://web.nfpa.com/education/learningresources-trainingprograms.aspx

One centre the author knows well is MSOE Fluid Power Institute, Milwaukee, WI, USA: https://www.msoe.edu/academics/how-we-teach/labs-and-research/engineering/fluid-power-institute

International Fluid Power Society – headquartered in the in USA, the society offers a comprehensive range of certified fluid power courses, including online training courses: http://www.ifps.org

FLUID POWER CERTIFICATION

Many companies are now placing increased emphasis on competency-based training, which involves third-party assessment of both knowledge and practical ability. The International Fluid Power Society (http://www.ifps.org) is the main organisation in North America to offer certification, while CETOP (https://www.cetop.org) offers a similar process in Europe. In the UK, the BFPA (http://bfpa.co.uk) has recently introduced a Training Passport scheme for individuals working in the fluid power industry, which documents their training achievements.

BESPOKE COURSES

Because hydraulic systems are applied to such a wide range of different machines, industries and applications, a user might prefer a bespoke course tailored to their needs. In this case it is worth approaching one of the large training establishments listed above and asking them to develop a machine-specific course.

Alternatively, search for a specialist company via your local trade association. Such companies often design, maintain or build hydraulic machinery for their customers and offer bespoke training as an add-on. An example of this is Bachy Soletanche (http://www.bacsol.co.uk) in the UK, which is a leading geotechnical specialist within the field of foundation and underground engineering. It operates a large fleet of drilling and boring machines made by many different manufacturers. Its training courses reflect the specific knowledge needed to maintain the hydraulic equipment on these machines.

In the USA, companies such as Eaton (https://www.eaton.com) and GPM Hydraulic Consulting Inc. (https://gpmhydraulic.com) supply a range of standard and bespoke training courses, as well as providing troubleshooting and reliability consulting services.

FLUID POWER RESOURCE WEBSITES

Webtec – engineer's reference guide, a host of useful formulae, white papers, and conversion factors related to hydraulics: https://en.webtec.com/education

EATON – useful hydraulic hints and troubleshooting guide:
http://www.eaton.com/ecm/groups/public/@pub/@eaton/@hyd/documents/content/ct_233701.pdf

ENG-TIPS – English-language engineering forum with a section on fluid power: http://www.eng-tips.com/threadminder.cfm?pid=1083

MP Filtri – *Fluid Condition Handbook*, useful guidance on contamination available as an e-book:
https://www.mpfiltri.co.uk/condition-monitoring-handbook-download

Gates – useful guidance on hoses and couplings:
http://www.gatesaustralia.com.au/~/media/files/gates-au/hydraulics/catalogues/safe-hydraulics-manual-us-2009.pdf

Insane Hydraulics – forum on hydraulic circuit design:
http://www.insanehydraulics.com/index.html

Engineer's Handbook – some formulae for fluid power:
http://www.engineershandbook.com/Tables/fluidpowerformulas.htm

AFS – hydraulic calculations, formulas and unit conversions:
http://advancedfluidsystems.com/tools-resources/fluid-power-calculator

FLUID POWER SOFTWARE AND APPS

Engineering ToolBox – various calculation apps for hydraulics and pneumatics: https://www.engineeringtoolbox.com/hydraulic-pneumatic-systems-t_59.html

Automation Studio – software for simulating hydraulic circuit designs and component sizing:
http://www.automationstudio.com

Webtec Hydraulic Oil Viscosity Calculator – online app for calculating oil viscosity at different temperatures and pressures:
http://www.webtec.com/education

CHAPTER

13 ABOUT THE AUTHORS

Steve Skinner started his career-long association with hydraulics as a student apprentice with the Dowty Group in the late 1960s. After graduating with a degree in mechanical engineering from the University of Bath, he joined hydraulics company Vickers Sperry Rand, first as a technical assistant, and subsequently as applications engineer and then customer engineer. During this time he became responsible for both the design, commissioning and troubleshooting of hydraulic systems, including a 5-month secondment to a large steelworks complex in Mexico.

In 1979 Steve took over responsibility for the Vickers' Hydraulic Training School, then located in Birmingham, UK, before moving to an international training role in 1985. He continued as European training manager when Eaton acquired Vickers in 1999, and in later years also took on product management responsibilities for the company's vane pump and heavy-duty transmission products.

He has been a member of several British Fluid Power Association committees, including being vice chairman of their education and training committee, and has authored several books on such subjects as hydraulic system troubleshooting, basic electronics, proportional valves, closed-loop control systems and variable-displacement pump controls. His recently published book on the history of hydraulic fluid power is the result of his keen interest in industrial history. He is now semi-retired, but continues to offer training services to the hydraulics industry and supports the excellent work carried out by the National Fluid Power Centre in Worksop, UK, where he remains a member of their Advisory Committee.

Martin Cuthbert grew up knowing about fluid power and the hydraulics industry thanks to his father, who ran the family company from the 1960s. Martin studied mechanical engineering at Sheffield, specialising in flow measurement in his final year. In 1997, after graduating with a master's degree in engineering, he joined Webtec and won the 1997 British Fluid Power Association's Prize for Young Engineers for his work on flow linearisation. He started out in the R&D department as a product engineer working on a team developing the next generation of turbine flow meters and test stand data-acquisition systems. By the late 1990s he had moved into a customer-facing role supporting UK customers.

In the early 2000s, Martin completed a postgraduate diploma in marketing and in 2003 was appointed commercial director, responsible for worldwide sales. On his father's retirement in 2006, he was appointed managing director. He regularly travels to North America and East Asia supporting Webtec's sales offices, visiting customers and giving training on selecting and using hydraulic test equipment.

He sits on the Advisory Committee for the National Fluid Power Centre in Worksop, UK.

CHAPTER

14 ABOUT WEBTEC

Webtec is a specialist manufacturer of hydraulic measurement and control products, helping to improve the productivity of heavy machinery.

For over 50 years it has been helping customers worldwide in the industrial, mobile, and agricultural sectors to diagnose hydraulic faults, verify hydraulic conditions and achieve repeatable hydraulic control.

It recognises that its customers' needs are unique, which is why it is proud of its ability to understand their requirements and then design and supply low-to-medium volume customised solutions on short lead times. It achieves this through technical innovation, investment in lean manufacturing and using cutting-edge IT solutions designed to maximise productivity.

Webtec's vision

> *We believe in being better than we are today, striving to constantly improve, and inspiring others to enjoy the challenges that engineering brings. We do this by focusing tirelessly on you, our customer, and your requirements when considering the products that we design, manufacture, supply, service and support.*

History

Webtec Products Ltd was founded in late 1964 as a joint venture between two American companies, Applied Power Inc. and Webster Electric Co. Inc., to manufacture the Webster range of hydraulic components. Roy Cuthbert started the business and eventually bought the company in 1970; he was MD for 36 years.

From 1971 the company diversified into flow measurement and related instrumentation. Webtec has continued to design and develop its own range of products both for the mobile and industrial machinery markets, and has also added complementary items from other manufacturers that it believes are useful to its customers.

A major part of Webtec's growth, since 1970, has been due to its investment in manufacturing and its focus on export business. Over the years, Webtec has purchased three machining companies, which have been consolidated in Webtec's modern machine shop in St Ives, UK.

In addition to the manufacturing and sales division in the UK, Webtec also has offices in the USA, France, Germany, Hong Kong and China; all operate primarily as sales and repair centres to support their worldwide distributor network. Webtec remains privately owned.

For further information visit www.webtec.com

INDEX

Note: Page references in *italics* refer to figures; those in **bold** refer to tables and definitions.

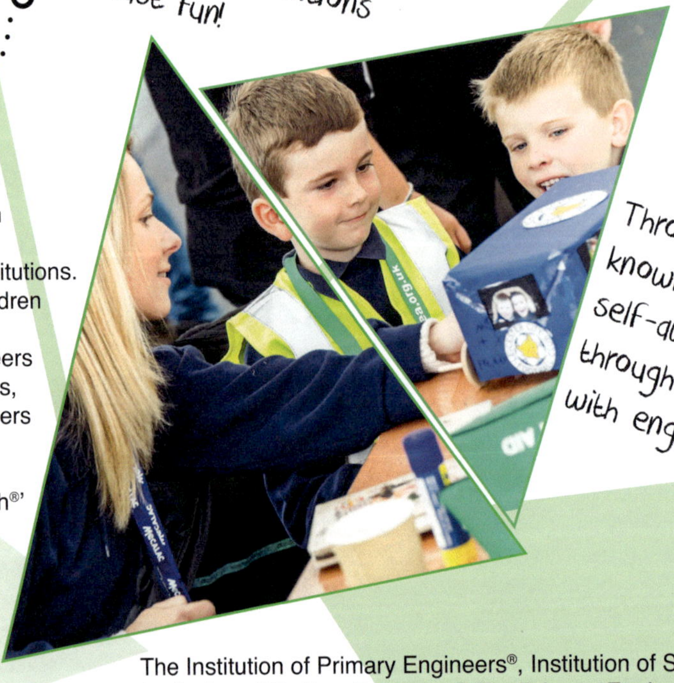

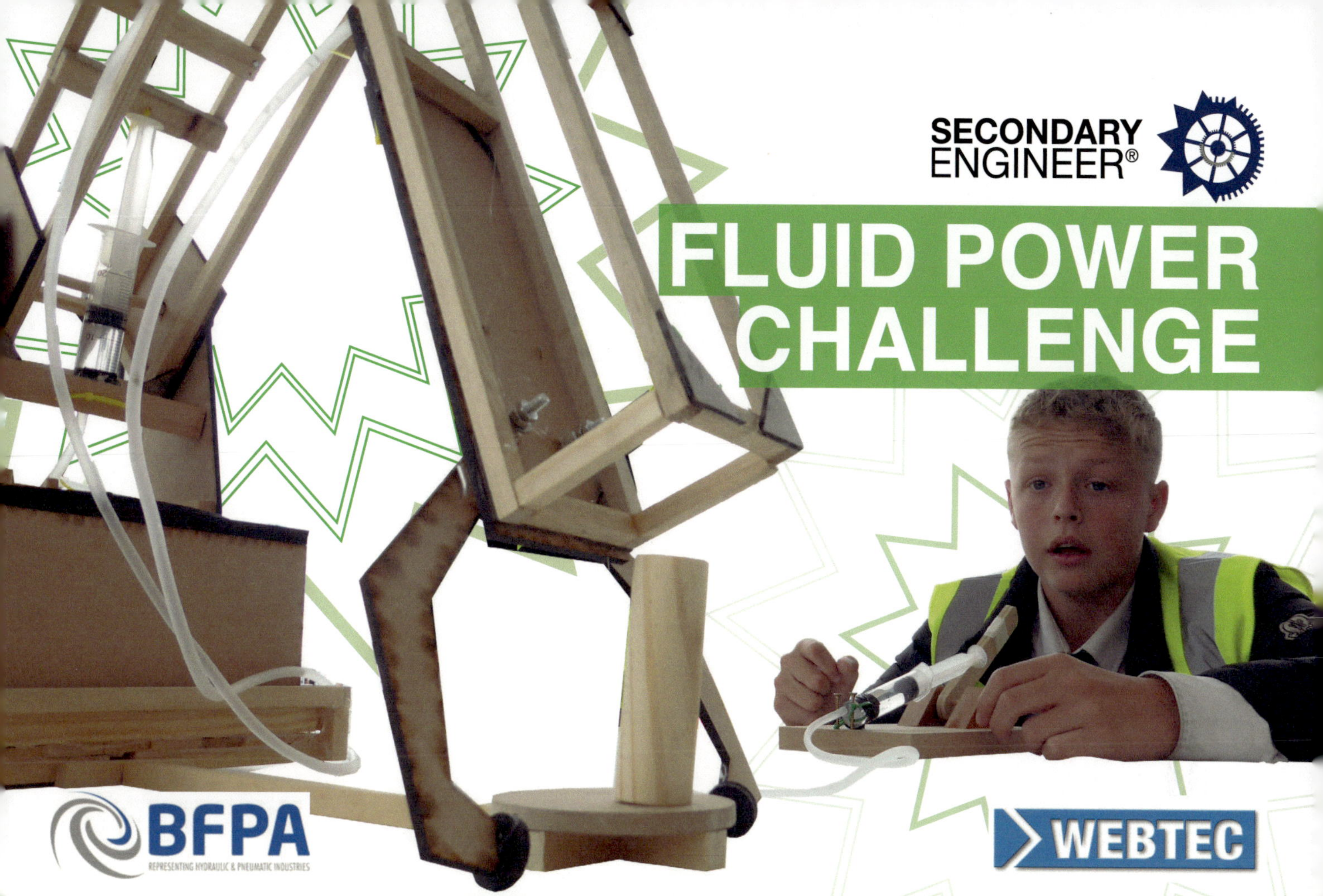

SECONDARY ENGINEER®

FLUID POWER CHALLENGE

BFPA
REPRESENTING HYDRAULIC & PNEUMATIC INDUSTRIES

WEBTEC

This course for secondary teachers of science, maths and technology and focuses on the dynamics of fluid power and the opportunities to explore hydraulics and pneumatics, science, maths and design technology. Fluid power plays a vital role in the creation of the built environment. This course investigates the technologies involved and is designed for pupils aged between 12-14 yrs. We invite engineers to join the teacher training and link with a school to support with the project by giving real world context to the engineering aspects of the challenge. On the course teachers are expected to engage with the practical aspects of the challenge designing their own Fluid Power Model to lift and move objects from one place to another. This course also highlights the competition element of the project and the requirements of the Fluid Power celebration event.

COURSE OUTCOMES

- Investigate the wider use of Fluid Power in engineering.
- Understand the maths and science involved in the design and manufacture of a practical solution.
- Create a working model in preparation for the celebration and teaching back in school.
- Teaching resources and consumables included.

If you would like to find out more visit our website: **www.secondaryengineer.com**
or contact us directly: **info@primaryengineer.com**

Primary Engineer
LEADERS AWARD
Secondary Engineer
LEADERS AWARD

ENGINEERING STEM
Professional Recognition
Postgraduate Certificate

Made in the USA
Las Vegas, NV
03 November 2026

57423763R00143